Numerical Modeling in Civil and Mining Geotechnical Engineering

Numerical Modeling in Civil and Mining Geotechnical Engineering

Editor

Li Li

MDPI • Basel • Beijing • Wuhan • Barcelona • Belgrade • Manchester • Tokyo • Cluj • Tianjin

Editor
Li Li
Department of Civil,
Geological and Mining
Engineering, Polytechnique
Montreal
Canada

Editorial Office
MDPI
St. Alban-Anlage 66
4052 Basel, Switzerland

This is a reprint of articles from the Special Issue published online in the open access journal *Processes* (ISSN 2227-9717) (available at: https://www.mdpi.com/journal/processes/special_issues/Modeling_Geotechnical).

For citation purposes, cite each article independently as indicated on the article page online and as indicated below:

LastName, A.A.; LastName, B.B.; LastName, C.C. Article Title. *Journal Name* **Year**, *Volume Number*, Page Range.

ISBN 978-3-0365-5441-9 (Hbk)
ISBN 978-3-0365-5442-6 (PDF)

Contents

About the Editor

Li Li

Li Li (Full Professor) is a registered engineer in Ordre des Ingénieurs du Québec (OIQ). He obtained a bachelor's degree and a master's degree, both in mining engineering, in China and a doctorate degree in France. Dr. Li has been working in geotechnical engineering for about 25 years in both academia and the industry. He was an Associate Professor at École de Technologie Supérieure (ÉTS) between 2010 and 2012. He was Associate Professor between 2012 and 2018 and became a Full Professor in 2018 at Polytechnique Montréal. Dr Li is interested in several fields in geotechnical engineering (rock, soil, and backfill mechanics). He is particularly active in research for estimating pressures and stresses in backfilled stopes, sizing barricades made of waste rocks, and assessing the required strength of side- or base-exposed backfill. He believes that a professor of engineering at the university level has several roles. As a researcher, he believes that improving our knowledge and improving the reliability of what we do have the same importance as originality and novelty. Simplification and step-by-step studies are his favorite methodologies in research. Any well-founded achievement or improvement should not be underrated. As engineers, we need to know what the value and application of our projects are in the industry. Our solutions and designs need to take into account the influencing factors as much as possible if it is impossible for all factors to be considered.

Preface to "Numerical Modeling in Civil and Mining Geotechnical Engineering"

Performing numerical modeling has become a routine task for academic researchers and consulting engineers in civil and mining engineering. Despite its great popularity and wide use in geotechnical engineering, one cannot hide the fact that a large portion of people who perform numerical modeling do not believe in their own numerical results—a phenomenon particular to geotechnical engineering. The crisis of confidence in numerical modeling in geotechnical engineering can be partly explained by the instability and unreliability of numerical results, the heterogeneity of field materials in space, and the unreliability or unrepresentativeness of experimental results. In the purpose of raising people's awareness on this critical issue and promoting people's confidence in the numerical modeling applied in civil and mining geotechnical engineering, this Special Issue (SI) was launched and paper invitations were sent to leading scholars in civil and mining geotechnical engineering. Fourteen articles were published not only for their originality and novelty but also for their details in numerical and/or physical models. These articles reflect the state-of-the-art of numerical modeling applied in geotechnical engineering and exhibit the useful process of numerical and physical modeling. The success of this SI would be impossible without the reviewers' significant efforts and contributions and the timely and constant support provided by the staff of the editorial office of Processes. The authors' efforts in accounting for the reviewers' comments and the academic editors' requirements of corrections equally deserve to be acknowledged.

Li Li
Editor

processes

Editorial

Special Issue on Numerical Modeling in Civil and Mining Geotechnical Engineering

Li Li

Research Institute on Mines and Environment (RIME UQAT-Polytechnique), Department of Civil, Geological and Mining Engineering, École Polytechnique de Montréal, C.P. 6079 Succursale Centre-Ville, Montréal, QC H3C 3A7, Canada; li.li@polymtl.ca

Numerical modeling is a widely used method in geotechnical engineering to understand the interactive responses of infrastructures with soils or/and rocks in both civil and mining engineering. Nowadays, computers are more and more powerful and commercialized software has become more and more easy to use because of graphic interfaces to facilitate the input and output parameters of numerical models. Numerical modeling has become much simpler and easier than it was decades ago. For some, numerical modeling can be carried out by anyone, including bachelor students. Ironically, the author's personal surveys indicate that 90% of the people who perform numerical modeling do not believe in their own numerical results. In addition, some believe that numerical modeling can only be used to provide qualitative information, such as a general idea or trend. This is another form of distrust in numerical modeling. How can we explain the crisis of confidence in numerical modeling, a phenomenon particular to geotechnical engineering?

Compared to other fields such as structural and mechanical engineering, one of the particularities of geotechnical engineering is the need to consider the ground. This is also one of the common points between civil geotechnical engineering and mining geotechnical engineering. Normally, the earth should be taken into account in numerical models of geotechnical engineering. However, the full consideration of the earth would result in an immense numerical model, requiring huge computing resource to simulate the earth. For most problems in geotechnical engineering, this is neither feasible nor necessary because the time and cost of calculations would be extremely high, and the uncertainties associated with the internal structures and properties of the earth could render the numerical results highly uncertain and even false. An alternative approach is to consider the studied structures with a semi-infinite space, as if the earth was considered as somehow flat without limit. This is not entirely false because the studied structures in civil engineering and mining engineering are for most cases very small compared to the size of the earth. It is, however, another source of problems in numerical modeling in civil and mining geotechnical engineering.

As most of the commonly used software in geotechnical engineering is based on continuum mechanics, discretization and meshing are necessary to represent the geometries of the ground and studied structures. Considering the whole semi-infinite space is impossible for most geotechnical engineering software developed on the finite element method or finite difference method. Cuts have to be made through the semi-infinite space to generate virtual boundaries, which, along with the ground surface and studied structure surfaces, constitute the domain of the numerical model. Boundary conditions are usually well-known along the ground surface and studied structure surfaces, and unknown along the virtual boundaries. A common method is to apply the initial in situ conditions (i.e., before any constructions or excavations) of the semi-infinite space along the virtual boundaries. The validity or representativeness of this hypothesis depends on the distances between the studied structures and the virtual boundaries. If the distances and the domain of the numerical model are too small, the virtual boundaries may fall within the influenced zones of the construction or excavation of the studied structures. Applying the initial in situ

Citation: Li, L. Special Issue on Numerical Modeling in Civil and Mining Geotechnical Engineering. *Processes* **2022**, *10*, 1571. https://doi.org/10.3390/pr10081571

Received: 27 July 2022
Accepted: 8 August 2022
Published: 11 August 2022

conditions of the semi-infinite space along the virtual boundaries is not representative of the field conditions. The constructed numerical model simulates a problem or situation different from the original one. The numerical results could be inaccurate and even false. If the distances between the virtual boundaries and studied structures and the domain of the numerical model are too large, the time of calculations could become uselessly long. Optimization of the domain is necessary to minimize the time of calculations and ensure stable and reliable numerical results. This is only possible through sensitivity analysis of the domain by considering a domain as small as possible to minimize the time of calculations, but large enough to ensure stable and reliable numerical results. The ensuing domain is called the optimal domain of the numerical model.

Another aspect affecting the time of calculations and quality of numerical results is the mesh sizes of the numerical model. Similar to the determination of the optimal domain, sensitivity analysis of mesh sizes is necessary in order to obtain the optimal meshes, which should be as coarse as possible to minimize the time of calculations, but fine enough to ensure stable and reliable numerical results.

Unstable and unreliable numerical results can result from a domain which is not large enough and/or meshes which are not fine enough. A good practice for minimizing the time of calculations and ensuring stable and reliable numerical results is to perform sensitivity analyses of the domain and meshes to obtain an optimal numerical model with an optimal domain and an optimal mesh. An example of this practice can be seen through the article of Zeng et al. [1], who present the validation of a newly implemented numerical model against analytical solutions. The method of obtaining stable and reliable numerical results of stresses in a backfilled stope through sensitivity analyses of the domain and meshes is also illustrated. When time is involved in numerical models, sensitivity analysis of time steps should be performed to obtain the optimal time step, which should be as large as possible to minimize the time of calculations, but small enough to ensure stable and reliable numerical results.

Normally, numerical results tend to become stable as long as the meshes of the numerical model are fine enough. It is, however, not always the case, especially when the default values of the controlling parameters given by commercialized software are used. The numerical results may become unstable, and the trend can even become irregular as the meshes of the numerical model are too fine. This is due to the fact that the default values of the controlling parameters given by commercialized software are valid for most of simple cases. When the number or/and size of the studied structures are large, accumulated errors associated with the approximation of numerical calculations could become large. The problem can be amplified by using meshes which are too fine. It is important to perform sensitivity analyses of all the controlling parameters. Examples can be found in Zhai [2] by increasing the number of iteration steps with FLAC3D [3], or in Jaouhar [4] through a reduction in the error tolerance with SIGMA/W [5].

In contrast to a belief that numerical modeling can be performed by anyone, the author believes that numerical modeling can only be performed by a qualified person. A minimum of training is necessary. The validation or verification of the used numerical code against closed-form (analytical) solutions should be the first step in the training. This is necessary for any new user, firstly to verify if the numerical code contains any errors or limitations, and secondly to verify if the new user can correctly use the numerical code to produce meaningful numerical results. It is very important for the trainee to understand that the most important aspects of the validation or verification step are the procedure of diverse sensitivity analyses, through which stable and reliable numerical results can be obtained. It is also very important for the trainee to apply the procedure of sensitivity analyses in all the numerical modeling with any specific projects.

A quite common and poor practice is to use an extremely large domain and very fine meshes through the whole domain during the step of validation or verification against a closed-form (analytical) solution. As the structure of closed-form solutions is usually simple and small, the time of calculation is not a problem. The trainee could quickly complete the validation step, but gains little knowledge on the limitation of the numerical models and on how to obtain stable and reliable numerical results. During the step of numerical modeling with a specific project, the number and size of studied structures become large. The power and availability of computing resources along with the time of calculations then become prevailing concerns, neglecting to examine whether the domain is large enough and whether the meshes are fine enough. The numerical model can be too small in domain and/or too coarse in meshes, resulting in unstable and unreliable numerical results. This is an importance source feeding people's distrust in numerical modeling in civil and mining geotechnical engineering.

Despite the importance of the step of validation or verification against closed-form (analytical) solutions, some people believe that this step is unnecessary, arguing that the numerical code employed is very popular and widely used. If one recognizes the fact that a good car tested by many people is not equivalent to a good driving by anyone, one would understand that a widely used numerical code does not automatically result in good numerical modeling by any new users. Subsequently, any new users must pass the step of validation, through which the new users can learn how to use the software, verify if the numerical code contains any errors and limitations, and learn how to obtain stable and reliable numerical results through the diverse sensitivity analyses.

Over the years, numerical codes based on the distinct element method or mesh-free methods are becoming more and more popular in geotechnical engineering to simulate the behavior of granular or fluid materials. While the domain and meshes are critical concerns of numerical modeling with numerical codes based on continuum mechanics, they are not applicable to numerical modeling with numerical codes based on distinct element method or mesh-free methods. However, the representativeness of the particles in numerical models based on the distinct element method and the density of nodes in numerical models based on the mesh-free methods become critical. How to obtain stable and reliable numerical results is still a critical concern. Again, their sensitivity analyses along with the sensitivity analyses of controlling parameters are necessary to obtain an optimal numerical model and ensure stable and reliable numerical results.

Another point of view quite common in geotechnical engineering is the need for validation of numerical models by experimental results. The numerical results are even considered unreliable and useless if there is no validation by experimental results. This point of view neglects the fact that experimental results can also involve numerous uncertainties and even errors due to human errors, instrumentation inaccuracy and flaws in testing norms [6,7]. A long discussion on the stability and reliability of experimental results is beyond the scope of this Special Issue (SI), but it is not uncommon to see geotechnical tests realized without calibrating all the testing instrumentation before and/or after the tests. This is particularly true in field measurements. The obtained experimental results can contain high uncertainties or even errors. In many cases, people make use of their own experimental results or published data to calibrate their numerical model by adjusting some model parameters to obtain good agreements between the numerical and experimental results. The ensued good agreement is usually called a "validation" of the numerical model or "prediction" of the experimental results. In the author's point of view, this process of calibration is neither a validation of the numerical model nor a prediction of the experimental results. If the experimental results are erroneous, it would be odd to conclude that the numerical model is validated by erroneous data and the numerical model successfully predicts wrong experimental results. If the experimental results are reliable, the process of calibration can then be considered as a test of the power or applicability of the numerical model. The numerical model along with the measured and calibrated model parameters can then be called "calibrated numerical model" [8–10]. The predictability of the calibrated numerical model needs to be verified against additional experimental data that are not used in the process of calibration.

The above-mentioned aspects are not the only sources feeding people's distrust in numerical modeling in geotechnical engineering. There exist other sources, such as the heterogeneity of material in space and poor investigation of sites. The reliability of numerical modeling diminishes as the reliability and representativeness of measured parameters decrease. This is, however, a problem of site investigation and parameter measurements, not a problem of numerical modeling. Here we understand that the reliability of a system needs the reliability of every element which constitutes the system.

Aiming to increase people's confidence in numerical modeling in civil and mining geotechnical engineering, the author accepted the journal invitation to host this SI. The collected articles not only present original and novel contributions to civil and mining geotechnical engineering, but also include sufficient details in order for readers to be able to reproduce the published results of the physical and numerical models. Comparisons between numerical and experimental results were encouraged, but not considered as mandatory. Emphasis was placed on the validation or verification of the used numerical model, including domain and mesh sensitivity analyses of numerical models as long as they are applicable.

Upon our invitations, more than 27 articles were submitted in this SI. A few had been rejected without being sent to the process of review as their contents were beyond the scope of this SI. Thirteen articles were rejected after evaluation by either academic editors or reviewers, while fourteen articles were published based on the positive recommendations of at least two reviewers for each article and the positive recommendations of academic editors. The SI contains seven articles in civil engineering [11–17] and seven articles in mining engineering [1,8,18–22]. Each of the published papers has its own merits and limitations (very normal in research). As the guest editor of this SI, the author hopes the readers enjoy the reading of all the 14 published articles. Any comments, suggestions and criticisms are welcome either in form of discussion articles or in form of personal communications.

The author takes this opportunity to thank Jim Wang, SI Managing Editor of MDPI, for his invitation and constant assistance from the beginning to the end of this SI. The SI would have been impossible without the efforts and significant contributions of the reviewers. Many thanks to Dr. Nevena Blagojev, Cady Chang, Susan Ji, Andjela Markovic, Dr. Iulian Patrascu, Amelia Qie, Melinda Simon-Varhelyi, Alfreda Song, Rich Tan, Frida Wang, Hana Wang, Frida Wang, Celine Xiao, Henry Zhang, Ella Zhang, Eileen Zhang, and Nan Zhang (I hope I didn't forget anyone; otherwise, please accept my apology) for their timely assistance. The author appreciates the efforts of academic editors Drs. Haiping Zhu, Blaž Likozar, Farhad Ein-Mozaffarie, Avelino Núñez-Delgado, and Orlando Vaselli. Prof. Giancarlo Cravotto, the Editor-in-Chief of *Processes*, is acknowledged for his official invitation to host this Special Issue and for his supports.

Funding: This research received no external funding.

Conflicts of Interest: The author declares no conflict of interest.

References

1. Zeng, F.; Li, L.; Aubertin, M.; Simon, R. implementation of the non-associated elastoplastic MSDP$_u$ model in FLAC3D and application for stress analysis of backfilled stopes. *Processes* **2022**, *10*, 1130. [CrossRef]
2. Zhai, Y. Design of Barricades Made of Waste Rocks for Backfilled Stopes. Ph.D. Thesis, École Polytechnique de Montréal, Montreal, QC, Canada, 2021.
3. Itasca. *FLAC3D—Fast Lagrangian Analysis of Continua in 3 Dimensions*; Version 7.0; User's Guide; Itasca Consulting Group: Minneapolis, MN, USA, 2019.
4. Jaouhar, E.M. Études Analytiques, Numériques et Expérimentales pour Évaluer les Contraintes et les Pressions dans les Chantiers Remblayés et sur les Barricades. Ph.D. Thesis, École Polytechnique de Montréal, Montreal, QC, Canada, 2019.
5. Geo-Slope. *Stress-Deformation Modeling with SIGMA/W 2007*, 3rd ed.; GEO-SLOPE International Ltd.: Calgary, AB, Canada, 2007.
6. Deiminiat, A.; Li, L.; Zeng, F.; Pabst, T.; Chiasson, P.; Chapuis, R. Determination of the shear strength of rockfill from small-scale laboratory shear tests: A critical review. *Adv. Civ. Eng.* **2020**, *2020*, 8890237. [CrossRef]
7. Deiminiat, A.; Li, L.; Zeng, F. Experimental study on the minimum required specimen width to maximum particle size ratio in direct shear tests. *CivilEng* **2022**, *3*, 66–84. [CrossRef]
8. Wang, R.; Zeng, F.; Li, L. Applicability of Constitutive Models to Describing the Compressibility of Mining Backfill: A Comparative Study. *Processes* **2021**, *9*, 2139. [CrossRef]

9. Wang, R.; Li, L.; Simon, R. A model for describing and predicting the creep strain of rocks from the primary to the tertiary stage. *Int. J. Rock Mech. Min. Sci.* **2019**, *123*, 104087. [CrossRef]
10. Wang, R.; Zeng, F.; Li, L. Stability analyses of side-exposed backfill considering mine depth and extraction of adjacent stope. *Int. J. Rock Mech. Min. Sci.* **2021**, *142*, 104735. [CrossRef]
11. Al-Atroush, M.E.; Hefny, A.M.; Sorour, T.M. A parametric numerical study for diagnosing the failure of large diameter bored piles using supervised machine learning approach. *Processes* **2021**, *9*, 1411. [CrossRef]
12. Bao, Y.; Jiang, B. Incompatible deformation model of rocks with defects around a thick-walled cylinder. *Processes* **2021**, *9*, 2215. [CrossRef]
13. Fatollahzadeh Gheisari, A.; Maghoul, P.; Holländer, H.M.; Kenyon, R.; Sinclair, R.; Saaly, M. The Long-Term Mitigating Effect of Horizontal Ground-Source Heat Exchangers on Permafrost Thaw Settlement. *Processes* **2021**, *9*, 1636. [CrossRef]
14. Granitzer, A.-N.; Tschuchnigg, F. Practice-Oriented Validation of Embedded Beam Formulations in Geotechnical Engineering. *Processes* **2021**, *9*, 1739. [CrossRef]
15. Karim, M.R.; Hughes, D.; Rahman, M.M. Unsaturated Hydraulic Conductivity Estimation—A Case Study Modelling the Soil-Atmospheric Boundary Interaction. *Processes* **2022**, *10*, 1306. [CrossRef]
16. Navarro, V.; Cabrera, V.; De la Morena, G.; González, D.; Asensio, L.; Yustres, Á. Precomputation of Critical State Soil Plastic Models. *Processes* **2021**, *9*, 2142. [CrossRef]
17. Wang, L.; Meguid, M.; Mitri, H.S. Impact of Ballast Fouling on the Mechanical Properties of Railway Ballast: Insights from Discrete Element Analysis. *Processes* **2021**, *9*, 1331. [CrossRef]
18. Chang, Q.; Yao, X.; Qin, J.; Li, M.; Wang, Y.; Zhou, H.; Xu, Y.; Sun, Y. Analysis on Water Inrush Prevention Mechanism of Paste-Filled Floor above Confined Water. *Processes* **2022**, *10*, 274. [CrossRef]
19. Rima, U.S.; Beier, N.; Abdulnabi, A. Modeling the Effects of Seasonal Weathering on Centrifuged Oil Sands Tailings. *Processes* **2021**, *9*, 1906. [CrossRef]
20. Wang, Q.; Feng, H.; Tang, P.; Peng, Y.; Li, C.; Jiang, L.; Mitri, H.S. Influence of Yield Pillar Width on Coal Mine Roadway Stability in Western China: A Case Study. *Processes* **2022**, *10*, 251. [CrossRef]
21. Zhao, Y.; Zhao, X.; Dai, J.; Yu, W. Analysis of the Surface Subsidence Induced by Mining Near-Surface Thick Lead-Zinc Deposit Based on Numerical Simulation. *Processes* **2021**, *9*, 717. [CrossRef]
22. Zhou, S.; Li, L. Numerical Investigation on the Impact of Tailings Slurry on Catch Dams Built at the Downstream of a Breached Tailings Pond. *Processes* **2022**, *10*, 898. [CrossRef]

processes

MDPI

Article

Impact of Ballast Fouling on the Mechanical Properties of Railway Ballast: Insights from Discrete Element Analysis

Luyu Wang [1], Mohamed Meguid [1] and Hani S. Mitri [2,*]

[1] Department of Civil Engineering, McGill University, Montreal, QC H3A 0C3, Canada; luyu.wang@mail.mcgill.ca (L.W.); mohamed.meguid@mcgill.ca (M.M.)
[2] Department of Mining and Materials Engineering, McGill University, Montreal, QC H3A 0E8, Canada
* Correspondence: hani.mitri@mcgill.ca; Tel.: +1-5-(14)-3984890

Abstract: Ballast fouling is a major factor that contributes to the reduction of shear strength of railway ballast, which can further affect the stability of railway supporting structure. The major sources of ballast fouling include infiltration of foreign fines into the ballast material and ballast degradation induced by train movement on the supported tracks. In this paper, a discrete element model is developed and used to simulate the shear stress–strain response of fouled ballast assembly subjected to direct shear loading. A simplified computational approach is then proposed to model the induced ballast fouling and capture the mechanical response of the ballast at various levels of contamination. The approach is based on the assumption that fine particles comprising the fouling material will not only change the interparticle friction angle, but also the contact stiffness between the ballast particles. Therefore, both the interparticle friction coefficient and effective modulus are adjusted based on a fouled ballast model that is validated using experimental results. The effect of ballast degradation is also investigated by gradually changing the particle size distribution of the ballast assembly in the discrete element model to account for the increased range of particle sizes. Using the developed model, the effect of ballast degradation on the shear strength is then evaluated. Conclusions are made to highlight the suitability of these approximate approaches in efficiently modeling ballast assemblies under shear loading conditions.

Keywords: railway ballast fouling; ballast degradation; micro-mechanical parameters; shear strength

Citation: Wang, L.; Meguid, M.; Mitri, H.S. Impact of Ballast Fouling on the Mechanical Properties of Railway Ballast: Insights from Discrete Element Analysis. *Processes* **2021**, *9*, 1331. https://doi.org/10.3390/pr9081331

Academic Editor: Li Li

Received: 26 June 2021
Accepted: 26 July 2021
Published: 30 July 2021

1. Introduction

Ballast is an essential structural component in most railway foundation systems. The primary function of the ballast layer is to resist the applied wheel loads from railway equipment through the tracks and to spread them uniformly to the underlying foundation material. Therefore, the mechanical behavior of the ballast is important for the stability and functionality of railway equipment. Voids usually exist within fresh, uniformly graded coarse aggregates that form a ballast layer. As ballast ages, such voids are filled with fine material, which is regarded as ballast fouling [1]. The major sources of ballast fouling include: (1) infiltration of foreign fine material, and (2) mechanical degradation due to the particle breakage [2,3]. Ballast fouling can cause unfavorable effects, most notably, shear strength reduction that further impacts the performance of the railway system [4].

The discrete element method (DEM) developed by Cundall and Strack [5] has been used to investigate the mechanical behavior of fresh ballast [6–10], as well as fouled ballast [3,4]. In previous studies, ballast fouling induced by infiltration of foreign fine material was simulated using two approaches. In the first approach, small particles are explicitly modeled and injected into the ballast assembly [3,11,12]. This approach is computationally demanding due to the fine particle sizes and the significant increase in the number of particles [5,7]. The second approach relies on reducing the ballast particle surface friction angle without injection of new fine particles [4]. In this case, the computational time is significantly reduced, and the influence of fine particles is reflected as lubricant between the ballast particles, thus affecting the interparticle friction angle [3,4]. The limitation of this approach is that the calibration process can be complex [11] and it is sometimes difficult to replicate the exact response [4].

Mechanical degradation, a major source of ballast fouling, produces approximately 75% of the fouling material under normal operating conditions [13]. In other studies [7,14], ballast degradation was simulated using breakable particles subjected to dynamic loading. As ballast degradation takes place, particle size distribution (PSD) changes from uniform to a broader grading [15,16]. The degraded ballast assembly is represented in DEM analysis by reducing the size of particles using either 2D or 3D models. Experimental and numerical investigations [3,14,15,17] that examined the effect of ballast degradation on the material shear strength revealed results that can differ from those obtained numerically. It is, therefore, necessary to further explore the numerical approaches to come up with a simplified method that would allow for the effect of ballast degradation to be investigated.

In this study, a discrete element model is developed using Particle Flow Code software in three dimensions (PFC3D) [18] to simulate the shear stress–strain behavior of conventional ballast material assembly under direct shear condition. A simplified computational approach is proposed to simulate foreign-material-induced ballast fouling. It captures the mechanical behavior of the ballast at various levels of void contamination using the so-called void contamination index (VCI). Model results are validated with experimental data under similar loading and boundary conditions. Empirical relationships between VCI and micro-mechanical parameters (effective modulus E^* and interparticle friction coefficient μ) are developed to efficiently determine the required E^* and μ for a given VCI ratio. The degraded particle assembly is tested using direct shear tests to obtain the shear strength properties. The impacts of degradation on the mechanical behavior of the ballast are discussed.

2. Modeling Ballast Particles

Modeling irregularly shaped particles in a discrete element model can realistically reflect the mechanical interaction between ballast particles, as it allows for sufficient interlocking to develop and minimizes excessive rolling [19]. Particle shape characteristics (i.e., angularity and flakiness) are known to affect the shear stress–strain behavior of ballast assembly [20–22]. In this study, a sample of railway ballast is obtained from Quebec, Canada, from which nine particles are randomly selected to represent typical particle size and shapes. The Bubble-Pack algorithm built in PFC3D is used to generate the irregular-shaped particle clumps based on the scanned three-dimensional ballast contours [23,24]. A clump is defined as a group of spheres of different sizes clustered into one unbreakable particle that acts like a single ballast particle in the DEM model [25].

Two parameters (φ and ρ) are usually specified in the Bubble-Pack algorithm to control the co-ordinates and the size of the spheres within the clump. The angle φ represents the maximum intersection angle between two pebbles, whereas ρ is the ratio of the smallest to the largest sphere size. The effect of varying φ and ρ on the shapes of the generated clumps is shown in Figure 1. As can be seen, increasing the interaction angle, φ, results in a much smoother clump surface, whereas decreasing the size ratio of the spheres, ρ, value results in a finer clump shape.

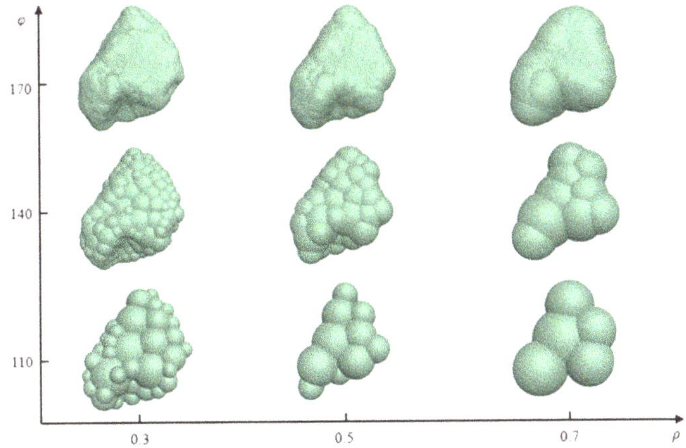

Figure 1. Clumps created using different sphere size ratios (ρ) and interaction angles (φ).

Normally, a large number of spheres is required for the algorithm to generate a realistic clump shape. However, the computational cost will proportionally increase. A balance is, therefore, needed between the number of particles required to sufficiently simulate the various ballast shapes and the cost of the analysis. In this study, the nine irregular-shaped clumps shown in Figure 2 are generated using several spherical particles that range between 10 and 16. These shapes approximately represent, to a large extent, the overall geometry of the investigated ballast particles and, at the same time, resulted in significant cost saving to the computation process.

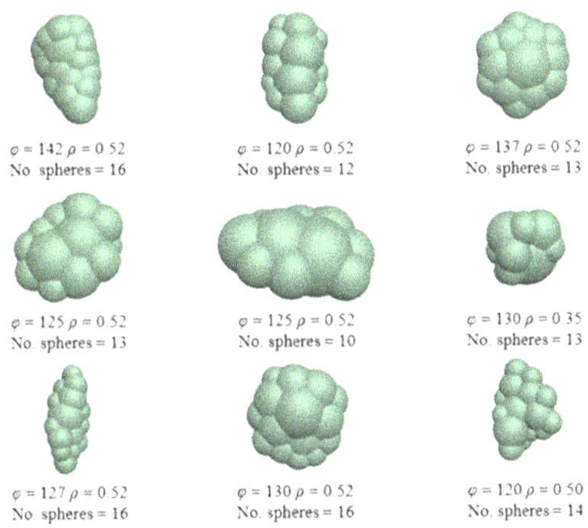

Figure 2. Irregular-shaped particles used in this study and their corresponding parameters.

3. Contact Model

The linear contact model (LCM) has been shown by researchers to be efficient in describing ballast interaction under different loading conditions [3,6,7,10,26]. The LCM, described in Figure 3, is usually used in discrete element analysis to represent particle–particle contact (PPC), as well as particle–wall contact (PWC).

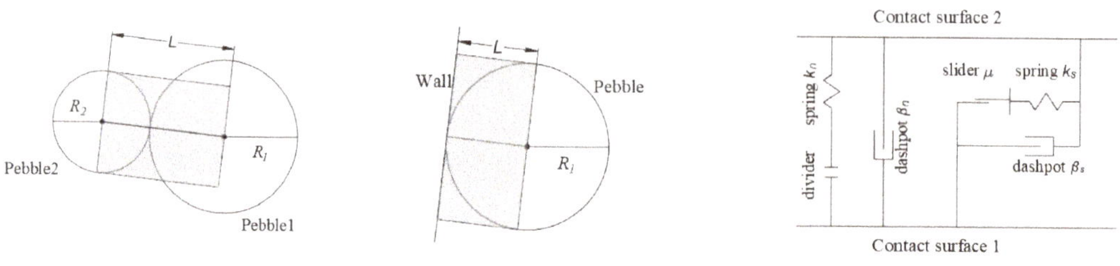

(a) (b) (c)

Figure 3. Geometric parameters and rheological components of the linear contact model. (**a**) Particle–particle contact; (**b**) particle–wall contact; (**c**) rheological model.

The micro-mechanical parameters that need to be defined in PFC3D include the effective modulus E^* and the normal-to-shear stiffness ratio k^*. These parameters are related to the normal and shear stiffness of the LCM as follows:

$$k_n = \frac{\pi R^2 E^*}{L} \tag{1}$$

$$k_s = \frac{k_n}{k^*} \tag{2}$$

$$R = \begin{cases} \min(R_1, R_2) & \text{PPC} \\ R_1, & \text{PWC} \end{cases} \tag{3}$$

$$L = \begin{cases} R_1 + R_2, & \text{PPC} \\ R_1, & \text{PWC} \end{cases} \tag{4}$$

where k_n is the contact stiffness in the normal direction and k_s is the contact stiffness in the tangential direction. R and L are geometry parameters of the contact described in Figure 3a,b. The contact stiffnesses are automatically scaled based on the individual particle size and the specified E^* and k^* values. Hence, the mechanical behavior of particle assemblies is associated with the elastic modulus and the Poisson's ratio at the macroscopic level [6].

The interparticle friction coefficient μ is a local contact surface property that controls the slip behavior between particles. Slip would occur when the contact shear force (F_i^s) is greater than the shear resistance. The shear resistance in the LCM is defined as:

$$F_{max}^s = \mu |F_i^n| \tag{5}$$

where F_{max}^s is the shear resistance and F_i^n is the normal contact force. A higher interparticle friction coefficient μ implies higher shear resistance and, hence, higher resistance to shear slip. This means that the overall stiffness and shear strength of the ballast assembly will increase [27].

Particle kinematic is calculated iteratively in the discrete element model for each time-step. The time-step should not generally exceed a critical value to ensure that disturbance only propagates from the particle to its immediate neighboring particle [28]. The critical time-step, Δt, used in PFC^{3D} is determined using Equation (6) below:

$$\Delta t = \sqrt{\frac{m}{K}} \tag{6}$$

where m is the smallest particle mass and K is the contact spring stiffness [29]. The critical time-step generally increases with the increase in particle sizes or the decrease in contact stiffness.

4. Model Setup and Calculation Procedure

Indraratna et al. [30] conducted a series of direct shear tests to examine the mechanical behavior of ballast fouled by coal dust at different contamination levels. The experiments were performed using a shear box measuring 300 mm × 300 mm × 195 mm. In this study, the direct shear tests are numerically simulated considering the dimensions and boundary conditions used in the experiments.

The numerical model used to simulate the direct shear test with the ballast assembly generated, using the irregular particle shapes presented earlier, is shown in Figure 4. The linear contact model is used to represent the contacts between particles, as well as between particles and the walls of the box, in both the normal and tangential directions. A total number of 8424 particles (density = 2700 kg/m^3) are randomly generated within the shear box. Typical ballast gradation, in practice, ranges from a maximum of 50–76 mm to a minimum of 12–20 mm, depending on the type of rail. However, in the experimental work of Indraratna et al. [30], the particle size ranges from 1.5 mm to 40 mm, as shown in Figure 5. Therefore, downscaling was deemed necessary to fit the PFC model with the experimental results by Indraratna et al. [30]. Considering that computing time is exponentially proportional with smaller particle size, it is impossible to replicate the lower limit of 1.5 mm in Figure 5. Therefore, it was rationalized that downscaling would apply to the $\frac{3}{4}$ inch sieve (19 mm) representing 0–10% ballast by weight in practice. This resulted in a particle size distribution that ranges from 7.5 to 40 mm. This is in line with the discrete element modeling by Indraratna et al. [3] for the same ballast sample. The randomly oriented particles are then cycled to satisfy the required porosity and to reach an equilibrium state inside the box. The porosity of the ballast assembly after initial packing reached 0.45, which is consistent with that reported in the experiment.

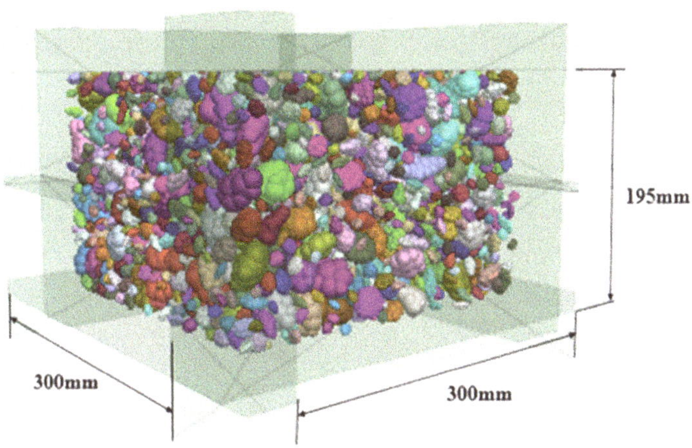

Figure 4. Direct shear box and ballast assembly in PFC^{3D}.

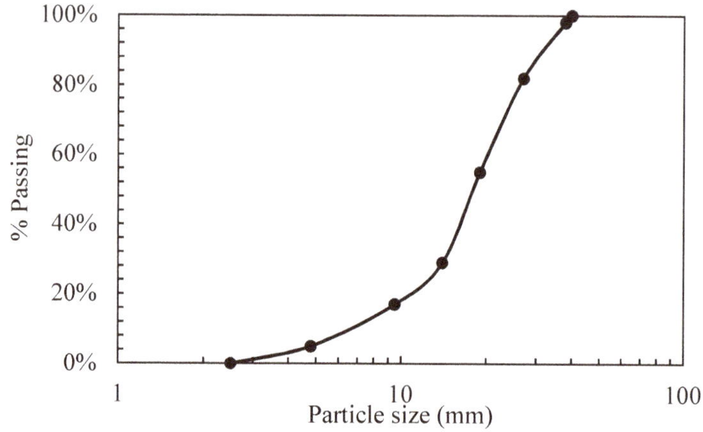

Figure 5. Particle size distribution used in the experiments (adapted from [30]).

After the system has reached equilibrium, normal stress is applied at the top and bottom of the assembly (see Figure 6a) and kept constant by adjusting the velocity of the top and bottom plates, utilizing the servo-control feature in PFC. The lower section of the box is then moved horizontally under a constant rate to apply the shear force on the ballast assembly [31]. A slow rate of shearing of 0.4 mm/s is used throughout the analysis. This rate was chosen by comparing the shear stress–strain results for different shear rates (from 0.04 mm/s to 40 mm/s) to avoid unduly sample disturbance during shear [3,27,32] and to save on computational time. At 40 mm shear displacement, the assembly would reach a maximum shear strain of about ε_s = 13.3%.

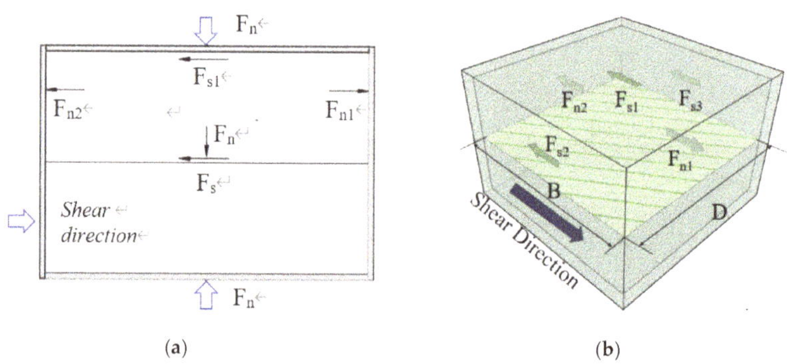

(a) (b)

Figure 6. Schematic of the forces acting on the direct shear box in (**a**) 2D and (**b**) 3D.

The forces acting on the shear box are shown in Figure 6b. The normal force (F_N) acting on the shear plane is equal to the normal load applied to the assembly, whereas the shear force (F_S) is equal to the sum of the horizontal forces in the upper section that can be calculated as follows.

$$F_s = F_{n1} + F_{n2} + F_{s1} + F_{s2} + F_{s3} \tag{7}$$

where F_{n1} and F_{n2} are the normal forces acting on the left and right walls of the box, F_{s1} is the shear force on the top plate, and F_{s2} and F_{s3} are the shear forces on the front and back walls, as shown in Figure 6b. The normal and shear stresses σ_n and σ_s acting on the shear plane are calculated using Equations (8) and (9) below:

$$\sigma_n = \frac{F_N}{D(B - vt)} \quad (8)$$

$$\sigma_s = \frac{F_S}{D(B - vt)} \quad (9)$$

where D and B are the length and width of the shear box (see Figure 4). The lower section moves at a constant rate v, which means that, after a time t has elapsed, the shear area would be $D(B - vt)$.

5. Modeling Fresh Ballast Assembly

The discrete element model is first validated using the results of the direct shear tests performed on fresh ballast under three different normal pressures, namely, 27 kPa, 51 kPa, and 75 kPa. To choose a suitable interparticle friction coefficient (μ) for the modeled material, sensitivity analysis was performed by modeling the shear stress–strain response under applied normal stress of 51 kPa. This is achieved by incrementally increasing μ value in five stages, 0.1, 1.1, 2.1, 3.1, and 4.1, respectively. Results showed that an interparticle friction coefficient of 2.1 or less could not capture the strain-softening behavior of the material. Some improvement is found when μ increased from 3.1 to 4.1. A value of 4.1 is, therefore, used in this study. The micro-mechanical parameters used in model validation are summarized in Table 1. The modeling results are presented in Figure 7, along with the experimental data.

Table 1. Micro-mechanical parameters for fresh ballast assembly.

Parameters	Value	
	PPC [1]	PWC [1]
Effective modulus, E^* (Pa)	1.8×10^7	3.6×10^7
Stiffness ratio, k^*	1.0	1.0
Interparticle friction coefficient, μ	4.1	0.0

[1] PPC—particle to particle contact; PWC—particle to wall contact.

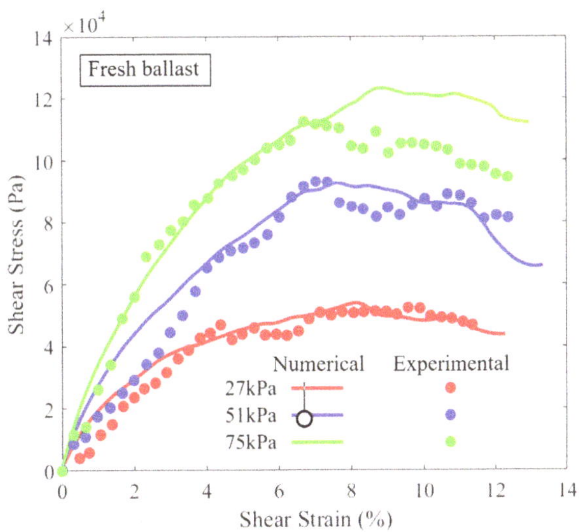

Figure 7. Shear behavior of fresh ballast under different confining pressures.

The trends of the computed responses are found to be in general agreement with the laboratory test results. It can also be seen that the DEM simulation is able to reproduce, to a large extent, the strain-softening behavior of the ballast material under applied normal stresses of 27 kPa and 51 kPa. Under normal stress of 75 kPa, however, the discrete element model overestimated the shear strength at shear strain of more than 7%. This may be attributed to possible ballast particle breakage in the shear box during the experiment under high normal stress (75 kPa). As particle breakage is not explicitly modeled in this study, the model may not capture the contribution of particle breakage to the softening response of the assembly [16,31].

In general, it can be concluded that the developed model performed reasonably well in capturing the overall shear stress–strain behavior of fresh ballast assembly under the investigated range of normal stresses. It is worth noting that, for typical 20 to 30-ton axle loads, the confining pressure within the ballast layer is usually in the range of 10–40 kPa and rarely exceeds 60 kPa [33].

6. Modeling Fouled Ballast Assembly

Ballast fouling resulting from mixing with fine material is first simulated under different VCI ratios. Instead of explicitly introducing discrete fine particles into the matrix of the ballast, fouled ballast is simulated in the following section by incrementally reducing the micro-mechanical parameters (E^* and μ) used in the fresh ballast model. By doing so, the number of particles in the discrete element model remains the same. Empirical relations are developed and used to determine the values of the reduced E^* and μ for a given VCI ratio. The shear strength envelopes for various VCI ratios are obtained using hyperbolic and power relationships based on the numerical results. Finally, ballast fouling due to degradation or particle breakage is also examined.

7. Modeling Fouled Ballast due to Particle Infiltration

The approach adopted in modeling fouled ballast due to the intrusion of foreign material into the ballast assembly is illustrated in Figure 8. Figure 8a shows that introducing external fines results in the interparticle force, F, changing to F', and the friction angle between particles $\varnothing$ changing to $\varnothing'$. In addition, as shown in Figure 8b, the interparticle contact stiffness, k_1 of fresh ballast changed to a lower stiffness, k_2. With this modeling approach, the interparticle friction coefficient μ and effective modulus E^* are also reduced [34].

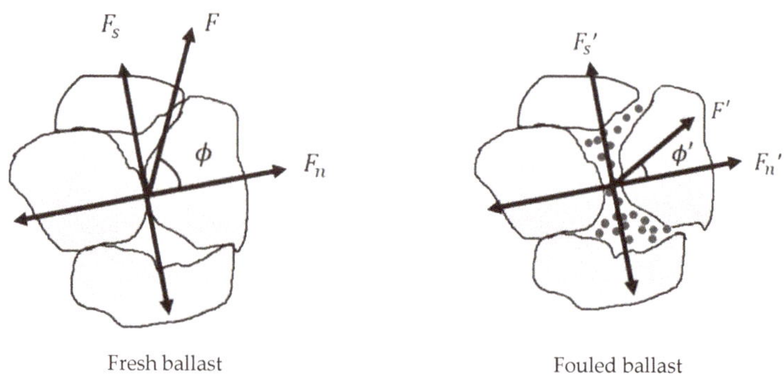

(a)

Figure 8. *Cont.*

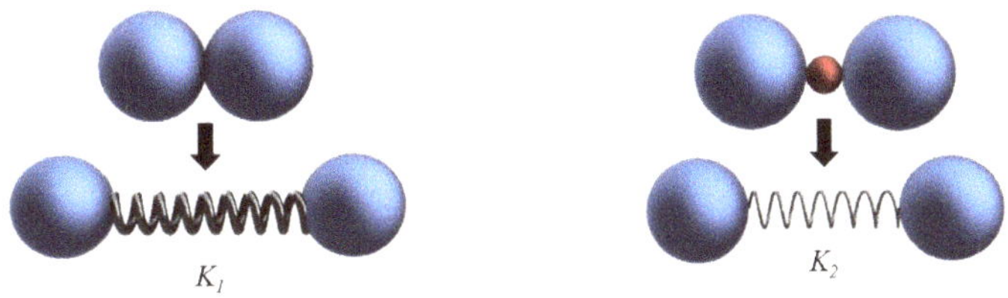

Fresh ballast, inter-particle contact stiffness K_1 Fouled ballast, inter-particle contact stiffness K_2

(**b**)

Figure 8. Adopted method for simulating fouled ballast; (**a**) changes to the interparticle friction coefficient due to the presence of fines; (**b**) changes to interparticle contact stiffness due to the presence of fines.

Void contamination index (VCI) represents the percentage of the voids within the fresh ballast assembly that is occupied by fouling material [35]. It is expressed as:

$$\text{VCI} = \frac{1+e_f}{e_b} \times \frac{G_{s\cdot b}}{G_{s\cdot f}} \times \frac{M_f}{M_b} \times 100 \quad (10)$$

where e_f is the void ratio of the fouling material; e_b is the void ratio of the fresh ballast; $G_{s\cdot b}$ is the specific gravity of the ballast; $G_{s\cdot f}$ is the specific gravity of the fouling material; M_f is the dry mass of the fouling material; and M_b is the dry mass of the fresh ballast. VCI can be used to effectively represent the level of the foreign material in the ballast.

Figure 9 compares the shear response of fouled ballast, calculated using the proposed model, and the experimental data for three different VCI ratios, 20%, 40%, and 95%, respectively. The effective modulus E^* and the interparticle friction coefficient μ for each VCI ratio are summarized in Table 2. The figures show typical strain-softening responses for the three investigated VCI ratios, which are reasonably captured by the model. The maximum shear stress values were found to slightly decrease with the increase in VCI ratio. For the investigated range of confining stress, the shear stress–strain model reasonably predicted the shear response of the fouled ballast assembly. In addition, the adopted approach used to model the different VCI levels by reducing the effective modulus E^* and the interparticle friction coefficient μ seem to be efficient and save about 40% of the computational time.

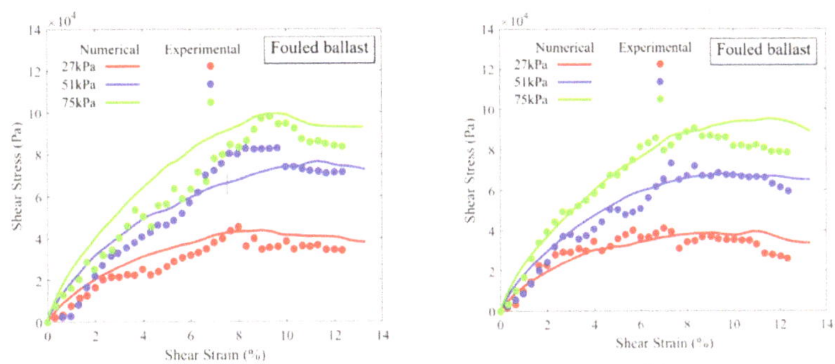

(**a**) (**b**)

Figure 9. *Cont.*

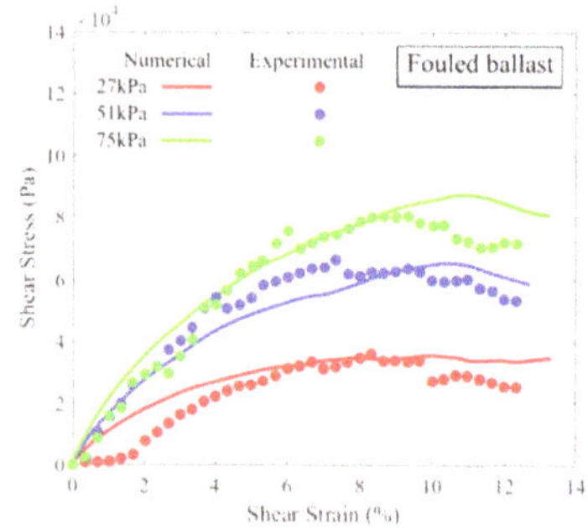

(c)

Figure 9. Shear response of fouled ballast for (**a**) VCI = 20%; (**b**) VCI = 40%; (**c**) VCI = 90%.

Table 2. Micro-mechanical parameters for fouled ballast assembly.

Parameters	Type of Contact	VCI		
		20%	40%	95%
Effective modulus, E^* (Pa)	PPC [1] PWC [1]	5.5×10^6 1.1×10^7	5.25×10^6 1.05×10^7	5.0×10^6 1.0×10^7
Interparticle friction coefficient, μ	PPC [1] PWC [1]	3.85 0.0	3.55 0.0	2.05 0.0

[1] PPC—particle to particle contact; PWC—particle to wall contact; k^* set as 1.0.

8. Estimating the Micro-Properties and Shear Strength of Fouled Ballast

Empirical correlations between the VCI and the micro-properties of the ballast assembly E^* and μ can be obtained by plotting the relationships depicted in Figure 10a.

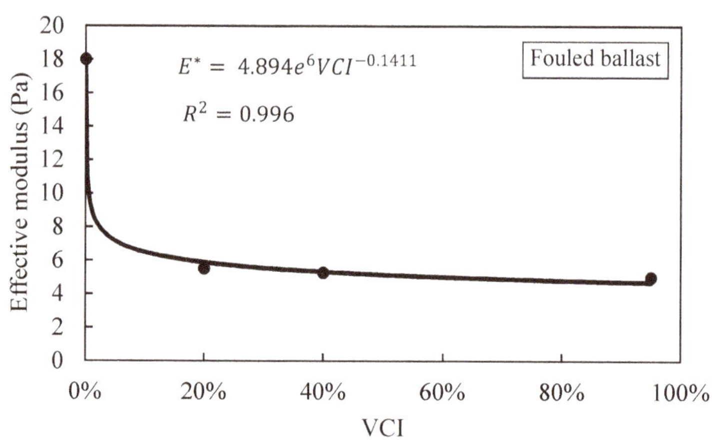

(**a**)

Figure 10. *Cont.*

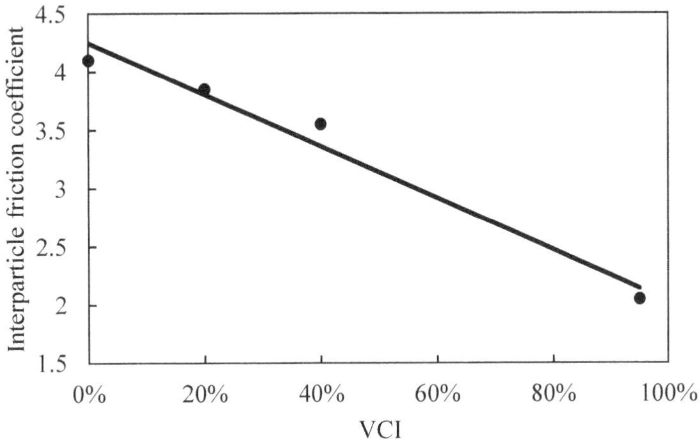

(b)

Figure 10. Changes in the micro-properties of the fouled ballast with the increase in VCI ratio: (**a**) change in effective modulus; (**b**) change in inter-particle friction coefficient.

The relationship between VCI and effective modulus is approximated using a power fit described by Equation (11). The effective modulus rapidly decreased from 18 Pa to about 5.5 Pa with the increase in VCI from 0% (fresh ballast) to about 20%. Little change in effective modulus was found for VCI of more than 20%.

$$E^* = 4.894e^6 \text{VCI}^{-0.1411} \tag{11}$$

The relationship between VCI and interparticle friction coefficient is characterized by a linear decrease with the increase in VCI, as shown in Figure 10b. The relationship can be described by Equation (12) below.

$$\mu = -2.218\text{VCI} + 4.247 \tag{12}$$

These relationships are useful for estimating the micro-parameters E^* and μ of a fouled ballast for a given VCI ratio. To validate these two expressions, a series of discrete element analyses was performed and used to calculate E^*, μ values for a given VCI ratio and compare the results with experimental data. The results are presented in Figure 11 for a range of applied normal stresses. For VCI = 70%, the estimated E^*, μ using the above expressions are found to be 5.15 MPa and 2.7, respectively. By inspecting the 45° line in Figure 11, the calculated strength was found to be in good agreement with the experimental data, which confirms the validity of the above expressions under direct shear conditions.

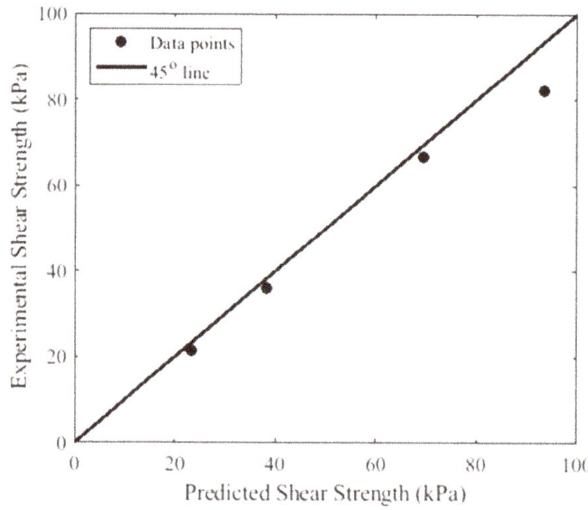

Figure 11. Comparison between measured and calculated shear strength result for VCI 70%.

Figure 12 shows the shear strength envelopes generated using the power-fitting expressions for different VCI ratios. In addition, both the experimental and numerical data are also presented. The power equations used to calculate the shear strength in each case are given below:

$$\tau_f = 4.427{\sigma_n}^{0.7692} \quad \text{Fresh ballast} \tag{13}$$

$$\tau_f = 2.762{\sigma_n}^{0.8429} \quad \text{VCI} = 20\% \tag{14}$$

$$\tau_f = 2.335{\sigma_n}^{0.8592} \quad \text{VCI} = 40\% \tag{15}$$

$$\tau_f = 2.304{\sigma_n}^{0.8540} \quad \text{VCI} = 70\% \tag{16}$$

$$\tau_f = 2.283{\sigma_n}^{0.8471} \quad \text{VCI} = 95\% \tag{17}$$

where the shear and normal stresses, τ_f and σ_n, in the above equations are expressed in kPa.

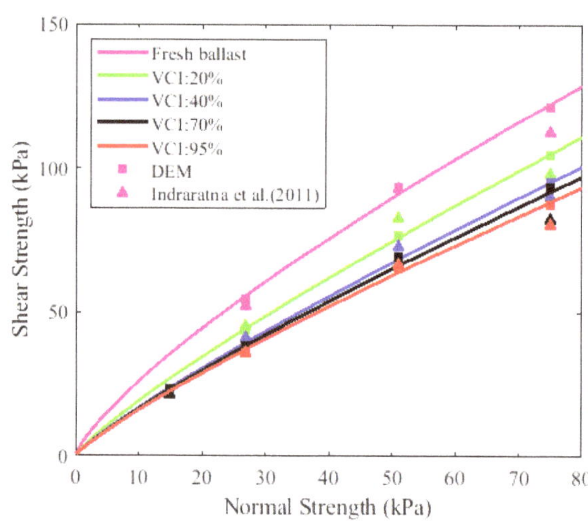

Figure 12. Normal stress versus shear strength results for different VCI ratios.

These envelopes suggest that the ballast shear strength is proportional to the applied normal stress and inversely proportional to the VCI ratio. For normal stress of 51 kPa, increasing VCI from 0% (fresh ballast) to 20% resulted in a significant decrease in the shear strength from about 90 kPa to about 75 kPa. An additional increase in VCI from 20% to 40% resulted in a moderate change in shear strength from 75 kPa to about 65 kPa. A further increase in VCI to more than 40% did not cause significant change in the shear strength of the fouled ballast.

The normalized shear strength of the fouled ballast can be related to VCI ratios and is determined using the hyperbolic relationship below [11]:

$$\frac{(\tau_p)_{Fouled\ ballast}}{\sigma_n} = \frac{(\tau_p)_{Fresh\ ballast}}{\sigma_n} - \frac{\mathrm{VCI}}{a \times \mathrm{VCI} + b} \tag{18}$$

where $(\tau_p)_{Fouled\ ballast}$ is the shear strength of fouled ballast sample, $(\tau_p)_{Fresh\ ballast}$ is the shear strength of fresh ballast sample, and a and b are hyperbolic constants that depend on the normal stress level. The linear relationship between the reduction in shear strength and VCI ratio can be obtained by rearranging Equation (18) as follows:

$$\frac{\mathrm{VCI}\cdot\sigma_n}{(\tau_p)_{Fresh\ ballast} - (\tau_p)_{Fouled\ ballast}} = a \times \mathrm{VCI} + b \tag{19}$$

The hyperbolic constants a and b are determined using linear regression, as illustrated in Figure 13, which are based on Equation (19). This expression is useful in estimating the shear strength reduction for a given VCI ratio.

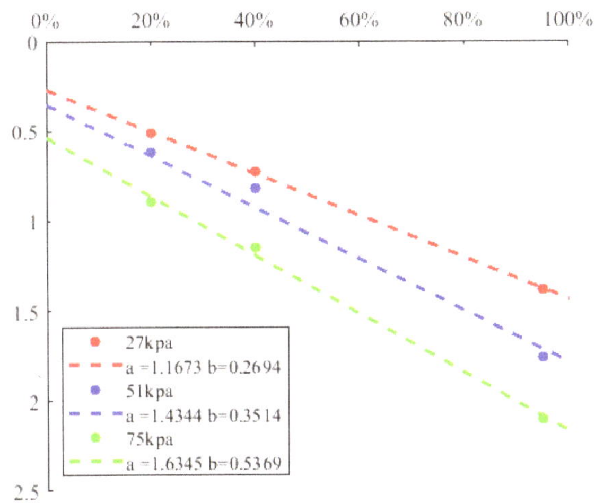

Figure 13. Hyperbolic constants a and b used to calculate the reduction in shear strength due to particle VCI.

9. Modeling the Effects of Ballast Degradation on Shear Strength

Ballast degradation generally causes the initial particle size distribution (PSD) to gradually shift to a degraded PSD with a wider range of particle sizes. The degree of ballast degradation is usually measured using the breakage (B) index [36], which is defined as the ratio between (i) the area between the fresh and degraded PSDs; and (ii) the area of the cumulative PSD. To simulate ballast degradation at different fine levels, the particle sizes are incrementally reduced, as summarized in Table 3. "It should be noted that this is an entirely different method from the one previously presented to simulate ballast fouling caused by foreign material injection whereby the microparameter properties were gradually decreased to simulate fouling. Therefore, in this section, a constant set of microparameters representing fresh ballast deemed sufficient." The ballast samples are sheared under normal stress of 51 kPa, and the corresponding shear strengths are determined.

Table 3. Particle size distributions of the 12 gradations.

Gradation	B (%)	C_u	D_M	D_{60}	D_{50}	D_{30}	D_{10}	D_m	Porosity	Co-Ordination No.
1	0.00	1.16	37.50	34.85	34.19	32.83	30.00	27.00	0.3872	3.3292
2	2.82	1.20	37.50	33.68	32.72	30.83	28.00	20.00	0.3780	3.5312
3	5.85	1.37	37.50	32.79	31.62	28.92	24.00	18.00	0.3624	3.2272
4	8.12	1.45	37.50	31.91	30.51	27.50	22.00	16.00	0.3585	3.0328
5	10.01	1.57	37.50	31.32	29.78	26.42	20.00	14.50	0.3600	2.9357
6	11.79	1.66	37.50	30.74	29.04	25.42	18.50	13.00	0.3472	2.7009
7	14.27	1.93	37.50	30.42	28.65	23.67	15.75	12.50	0.3326	2.8360
8	16.55	2.10	37.50	30.08	28.23	22.27	14.31	10.00	0.3498	2.1972
9	18.97	2.37	37.50	29.64	27.67	20.97	12.50	9.00	0.3220	2.1343
10	22.46	2.63	37.50	28.98	26.85	17.79	11.00	8.50	0.3131	2.0202
11	26.37	2.91	37.50	28.30	26.00	16.94	9.71	4.50	0.2897	1.7487
12	28.72	3.12	37.50	28.11	25.36	15.58	9.00	4.00	0.3050	1.5823

Note: B: breakage index (%); C_u: coefficient of uniformity, determined by: $C_u = D_{60}/D_{10}$. D_M: maximum ballast size used in this study; D_m: minimum ballast size used in this study; D_{10}, D_{30}, D_{50}, D_{60}: diameters in millimeters at which 10%, 30%, 50%, and 60% by weight of ballast passes through the sieve.

The shear strength results corresponding to different ballast degradation levels are plotted in Figure 14a, along with the best fit line. The shear strength is found to gradually decrease from 140 kPa to about 80 kPa as the breakage index increased from 0% to 20%. No significant change in shear strength occurs as a result of further increase in the breakage index. Figure 14b shows the change in co-ordination number with the increase in the breakage index. The co-ordination number was found to decrease to about half of its value when the breakage index reached 30%. Similarly, the porosity of the ballast assembly (seen in Figure 14c) was found to decrease from about 0.39 to about 0.29 as the breakage index increased from 0% to about 30%.

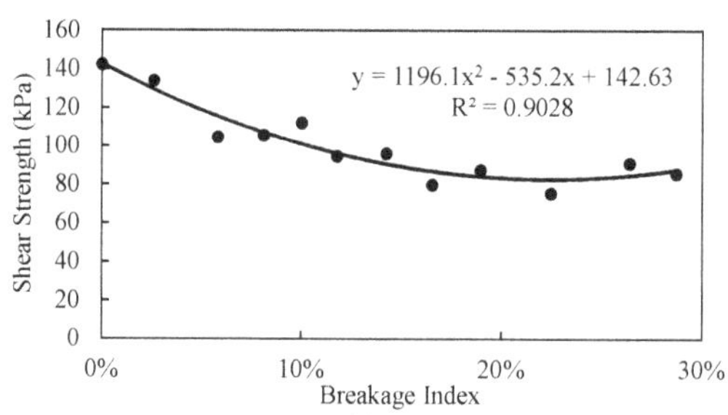

(a)

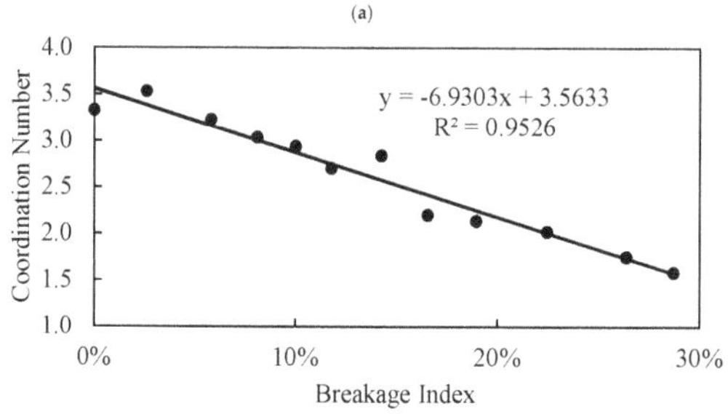

(b)

Figure 14. *Cont.*

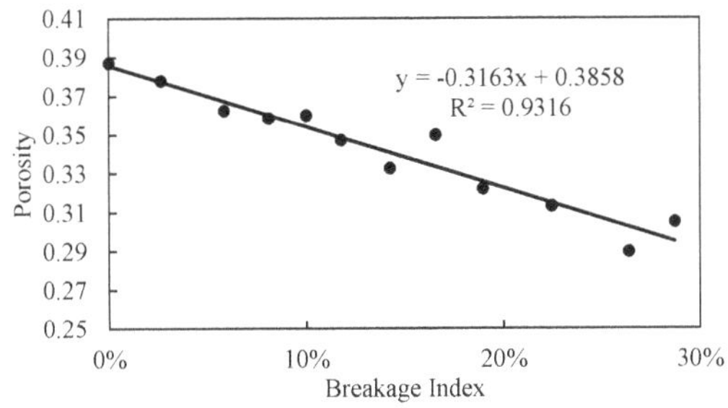

(c)

Figure 14. Relationships between breakage index and: (**a**) shear strength; (**b**) co-ordination number; (**c**) porosity.

10. Conclusions

In this study, the shear stress–strain behavior of both fresh and fouled ballast subjected to direct shear loading is investigated using the discrete element method. Simplified models that account for the intrusion of fine material and particle degradation are proposed. Empirical relationships between the micro-mechanical parameters and the void contamination index (VCI) are presented to help calculate the parameters needed for the analysis. The results are validated by comparing the numerical results with published experimental data. Shear strength envelopes for different levels of VCI are generated to examine the relation between the shear strength of the fouled ballast assembly and the VCI values. Increasing the range of particle size distribution is adopted as a tool to simulate the effect of ballast degradation on the shear strength of the assembly.

The analysis shows that reducing the effective modulus and interparticle friction coefficient can be used to numerically represent the condition where fine particles intrude fresh ballast. The modeling results were found to produce shear stress–strain responses that are consistent with the experimental data.

Finally, it is evident from the reported results that both infiltration of fines and particle degradation can negatively impact the shear strength of the ballast. That said, particle degradation alone is found to have significant effects on the strength of the ballast assembly that are comparable to those of fine particle infiltration.

It should be noted that, while the findings of this study are specific to the particle shapes and sizes investigated, the methodology presented herein to examine the influence of ballast fouling should be equally applicable to other ballast particle configurations.

Author Contributions: L.W.: Conceptualization, Software, Data curation, Writing—original draft, Visualization. M.M.: Methodology, Supervision. H.S.M.: Supervision, Writing—review & editing. All authors have read and agreed to the published version of the manuscript.

Funding: This work was financially supported by scholarships from Henan Polytechnic University, China and McGill University, Canada. The authors are grateful for their support.

Institutional Review Board Statement: Not applicable.

Data Availability Statement: Data is contained within the article.

Acknowledgments: This work was financially supported by scholarships from Henan Polytechnic University, China and McGill University, Canada. The authors would like to express their appreciation for those who supported this research project.

Conflicts of Interest: The authors declare no conflict of interest.

References

1. Selig, E.T.; Waters, J.M. *Track Geotechnology and Substructure Management*; Thomas Telford: London, UK, 1994; pp. 8.18–8.32.
2. Dombrow, W.; Huang, H.; Tutumluer, E. Comparison of coal dust fouled railroad ballast behavior-granite vs. limestone. In Proceedings of the 8th International Conference on the Bearing Capacity of Roads, Railways and Airfields, BCR2A'09, Champaign, IL, USA, 29 June–2 July 2009; pp. 1349–1357.
3. Indraratna, B.; Ngo, T.; Rujikiatkamjorn, C.; Vinod, J.S. Behavior of fresh and fouled railway ballast subjected to direct shear testing: Discrete element simulation. *Int. J. Géoméch.* **2014**, *14*, 34–44. [CrossRef]
4. Huang, H.; Tutumluer, E. Discrete element modeling for fouled railroad ballast. *Constr. Build. Mater.* **2011**, *25*, 3306–3312. [CrossRef]
5. Cundall, P.A.; Strack, O.D. A discrete numerical model for granular assemblies. *Geotechnique* **1979**, *29*, 47–65. [CrossRef]
6. Gao, G.; Meguid, M. Effect of particle shape on the response of geogrid-reinforced systems: Insights from 3D discrete element analysis. *Geotext. Geomembr.* **2018**, *46*, 685–698. [CrossRef]
7. Hossain, Z.; Indraratna, B.; Darve, F.; Thakur, P.K. DEM analysis of angular ballast breakage under cyclic loading. *Géoméch. Geoengin.* **2007**, *2*, 175–181. [CrossRef]
8. McDowell, G.; Harireche, O.; Konietzky, H.; Brown, S.; Thom, N. Discrete element modelling of geogrid-reinforced aggregates. *Proc. Inst. Civ. Eng. Geotech. Eng.* **2006**, *159*, 35–48. [CrossRef]
9. O'Sullivan, C.; Cui, L.; O'Neill, S.C. Discrete element analysis of the response of granular materials during cyclic loading. *Soils Found.* **2008**, *48*, 511–530. [CrossRef]
10. Stahl, M.; Konietzky, H. Discrete element simulation of ballast and gravel under special consideration of grain-shape, grain-size and relative density. *Granul. Matter.* **2011**, *13*, 417–428. [CrossRef]
11. Xu, Y.; Gao, L.; Zhang, Y.-r.; Yin, H.; Cai, X.-p. Discrete element method analysis of lateral resistance of fouled ballast bed. *J. Cent. South Univ.* **2016**, *23*, 2373–2381. [CrossRef]
12. Yanli, Y. Experimental study on design parameters of longitudinal and lateral resistance of ballast bed for III-type concrete sleeper. *J. Railw. Eng. Soc.* **2010**, *10*, 49–51.
13. Al-Qadi, I.L.; Xie, W.; Roberts, R. Scattering analysis of ground-penetrating radar data to quantify railroad ballast contamination. *NDT E Int.* **2008**, *41*, 441–447. [CrossRef]
14. Indraratna, B.; Thakur, P.K.; Vinod, J.S. Experimental and numerical study of railway ballast behavior under cyclic loading. *Int. J. Géoméch.* **2010**, *10*, 136–144. [CrossRef]
15. Danesh, A.; Palassi, M.; Mirghasemi, A.A. Evaluating the influence of ballast degradation on its shear behaviour. *Int. J. Rail Transp.* **2018**, *6*, 145–162. [CrossRef]
16. Indraratna, B.; Lackenby, J.; Christie, D. Effect of confining pressure on the degradation of ballast under cyclic loading. *Geotechnique* **2005**, *55*, 325–328. [CrossRef]
17. Qian, Y.; Tutumluer, E.; Hashash, Y.; Ghaboussi, J. Effects of ballast degradation on permanent deformation behavior from large-scale triaxial tests. In Proceedings of the 2014 Joint Rail Conference, Colorado Springs, CO, USA, 2–4 April 2014.
18. Itasca. *Particle Flow Code in Three Dimensional (PFC3D)*; Itasca Consulting Group, Inc.: Minneapolis, MN, USA, 2019.
19. Guo, Y.; Ji, Y.; Zhou, Q.; Markine, V.; Jing, G. Discrete element modelling of rubber-protected ballast performance subjected to direct shear test and cyclic loading. *Sustainability* **2020**, *12*, 2836. [CrossRef]
20. Bian, X.; Li, W.; Qian, Y.; Tutumluer, E. Micromechanical particle interactions in railway ballast through DEM simulations of direct shear tests. *Int. J. Géoméch.* **2019**, *19*, 04019031. [CrossRef]
21. Danesh, A.; Mirghasemi, A.A.; Palassi, M. Evaluation of particle shape on direct shear mechanical behavior of ballast assembly using discrete element method (DEM). *Transp. Geotech.* **2020**, *23*, 100357. [CrossRef]
22. Kozicki, J.; Tejchman, J.; Mroz, Z. Effect of grain roughness on strength, volume changes, elastic and dissipated energies during quasi-static homogeneous triaxial compression using DEM. *Granul. Matter* **2012**, *14*, 457–468. [CrossRef]
23. Ferellec, J.-F.; McDowell, G.R. A method to model realistic particle shape and inertia in DEM. *Granul. Matter.* **2010**, *12*, 459–467. [CrossRef]
24. Taghavi, R. Automatic clump generation based on mid-surface. In Proceedings of the 2nd International FLAC/DEM Symposium, Melbourne, Australia, 14–16 February 2011; pp. 791–797.
25. Gao, G.; Meguid, M. On the role of sphericity of falling rock clusters—Insights from experimental and numerical investigations. *Landslides* **2018**, *15*, 219–232. [CrossRef]
26. Zhang, X.; Zhao, C.; Zhai, W. Dynamic behavior analysis of high-speed railway ballast under moving vehicle loads using discrete element method. *Int. J. Géoméch.* **2017**, *17*, 04016157. [CrossRef]
27. Wang, X.; Li, J. Simulation of triaxial response of granular materials by modified DEM. *Sci. China Ser. G Phys. Mech. Astron.* **2014**, *57*, 2297–2308. [CrossRef]
28. Djordjevic, N. Influence of charge size distribution on net-power draw of tumbling mill based on DEM modelling. *Miner. Eng.* **2005**, *18*, 375–378. [CrossRef]
29. O'Sullivan, C.; Bray, J.D. Selecting a suitable time step for discrete element simulations that use the central difference time integration scheme. *Eng. Comput.* **2004**, *21*, 278–303. [CrossRef]
30. Indraratna, B.; Ngo, T.; Rujikiatkamjorn, C. Behavior of geogrid-reinforced ballast under various levels of fouling. *Geotext. Geomembr.* **2011**, *29*, 313–322. [CrossRef]

31. Jiang, M.; Konrad, J.; Leroueil, S. An efficient technique for generating homogeneous specimens for DEM studies. *Comput. Geotech.* **2003**, *30*, 579–597. [CrossRef]
32. Liu, S.H. Simulating a direct shear box test by DEM. *Can. Geotech. J.* **2006**, *43*, 155–168. [CrossRef]
33. Ngo, N.T.; Indraratna, B.; Rujikiatkamjorn, C. Micromechanics-based investigation of fouled ballast using large-scale triaxial tests and discrete element modeling. *J. Geotech. Geoenviron. Eng.* **2017**, *143*, 4016089. [CrossRef]
34. Mishra, B.; Murty, C. On the determination of contact parameters for realistic DEM simulations of ball mills. *Powder Technol.* **2001**, *115*, 290–297. [CrossRef]
35. Tennakoon, N.; Indraratna, B.; Rujikiatkamjorn, C.; Nimbalkar, S.; Neville, T. The role of ballast-fouling characteristics on the drainage capacity of rail substructure. *Geotech. Test. J.* **2012**, *35*, 629–640. [CrossRef]
36. Einav, I. Breakage mechanics—Part I: Theory. *J. Mech. Phys. Solids* **2007**, *55*, 1274–1297. [CrossRef]

processes

MDPI

Article

The Long-Term Mitigating Effect of Horizontal Ground-Source Heat Exchangers on Permafrost Thaw Settlement

Amir Fatollahzadeh Gheisari [1], Pooneh Maghoul [1,2,*], Hartmut M. Holländer [1], Rob Kenyon [3], Rob Sinclair [3] and Maryam Saaly [3]

1 Department of Civil Engineering, University of Manitoba, Winnipeg, MB R3T 5V6, Canada; fatollaa@myumanitoba.ca (A.F.G.); Hartmut.Hollaender@umanitoba.ca (H.M.H.)
2 Department of Civil, Geological and Mining Engineering (CGM), Polytechnique Montréal, Montréal, QC H3C 3A7, Canada
3 KGS Group, Winnipeg, MB R3T 5P4, Canada; rkenyon@kgsgroup.com (R.K.); rsinclair@kgsgroup.com (R.S.); MSaaly@kgsgroup.com (M.S.)
* Correspondence: Pooneh.Maghoul@polymtl.ca or Pooneh.Maghoul@umanitoba.ca

Abstract: This study investigated the long-term effect of horizontal Ground-Source Heat Exchangers (GSHEs) on mitigating permafrost thaw settlement. In the conceptual system, a fan coil was used to chill the recirculating fluid in the linear High-Density Polyethylene (HDPE) ground loop system. A fully coupled thermo-hydro-mechanical finite element framework was employed to analyze multiphysics processes involved in the thaw settlement phenomenon. To investigate the sustainability of the system, a period of 50 years was simulated. Two operational modes were defined: one without and the other with HDPE. Different heat carrier velocities and inlet temperatures, and heat exchanger depths were examined to explore their effects on the thaw settlement rate. It was concluded that the proposed system can effectively alleviate the predicted permafrost thaw settlement over the study period. Moreover, the heat carrier temperature was found to have a prominent impact on the thaw settlement rate amongst other parameters.

Keywords: geothermal heat exchangers; permafrost; thaw settlement; sustainability

Citation: Fatollahzadeh Gheisari, A.; Maghoul, P.; Holländer, H.M.; Kenyon, R.; Sinclair, R.; Saaly, M. The Long-Term Mitigating Effect of Horizontal Ground-Source Heat Exchangers on Permafrost Thaw Settlement. *Processes* **2021**, *9*, 1636. https://doi.org/10.3390/pr9091636

Received: 23 June 2021
Accepted: 28 August 2021
Published: 10 September 2021

1. Introduction

Permafrost is ground (solid, sediment, rock) that remains frozen for at least two consecutive years [1]. Over the last 30 years, the mean annual temperature in the Arctic has risen at a rate of 0.54 °C per decade, a triple increase compared to the global average temperature [2]. This rapid increase in air temperature is leading the permafrost to thaw at an unprecedented rate. Permafrost thaw settlement as a consequence of climate change causes considerable damages to the northern infrastructures. Therefore, the assessment of permafrost thaw settlement is of paramount importance in investigating the resiliency of northern infrastructures and selecting the most sustainable mitigation strategies.

Permafrost thaw settlement involves heat and mass transfer, pore water pressure dissipation, deformation, and strength evolution. It is indeed a complex multiphysics process involving thermal (T), hydraulic (H), and mechanical (M) analyses, which is hereafter referred to as thermo-hydro-mechanical (THM) analysis [3]. The first complete thaw consolidation theory was introduced and developed by Morgenstern and Nixon [4], where they formulated the settlement based on the small-strain consolidation theory. However, the proposed model cannot be applied to an ice-rich permafrost where the soil undergoes large strain settlements [5]. Foriero and Ladanyi [6] overcame this shortcoming by developing a large-strain consolidation theory which allows finite strains and considers the variation of hydraulic conductivity and compressibility during the consolidation process.

Besides the effort in modelling the permafrost thaw settlement process, numerous climate change adaptation strategies and mitigation solutions have been developed in the last few decades. These techniques are generally classified into three main groups: (i) limiting the heat transfer from the atmosphere into permafrost (e.g., providing shades, vegetation, etc.), (ii) extracting heat to cool down permafrost, and (iii) reinforcing the infrastructure against settlement [7]. Among a wide range of permafrost stabilization methods, thermosiphons have been most widely used due to their simple structure, high efficiency, and low fabrication costs [8]. This technology comes under the second category that aims to extract heat out the system and lessen the temperature of degrading permafrost. Chen et al. [9] studied the effect of thermosiphons on reducing thaw settlement of an embankment built on a sandy permafrost. They found that the thermosiphon performance in mitigating thaw settlement highly depends on the mean annual air temperature.

Ground-Source Heat Exchangers (GSHEs) with a similar function of heat extraction to thermosiphons have been traditionally used for heat energy supply and the heating/cooling demand of buildings. However, their effects on permafrost stabilization have rarely been studied. Fontaine et al. [10] presented a new analytical model to study the effect of a spiral-shaped horizontal ground heat exchanger on permafrost stabilization. However, their model overestimated the summer ground temperature and thaw depth. Only a short-term (5 years) analysis was carried out and the long-term performance of the GSHE system was not investigated.

In this paper, the effectiveness of a horizontal GSHE system as a long-term and sustainable solution for permafrost stabilization was evaluated. The study site was located in a wastewater lagoon facility in Ross River, Yukon, Canada, where the huge heat leakage from the wastewater was expected to induce permafrost degradation and significant thaw settlement. The proposed system consisted of a horizontal closed-loop system that circulates an anti-freeze coolant to cool down the ground and mitigate the thawing permafrost beneath the base of the lagoon.

2. Materials and Methods

2.1. Study Area

This research was carried out on a freeze-back system at the Ross River's wastewater lagoon in a permafrost rich area, Yukon, Canada (Figure 1a). The Ross River community (61°32′, 132°35′) is located within the zone of extensive discontinuous and mostly warm permafrost. The present wastewater lagoon for the community is located to the west of the community, approximately 700 m south of the Pelly River. The lagoon system was constructed in fall 2017 and consisted of two cells—a primary cell and a secondary cell with the base dimensions of 75 m × 15 m and 75 m × 35 m, respectively, to provide continuous treatment efficiency and ensure no raw, untreated influent discharges. The lagoon base was placed 3 m below the original ground level through excavation, and the maximum lagoon depth was 3 m, as well. The wastewater lagoon system was designed for the wastewater production per capita of 110 L/day/person, the annual wastewater production of 17,000 m^3/year, and 10-month storage volume of 14,000 m^3/year [11]. The 3 m depth of excavation for the construction of the wastewater lagoon exposed the underlying permafrost to a downward heat flux from the lagoon base. The lagoon heat regime comes from the solar radiation into the wastewater in the lagoon, leading the lagoon to maintain a temperature above the freezing temperature all year round. Lagoon operation will also impede seasonal freeze back within the active layer in the winter.

A geotechnical investigation [12] reported that the site stratigraphy consists of approximately 2 to 3 m of sandy gravel overlying a 3.2 to 4 m thick clayey silt layer. Ice lensing with a maximum thickness of 20 mm had been detected within the clayey silt. According to the field observation from the two boreholes of BH18-05 and BH18-06 (Figure 1b), the average permafrost layer started at 8 m below the original ground level and 5 m below the lagoon base, extending to 22 m in depth (Figure 1c). The investigation also reported that the water content was 8% and 30% in the sandy gravel and clayey silt layer, respectively.

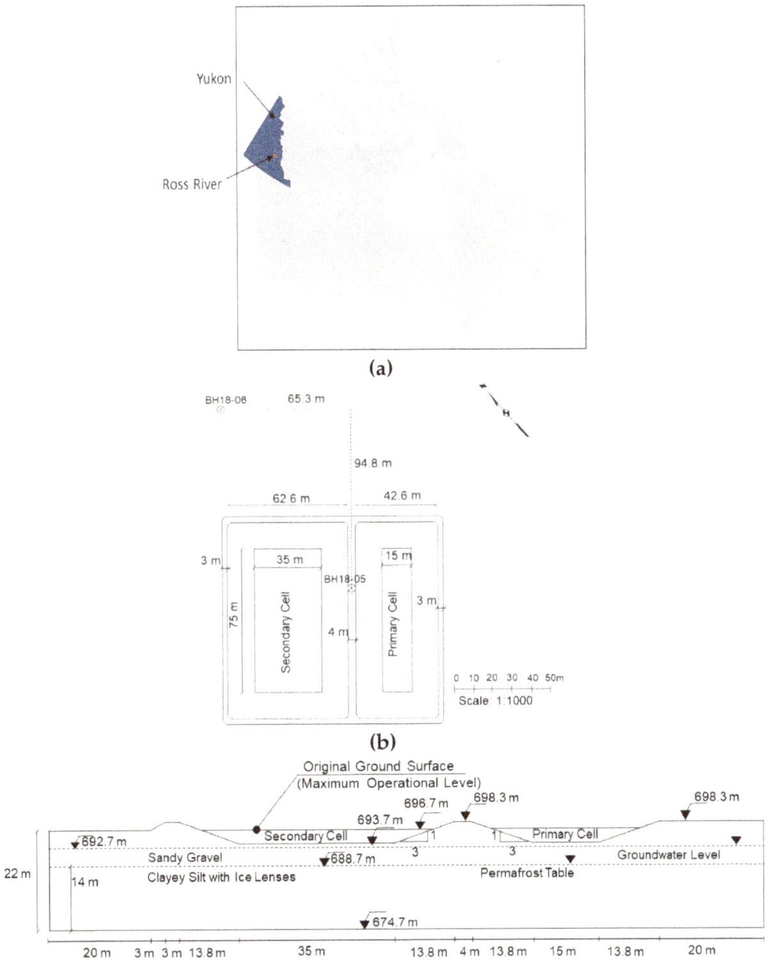

Figure 1. (**a**) Ross River, Yukon, (**b**) site plan of the Ross River wastewater lagoon and the borehole locations, and (**c**) site stratigraphy, elevation level from average sea level and permafrost table [8].

2.2. Specification and Conceptual Model of the Proposed GSHE System

The studied GSHE system consisted of horizontal HDPE pipes, each 50 mm in diameter and 45 and 70 m long, which were horizontally placed 2 m away and 1.5 m below the lagoon's base (4.5 m from the original ground level) under the primary and secondary cells, respectively (Figure 2a). Using the symmetry in the layout of the HDPE pipes, only one repeated block of the soil domain with five embedded heat exchangers (Figure 2b) was considered to decrease the computational time [13]. The coolant fluid was a 30% ethanol-water blend with a velocity and inlet temperature of 0.4 m/s and −5 °C, respectively. Different pipe burial depths, heat exchanger velocities, and temperatures were modelled to understand their effects on the efficiency of system. For this purpose, 3 different pipe depths (1.5, 2.5, and 3.5 m), 3 heat exchanger inlet velocities (0.2, 0.4, and 0.6 m/s) and 4 heat exchanger inlet temperatures (−1, −5, −10, and −15 °C) were tested. Due to the considerable lagoon heat flux and significant permafrost thaw rate, the GSHE system operated year round. Although the extracted geothermal heat can be pumped to nearby buildings, the heat recovery from the Ross River Lagoon was not used for residential or industrial buildings as they were not located within close distance to the lagoon. According to the field observation from BH18-05, the initial soil temperature (2019) was rather stable in permafrost at −0.2 °C and gradually increased in the shallower areas, reaching 8 °C at the ground surface.

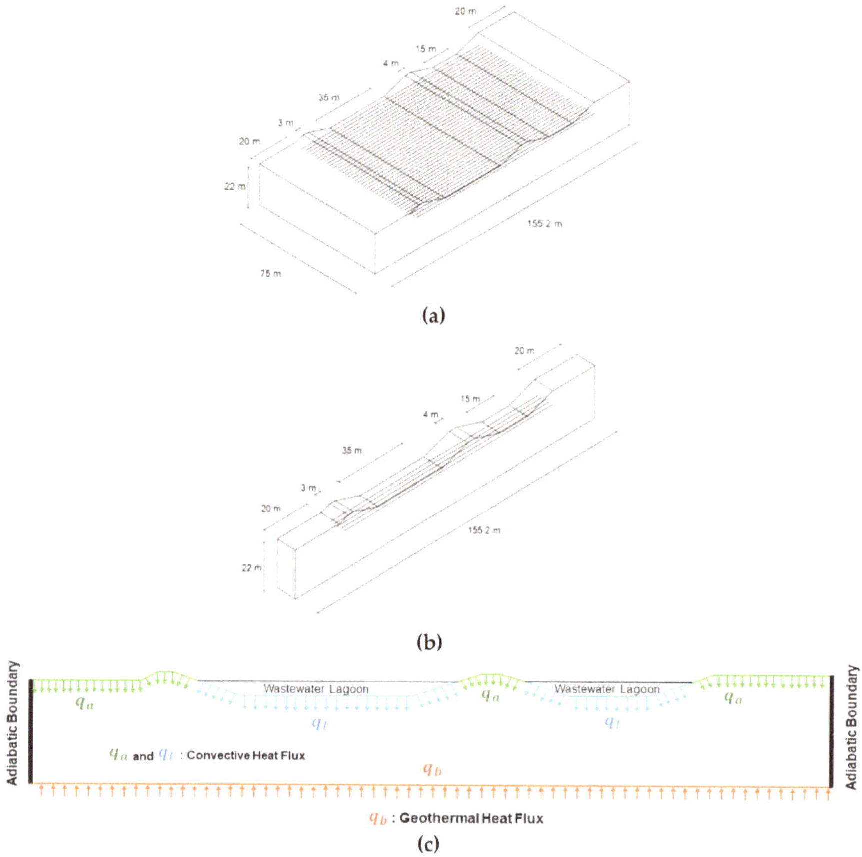

Figure 2. (**a**) The whole domain, (**b**) the repeated block of porous matrix and the embedded GSHEs, and (**c**) model thermal boundaries.

The top boundary of the domain was in contact with the surrounding air and the lagoon base. Because of the exposure to the ambient air, a forced convective heat flux (q_a) was applied to the parts of the top boundary exposed to the air. The boundary condition at the interface between the lagoon base and the soil was also defined by a convective heat flux (q_l) [14]. The lagoon temperature varies with a rather sinusoidal pattern, starting from 4 °C in mid-winter to 20 °C in mid-summer [11].

In addition, a constant upward heat flux (q_b) was assigned to the bottom of the domain to represent the geothermal heat flux that exists at the site. The sides of the domain were defined as adiabatic boundaries as shown in Figure 2c.

Mechanical boundary condition consisted of an overburden pressure equivalent to the weight of the wastewater under the lagoon's full operational mode, applied on the lagoon's bottoms and walls. To model the elastoplastic soil behavior, the initial stress was defined as the sum of the overburden pressure and the soil total stress (multiplied by lateral pressure coefficient at rest (k_0) for the initial stress in the X-direction). The bottom of the domain was set to the fixed support and the domain's sides were defined as roller support with a degree of freedom in the vertical direction. Darcy's law was utilized to model the hydraulic boundaries and initial conditions. The bottom and sides of the domain were defined as no-flow boundaries and an initial hydrostatic pressure was assigned to the entire domain.

A series of 2D multiphysics finite element (FE) simulations of a GSHE system, including thermal analyses of heat exchangers and the THM process in foundation permafrost, was conducted in COMSOL Multiphysics v5.4 (COMSOL INC, Stockholm, Södermanland and Uppland, Sweden). Thaw settlement and possible frost heave (due to the seasonal freezing) were calculated based on a THM model which accounts for the evolution of strength, porosity, hydraulic conductivity, and pore water pressure during the freeze–thaw cycles. To eliminate unwanted effects of boundaries on the analysis and avoid erroneous results, different distances from the outermost dike edge, including 10, 20, and 30 m, and from the permafrost bottom level, including 0, 10 m, and 20 m, were examined (Figure 3). The thermal analysis showed that the model with a 20 m horizontal distance from the outermost dike edge and 0 m vertical distance from the permafrost bottom level resulted in consistent results in the permafrost temperature distribution even if larger distances were selected. The study period was 50 years, starting from 2019 and ending in 2069, with 15-day time intervals. To study the remedial effect of the proposed GSHE system in reducing thaw settlement, two scenarios were introduced: one without GHSEs and the other with GSHEs, and the results were compared.

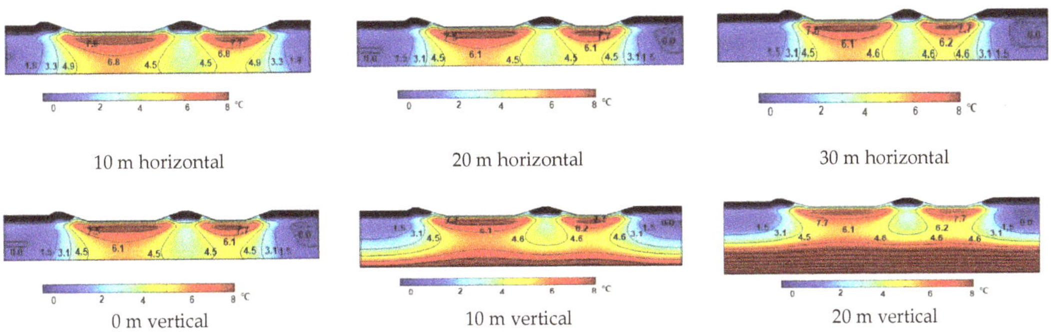

Figure 3. Model different dimensions from the outermost dike edges and the permafrost bottom level.

2.3. Governing Equations

The thaw settlement phenomenon consists of different physics interacting with each other, including: (1) the porous nature of soil and its temperature-dependent physical and mechanical properties, (2) heat transfer mechanism in porous soil, (3) mechanical behavior of the thawing soil, and (4) heat and mass transfer mechanism of the pore fluid. To study the effectiveness of the GSHE on limiting the thaw settlement, the pipe flow and heat exchange (through its walls) must be coupled with the above-mentioned physics. Described in the following subsection.

2.3.1. Heat Transfer within Degrading Permafrost during Freeze–Thaw Cycle

Conductive Heat Transfer Considering Pore Water Phase Change

The transient heat conduction in the soil by considering the latent heat of pore water can be described as [13]:

$$C_{app}\frac{\partial T}{\partial t} + \nabla \cdot q = Q \tag{1}$$

where T (K) is the soil temperature, Q (W/m^3) is the heat source/sink, $\nabla = \left(\frac{\partial}{\partial x}, \frac{\partial}{\partial y}, \frac{\partial}{\partial z}\right)$ is the gradient operator, t (s) is time, and q (W/m^2) is the conductive heat flux defined by Fourier's law:

$$q = \lambda \nabla T \tag{2}$$

where (W/(m. K)) is the thermal conductivity of the saturated frozen medium. The term C_app is defined as the apparent heat capacity (J/(kg. K)), expressed as:

$$C_{app} = \frac{1}{\rho}\left(\rho_{ph}C_{ph} - L_f\rho_i\frac{\partial\theta_i}{\partial t}\right) \tag{3}$$

where L_f (kJ/kg) is the latent heat per unit mass of water, ρ (m/kg^3) is the bulk density of the porous medium, ρ_i (m/kg^3) is the density of ice, and θ_i is the volumetric fraction of ice in pores. The term $\rho_{ph}C_{ph}$ (J/(m^3. K)) denotes the volumetric heat capacity of the soil mixture which can be calculated by the sum of the volumetric heat capacity of each constituent of the saturated freezing medium (solid skeleton, water, and ice) multiplied by its volumetric fraction (θ) as follows:

$$\rho_{ph}C_{ph} = \rho_s C_s \theta_s + \rho_w C_w \theta_w + \rho_i C_i \theta_i \tag{4}$$

where s, w, and i denote solid skeleton, water, and ice, respectively. Similarly, thermal conductivity (λ) and the bulk density (ρ) of the saturated frozen soil mixture can be defined as Equations (5) and (6), respectively:

$$\lambda = \rho_s\theta_s\lambda_s + \rho_w\theta_w\lambda_w + \rho_i\theta_i\lambda_i \tag{5}$$

$$\rho = \rho_s\theta_s + \rho_w\theta_w + \rho_i\theta_i \tag{6}$$

where ρ_s, ρ_w, and ρ_i and λ_s, λ_w, and λ_i are are the density and thermal conductivity of soil, water, and ice, respectively.

2.3.2. Mechanical Behavior of the Soil during Freeze–Thaw Cycles

Kinematic Formulation

The linearized form of Green–Lagrange strain tensor (ε_{ij}) for infinitesimal deformations is formulated as:

$$\varepsilon_{ij} = \frac{1}{2}(u_{i,j} + u_{j,i}) \tag{7}$$

in which u_i and u_j are the displacement vector in the *i* and *j* direction, respectively.

Constitutive Model

The effective stress is defined as the difference between the total stress (σ'_{ij}) and the pore water pressure (u):

$$\sigma'_{ij} = \sigma_{ij} - \alpha u \delta_{ij} \tag{8}$$

where α is the Biot coefficient.

The effective stress is averaged over the solid mineral particles and ice. The equation of equilibrium is formulated as:

$$\sigma_{ji,j} + F_i = 0 \tag{9}$$

where F_i is the body force.

The mass conservation principle is:

$$\frac{\partial\theta_w}{\partial t} + \frac{\rho_i}{\rho_w}\frac{\partial\theta_i}{\partial t} + \nabla q_w + \frac{\partial\varepsilon_v}{\partial t} = 0 \tag{10}$$

where ε_v is the volumetric strain and q_w is water flux governed by Darcy's law.

Pore ice ratio (e_{ip}) as a key parameter in describing the behavior of the frozen soil [3] is introduced as:

$$e_{ip} = \frac{V_i}{V_s} \tag{11}$$

in which V_i and V_s are the pore ice and soil skeleton volume, respectively. Moreover, to describe the volumetric changes of the soil, specific volume is defined as follows:

$$v = \frac{V}{V_s} = 1 + e \tag{12}$$

where, υ is the total volume of the soil and e is the void ratio. The mean effective stress (p') is defined as:

$$p' = \frac{1}{3}\sigma'_{kk} \quad k = 1,2,3 \tag{13}$$

where σ'_{kk} is the principal stresses of the effective stress tensor.

The changes in the void ratio (and thus specific volume) versus effective stress is non-linear over the range of stresses in practice [5,14]. Experimental studies demonstrated that this behavior can truly be captured by a semi-logarithm relationship, formulated as follows [5]:

$$v = v_0 - \lambda' \ln \frac{p'}{p^r} \tag{14}$$

where v_0 is the specific volume at the reference pressure p^r (MPa) and λ' is the slope of the normally consolidation line (NCL) for unfrozen soil. The elastic changes of the specific volume in the unloading-reloading line (URL) is:

$$dv^e = -\kappa \frac{dp'}{p'} \tag{15}$$

where k is the slope of URL.

The increase in the soil strength is a function of pore ice ratio (e_{ip}). The subsequent relationships for the slope of NCL and URL for frozen soil are:

$$\lambda'_f = \lambda' \exp(-\alpha_1 e_{ip}) \tag{16}$$

$$\kappa_f = \kappa \exp(-\alpha_2 e_{ip}) \tag{17}$$

where α_1 and α_2 are the soil constants [3]. Slopes λ' and k for the unfrozen soil and λ'_f and k_f for the frozen soil can be obtained from isotropic compression tests in a triaxial apparatus.

The pre-consolidation pressure increases during freezing process and reaches p_0^f for frozen soil which can be obtained from the following equation:

$$\frac{p_0^f}{p^r} = \exp\left(\frac{\beta e_{ip}}{\lambda'_f - \kappa_f}\right)\left(\frac{p_0}{p^r}\right)^{\left(\frac{\lambda' - \kappa_f}{\lambda'_f - \kappa_f}\right)} \tag{18}$$

where β is a function of p^r which can be obtained from isotropic compression test in a triaxial apparatus.

Non-Isothermal Heat Transfer from Pipe Flow

In order to couple the heat transfer in a solid with the heat transfer from the pipe flow, the following differential energy equation must be satisfied [15]:

$$\rho A C_{app} \frac{\partial T}{\partial t} + \rho_f A C_{pf} \overline{v} \nabla T = \Delta A \lambda_f \nabla T + \frac{\rho}{2d_h} |\overline{v}|^3 + Q_{wall} \tag{19}$$

where A is pipe cross section area (m^2), C_{pf} (J/kg. K) is the heat capacity of the fluid at constant pressure, $\overline{v}$ (m/s), T (K), and λ_f (W/(m·K)) are the fluid velocity, temperature, and thermal conductivity, and d_h (m) is the hydraulic diameter. The term Q_{wall} (W/m^3) is the external heat source/sink term through the pipe wall.

Convective Heat Flux

To model the seasonal temperature variation on the soil temperature, the below convective heat flux was assigned to the area of the top domain boundary with a contact to the ambient air:

$$\text{Heat Flux}: q_t(t) = h_{conv(a)}(T_a(t) - T_s(t)) \tag{20}$$

where $h_{conv(a)}$ (W/m^2°C) is the convection heat transfer coefficient for the domain area exposed to the ambient air. T_s (°C) is the ground surface temperature (calculated in each time step) and T_a (°C) is the air temperature at the ground surface. The boundary conditions at the interface between the lagoon base and the ground surface are also defined by a convective heat flux, formulated as follows:

$$\text{Heat Flux}: q_l(t) = h_{conv(l)}(T_l(t) - T_s(t)) \tag{21}$$

where T_l (°C) is the lagoon temperature and $h_{conv(l)}$ is the lagoon heat transfer coefficient. The wastewater lagoon bottom and walls are made of concrete. Since the value of the heat transfer coefficient for the concrete between wastewater and soil does not exist in the literature, in this study this value was quantified numerically using finite element analysis. The result of the stationary heat transfer analysis yielded 4.0 and 5.0 (W/m^2°C), for the heat transfer coefficient of the inclined sides and horizontal bottom boundary, respectively.

2.4. Numerical Modeling

2.4.1. Soil Properties

Since no geotechnical laboratory tests had been conducted to determine the in situ soil properties of the site, the average values for clay/sandy soils [16] were considered, as given in Table 1. Other parameters including water, ice, and coolant fluid thermal properties were taken from previous studies [3,16]. The required parameters for modelling the heat transfer in a porous matrix and non-isothermal pipe flow are given in Table 2. In addition, the associated parameters of the thaw settlement for clayey soil are as follows: λ' = 0.35, k = 0.07, p_0 = 650 MPa, α_1 = 0.4, α_2 = 1.8, p^r = 0.1 MPa, and β = 0.18 [3].

Table 1. Properties of the soils and fluids [16].

Material	Thermal Conductivity (W/(m. K))	Mass Heat Capacity (J/kg. K)	Density (kg/m^3)
Sandy Gravel	2.3	1255	2300
Clayey Silt	1.25	942	1900
Water	0.56	4188	1000
Ice	2.2	2093	917
Coolant	0.41	4250	955

Table 2. Properties of the soils and fluids [3,17].

Property	Value
Latent heat of water (L_f) (kJ/kg)	334
Phase change temperature (T_{pc}) (°C)	0
Upward heat flux (W/m^2)	0.075
HDPE pipe wall thickness (mm)	2
HDPE pipe wall thermal conductivity (W/(m. K))	0.46

2.4.2. Mesh Sensitivity Analysis

The triangular Lagrange finite elements were applied for the heat transfer, non-isothermal pipe models and solid mechanics [18]. The mesh consistency testing showed that the nodal distance associated with "extra fine" mesh in COMSOL Multiphysics resulted in consistent results in the maximum temperature and surface thaw settlement even if a smaller nodal distance was selected (Figure 4). Moreover, "extremely fine" meshing was assigned to the area close to the lagoon base and GSHEs to increase the precision in the analysis (Figure 5).

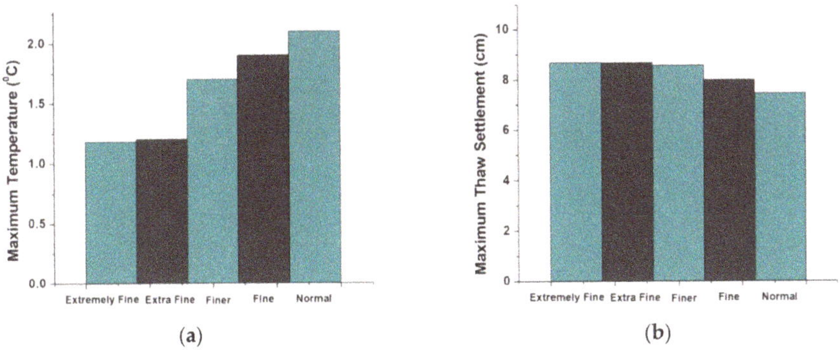

Figure 4. The influence of mesh sizes on (**a**) temperature of permafrost top surface and (**b**) lagoon thaw settlement.

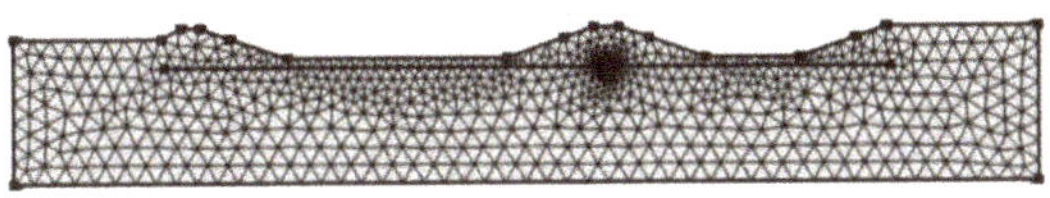

Figure 5. Mesh distribution.

2.5. Model Validation

A laboratory test conducted by Wang et al. [19] on the settlement of the sandy clay, similar to Ross River soil stratigraphy, was used to verify the results of the numerical simulation. In this experiment, a cylindrical sandy clay sample with 100 mm diameter and 112.8 mm height was tested with a dry density of 1400 kg/m^3 and 20.9% water content. The initial temperature of the sample was set to −1 °C and a simplified thermal boundary was defined on the top boundary as:

$$T_{top}(^\circ C) = 5.0 \sin(1.047t + 1.57) \tag{22}$$

where t (h) is time. In addition, a constant temperature of 1 °C was applied on the bottom boundary. The study period was 100 h with 30-min time intervals. Further information regarding the material, hydraulic, and mechanical properties of the tested frozen soil can be found in [19].

In a similar manner to the above laboratory experiment, a finite element simulation was carried out. The experimental and numerical results of the thaw settlement at the top domain and the temperature variation of a point at the depth of 6 cm, both located on the centerline, are plotted in Figure 6.

The maximum difference in the vertical thaw settlement between the experimental measurement and numerical results (Figure 6a) was 2.68 mm, which accounts for 2.68% of the initial height of the sample. In addition, the final settlement value (8 mm) was consistent in both numerical analysis and the lab test. The maximum difference in the temperature of the selected point was 0.54 °C after the 45th hour (Figure 6b). Given the marginal difference between the experimental measurement and the numerical model, a satisfactory level of accuracy was obtained. Therefore, the same numerical modelling procedure was implemented for estimating the long-term thaw settlement under the Ross River wastewater lagoon. The soil properties, model geometry, thermal boundaries, and study time were adjusted accordingly.

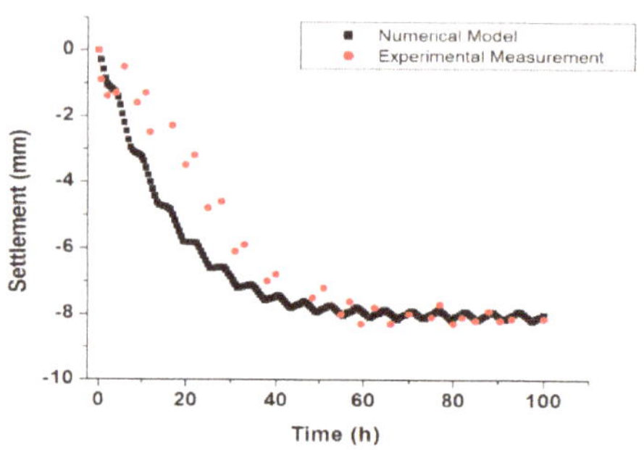

(a)

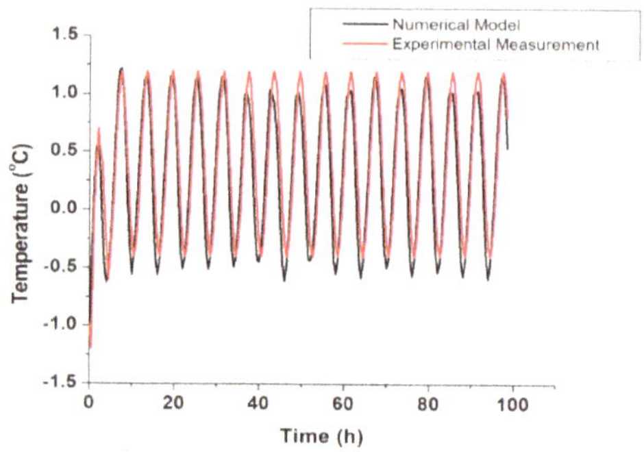

(b)

Figure 6. (**a**) Thaw settlement and (**b**) temperature variation of numerical results versus experimental results of [18].

3. Results and Discussion

3.1. Thaw Settlement Due to the Lagoon Heat Flux and Climatic Conditions without Embedded GHSE System

First, the temperature variation of different points at three depths were studied to investigate the thaw settlement. These three points were chosen at the depth of −5 (point C), −12 (point D), and −18 m (point E), which were representative of the top, middle, and bottom of the permafrost, respectively (Figure 7). The 50-year temperature variation of the selected points (Figure 8a) indicates that Point C, D, and E reached the above-freezing temperature after 1, 10, and 7 years, respectively. Thaw at point E took place earlier than point D due to the upward geothermal heat flux which existed at the model bottom.

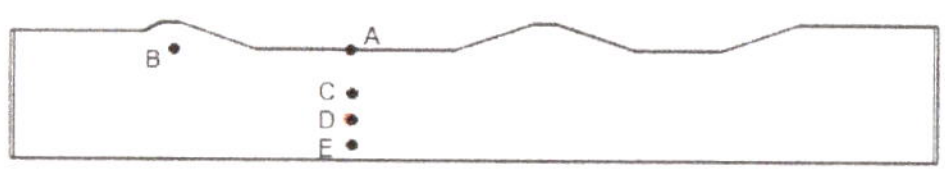

Figure 7. Point locations.

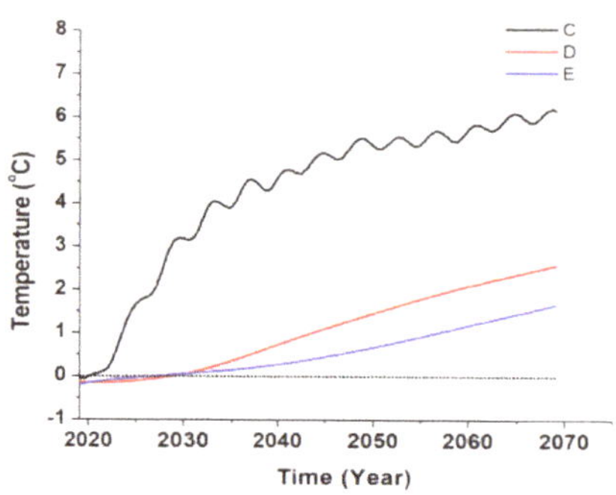

(a)

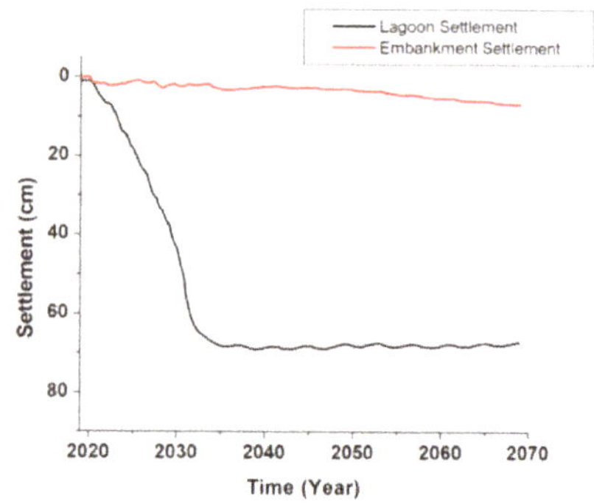

(b)

Figure 8. (**a**) Temperature variation and (**b**) thaw settlement without embedded GSHE system.

Thaw settlement at the lagoon elevation and the embankment was studied comparing points A and B, respectively (Figure 8b). Uneven settlement is one of the main contributing factors in damaging the functionality of the wastewater lagoon. The results indicate that the thaw settlement under the wastewater lagoon was initiated immediately after the lagoon operation due to its heat leakage and reached 69 cm in 2034, a year in which all permafrost thawed.

The permafrost thaw settlement of the embankment gradually increased over the 50-year study time. Unlike the permafrost underneath the lagoon which would disappear by 2031, permafrost under the embankment thawed at a slower rate and some permafrost existed even after 50 years. The maximum embankment settlement was 6 cm in 2069, which was 63 cm less than that of lagoon at the end of the 50-year study period. The maximum difference in the settlement under the lagoon and the embankment was 66 cm in 2039. However, the difference in the thaw settlement of the lagoon and embankment decreased over time.

3.2. Effect of Horizontal GSHE System on Permafrost Preservation

The horizontal HDPE heat exchangers were embedded within the domain to see the effect of GSHEs on the thaw settlement. Temperature variation of the three points illustrate that the GSHEs can considerably reduce the permafrost temperature at different points (Figure 9a). Using horizontal GSHEs maintained the temperature of points C and D in the freezing range throughout the 50-year study time. However, the temperature of point E (the furthest point to the pipes) rose to above zero by 2021 and it took the freezing effect of the pipes 2 years to neutralize the upward geothermal heat flux from the model bottom. As a result, by 2021 the lagoon and embankment underwent 2.4 and 4.4 cm thaw settlement, respectively, a 96% and 28% reduction in the maximum settlement compared to the case when no GSHE was used. The maximum difference in the thaw settlement between the lagoon and embankment also decreased by 97%. It should be noted that the thaw settlement of the embankment was greater than the lagoon due to being away from the GSHEs. Following that, the freezing effect of the pipes induced some level of heave in both lagoon and embankment. The heave in the lagoon was 0.8 cm by 2037, followed by a growth to a maximum of 1.3 cm by 2069 due to the combined effect of the lagoon temperature and GSHEs. The heave in the embankment surface was 2.3 cm and stayed almost constant with the exception of some fluctuations in 2041 and 2058. The induced heave compensated for the thaw settlement to some extent, resulting in the net settlement reaching 1.3 and 2.6 cm in the lagoon and embankment, respectively.

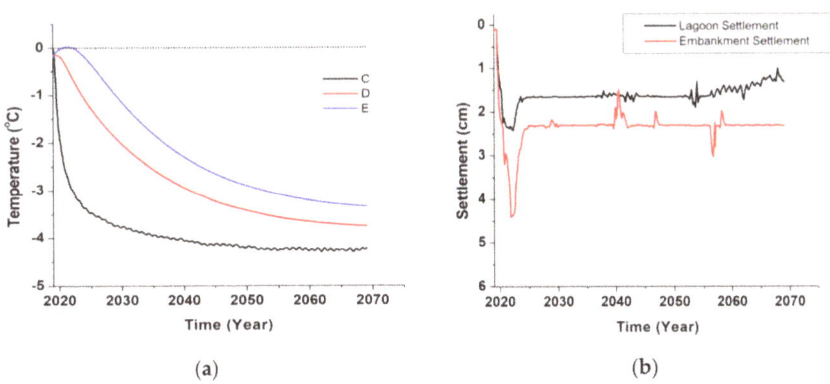

Figure 9. (**a**) Temperature variation and (**b**) thaw settlement in the presence of the embedded GSHE system.

3.3. Effect of Operational Parameters of the Horizontal GSHE System on the Heat Extraction Power and Thaw Settlement

Analyzing different operational parameters given in Table 3 proved that these parameters have a significant impact on the heat extraction power and permafrost thaw settlement.

Table 3. The effect of operational parameters on the heat extraction power permafrost thaw settlement.

Operational Parameter	Value	Heat Extraction [1] (MW·h)	Final Thaw Settlement (cm)	
			Lagoon	Embankment
Fluid Temperature (°C)	−1	328	5.1	7.4
	−5	573	2.2	2.4
	−10	954	−0.5 (Heave)	3.5
	−15	1339	−1.9 (Heave)	2.3
Fluid Velocity (m/s)	0.2	547	2.5	28.9
	0.4	573	2.2	2.4
	0.6	565	2.4	9.8
Burial Depth (m)	1.5	573	2.2	2.4
	2.5	387	2.1	16.3
	3.5	336	2.2	10.9

[1] Heat extraction power of a set of GSHP (one below the primary and one below the secondary cell) in their 50-year lifetime.

As expected, the heat extraction power increased when lower fluid inlet temperatures were used. Based on Equation (2), larger temperature gradient between the fluid coolant and the surrounding soil leads to higher heat transfer quantities. Conversely, the fastest coolant did not contribute to the highest heat extraction power and 1.4% decrease was observed when the fluid inlet velocity changed from 0.4 to 0.6 m/s. Similar to the inlet temperature, changing the burial depth resulted in significant variation of the heat extraction power and compared to 1.5 m burial depth, 32% and 41% reduction was observed with 2.5 and 3.5 m pipe burial depths. Deeper operational pipes were further from the lagoon heat flux and therefore they were exposed to smaller temperature gradient.

With regard to settlement, using a −5 °C coolant instead of −1 °C reduced the lagoon thaw settlement by 57%. Similarly, the embankment settlement decreased from 7.4 to 2.4 cm, and the maximum difference in the lagoon and embankment declined by 91%. Colder fluids with a temperature of −10 and −15 °C resulted in larger heave in the lagoon bottom which could initiate cracks in the wastewater facility. In the embankment, using −10 °C fluid temperature increased the settlement by 1.1 cm compared to −5 °C fluid. The smallest embankment settlement occurred when −15 °C fluid temperature was used. In light of the above, a heat carrier temperature of −5 °C provided the best results.

Changing the heat carrier inlet velocity from 0.4 to 0.2 m/s and from 0.4 to 0.6 m/s resulted in 0.3 and 0.2 cm increase in the final thaw settlement of the lagoon, respectively. Greater differences emerged in the embankment thaw settlement, where changing the fluid inlet velocity from 0.4 to 0.2 m/s resulted in a roughly 12-fold increase in the thaw settlement. Changing this parameter to 0.6 m/s resulted in a fourfold increase in the thaw settlement. A faster heat carrier passes the domain in a shorter time, so it loses the cold much less. Therefore, the passing fluid stays cold enough along the system to freeze back the permafrost. However, when this parameter reaches a certain value, no improvement is achieved as the friction forces produce thermal energy in the pipes. Therefore, 0.4 m/s fluid velocity was the optimum value for permafrost stabilization in this research.

Pipe burial depth variation only caused marginal differences in the lagoon thaw settlement. The embankment thaw settlement, on the other hand, increased to 16.3 cm with GSHE located at the depth of 2.5 m. Burial depth of 3.5 m caused an 8.5 cm growth in the embankment thaw settlement, a smaller increase compared to 2.5 m burial depth. While the deeper pipes can still maintain the soil below the lagoon frozen, their influence becomes trivial for the areas underlying the embankment as they go deeper.

It can be concluded that no predictable relationship exists between the increasing/decreasing GSHE operational parameters and the GSHE performance. In other words, the optimum value for each operational parameter should be investigated separately and through trial and error. These optimum values can also vary from project to project, depending upon soil thermal, hydraulic, and mechanical properties, as well as permafrost depth and temperature.

4. Conclusions

This research investigated the long-term performance of the horizontal GSHE system and the effect of some of its operational parameters on mitigating permafrost thaw settlement. The study site was Ross River, Yukon, Canada, where the huge heat leakage in a conventional lagoon was expected to induce permafrost degradation and significant thaw settlement. The results indicated that the entire permafrost would thaw after twelve years under the climatic and lagoon heat fluxes. The thaw rate was more significant under the lagoon in the long term, resulting in an uneven settlement in the subsurface. Using the GSHE system proved to efficiently maintain the permafrost temperature within the freezing temperature and significantly reduce the thaw settlement in both lagoon and embankment. GSHE operational parameters marginally differentiated the final lagoon settlement value. However, these parameters contributed to resulting in more notable changes in the embankment final settlement. Heat extraction power is greatly affected due to the changes in the fluid inlet temperature and the pipe burial depth. It should be noted that the relationship between increasing/decreasing the operational parameters and the heat extraction power/final thaw settlement cannot be predicted without analyzing them and the optimum values of these parameters should be found through trial and error.

Author Contributions: Conceptualization, A.F.G., P.M., H.M.H. and R.S.; methodology, A.F.G., P.M. and H.M.H.; software, A.F.G.; validation, A.F.G.; formal analysis, A.F.G.; investigation, A.F.G.; resources, P.M. and H.M.H.; data curation, A.F.G.; writing—original draft preparation, A.F.G.; writing—review and editing, A.F.G., P.M., H.M.H., R.S., R.K. and M.S.; supervision, P.M. and H.M.H.; project administration, P.M. and H.M.H.; funding acquisition, P.M. and H.M.H. All authors have read and agreed to the published version of the manuscript.

Funding: This research was funded by Mitacs [IT19450] and KGS Group.

Institutional Review Board Statement: Not applicable.

Informed Consent Statement: Not applicable.

Acknowledgments: The authors acknowledge the support of KGS Group for providing data on Ross River wastewater lagoon.

Conflicts of Interest: The authors declare no conflict of interest.

References

1. Subcommittee, P. Glossary of permafrost and related ground-ice terms. In *Associate Committee on Geotechnical Research*; National Research Council of Canada: Ottawa, ON, Canada, 1988; p. 156.
2. GISTEMP Team. G. S. T. A. G., version 4. NASA Goddard Institute for Space Studies. Available online: https://data.giss.nasa.gov/gistemp/ (accessed on 25 January 2021).
3. Zhang, Y.; Michalowski, R.L. Thermal-hydro-mechanical analysis of frost heave and thaw settlement. *J. Geotech. Geoenviron. Eng.* **2015**, *141*, 04015027. [CrossRef]
4. Morgenstern, N.T.; Nixon, J. One-dimensional consolidation of thawing soils. *Can. Geotech. J.* **1971**, *8*, 558–565. [CrossRef]
5. Dumais, S.; Konrad, J.-M. One-dimensional large-strain thaw consolidation using nonlinear effective stress–void ratio–hydraulic conductivity relationships. *Can. Geotech. J.* **2018**, *55*, 414–426. [CrossRef]
6. Foriero, A.; Ladanyi, B. FEM assessment of large-strain thaw consolidation. *J. Geotech. Eng.* **1995**, *121*, 126–138. [CrossRef]
7. Doré, G.; Niu, F.; Brooks, H. Adaptation methods for transportation infrastructure built on degrading permafrost. *Permafr. Periglac. Process.* **2016**, *27*, 352–364. [CrossRef]
8. Jiao, B.; Qiu, L.; Zhang, X.; Zhang, Y. Investigation on the effect of filling ratio on the steady-state heat transfer performance of a vertical two-phase closed thermosyphon. *Appl. Therm. Eng.* **2008**, *28*, 1417–1426. [CrossRef]
9. Chen, L.; Yu, W.; Lu, Y.; Liu, W. Numerical simulation on the performance of thermosyphon adopted to mitigate thaw settlement of embankment in sandy permafrost zone. *Appl. Therm. Eng.* **2018**, *128*, 1624–1633. [CrossRef]
10. Fontaine, P.-O.; Marcotte, D.; Pasquier, P.; Thibodeau, D. Modeling of horizontal geoexchange systems for building heating and permafrost stabilization. *Geothermics* **2011**, *40*, 211–220. [CrossRef]
11. Sinclair, R.; Kenton, T. *Ross River Wastewater Treatment System, Ross River, Yukon*; Final Design Report; KGS Group: Winnipeg, MB, USA, 2014.

12. Tetra Tech EBA. *Geothermal Evaluation-Proposed Wastewater Disposal Facilities*; Tetra Tech EBA: Ross River, YT, Canada, 2014.
13. Fatolahzadeh Gheysari, A.; Holländer, H.M.; Maghoul, P.; Shalaby, A. Sustainability, climate resiliency, and mitigation capacity of geothermal heat pump systems in cold regions. *Geothermics* **2021**, *91*, 101979. [CrossRef]
14. Konrad, J.-M. Hydraulic conductivity changes of a low-plasticity till subjected to freeze–thaw cycles. *Géotechnique* **2010**, *60*, 679–690. [CrossRef]
15. Dacquay, C.; Holländer, H.M.; Kavgic, M.; Maghoul, P.; Liu, H.; Fujii, H. Evaluation of an integrated sewage pipe with ground heat exchanger for long-term efficiency estimation. *Geothermics* **2020**, *86*, 101796. [CrossRef]
16. Bergman, T.L.; Lavine, A.; Incropera, F.P.; Dewitt, D.P. *Fundamentals of Heat and Mass Transfer*; John Wiley Sons: New York, NY, USA, 2017.
17. Saaly, M.; Sinclair, R.; Kurz, D.; Maghoul, P.; Holländer, H.; Gheisari, A.F. Assessment of a Closed-Loop Geothermal System for Seasonal Freeze-Back Stabilization of Permafrost. *GeoVirtual* **2020**. Available online: https://geovirtual2020.ca/wp-content/files/343.pdf (accessed on 15 July 2021).
18. Liu, H.; Maghoul, P.; Shalaby, A.; Bahari, A. Thermo-hydro-mechanical modeling of frost heave using the theory of poroelasticity for frost-susceptible soils in double-barrel culvert sites. *Transp. Geotech.* **2019**, *20*, 100251. [CrossRef]
19. Wang, S.; Qi, J.; Yin, Z.; Zhang, J.; Ma, W. A simple rheological element based creep model for frozen soils. *Cold Reg. Sci. Technol.* **2014**, *106*, 47–54. [CrossRef]

processes

MDPI

Article

Incompatible Deformation Model of Rocks with Defects around a Thick-Walled Cylinder

Yingji Bao [1,2] and Binsong Jiang [1,*]

1 State Key Laboratory for Geomechanics and Deep Underground Engineering, China University of Mining and Technology, Xuzhou 221116, China; 10704@jsjzi.edu.cn
2 School of Transportation Engineering, Jiangsu Vocational Institute of Architectural Technology, Xuzhou 221116, China
* Correspondence: 10388@jsjzi.edu.cn

Abstract: Before the excavation of underground engineering, joints, fissures, and voids already exist in the rock—that is, there are defects in the rock. Due to the existence of these defects, the rock produces plastic deformation, which can lead to incompatible deformation. Therefore, the classic continuum theory cannot accurately describe the deformation of the rock. In this paper, a relationship between the strain tensor and metric tensor was studied by analyzing the three states of elastic plastic deformation, and the elasto-plastic incompatible model was built. Additionally, the stress and deformation of a thick-walled cylinder under hydrostatic pressure was investigated by using a finite element program written in the FORTRAN language. The results show that the plastic strain is associated with not only deviator stress but also the distribution of defects (represented by the incompatible parameter R). With the value of R increasing, the defects in the rock increased, but the elastic plastic stiffness matrix decreased. Thus, as more rock enters the plastic state, the deformation of the surrounding rock is enlarged.

Keywords: incompatibility; rock; plastic deformation; finite element

Citation: Bao, Y.; Jiang, B. Incompatible Deformation Model of Rocks with Defects around a Thick-Walled Cylinder. *Processes* **2021**, *9*, 2215. https://doi.org/10.3390/pr9122215

Academic Editors: Li Li and Haiping Zhu

Received: 21 October 2021
Accepted: 3 December 2021
Published: 8 December 2021

1. Introduction

The size and shape of an object will change under external forces, namely the deformation of objects. After the external force is removed, the part of the deformation that has disappeared is called elastic deformation, and the remaining deformation is plastic deformation. The important characteristic of elastic deformation is reversibility—that is, deformation occurs after the force is applied and disappears after the force is removed. This indicates that the elastic deformation is determined by the binding force between atoms. If an external force overcomes the gravitational pull between the atoms and pulls them apart, the object will break, and its strength is called breaking strength. The actual object contains defects such as dislocations and disclinations; with a small elastic deformation, the stress is capable of activating the dislocation and making it move, resulting in plastic deformation. For brittle materials, as they are more sensitive to stress concentration, when the stress is slightly higher, the concentrated stress can lead to the movement and proliferation of these defects, resulting in incompatible deformation. Therefore, the classical continuum theory is not suitable for analyzing the defected rocks, and the incompatible deformation theory can be used.

The incompatible deformation theory originates from the defect theory and has been studied by many researchers [1–3]. Kondo first introduced differential geometry into the defect theory. He established the relationship between the dislocation theory and non-Riemannian geometry [4]. Later on, Anthony pointed out the relationship between differential geometry and disclinations [5]. Then, Kroner completed the underlying theory for defects and differential geometry including dislocations and extra-matter [6]. These investigations were conducted to establish a relation between the parameters of non-Euclidean geometry and the elastic strain, bend-twist, and quasi-plastic strain of the defect theory [7]. M.A. Guzev first introduced the incompatible deformation theory to geotechnical engineering and assumed that the deformed body is in a Riemannian space. The curvature tensor R is not zero, so it is treated as an independent thermodynamic variable. As a result, a Riemann continuous model is established by using unbalanced thermodynamics. The stress distribution of a deep circle tunnel under plane strain was analyzed [8–13]. Zhou Xiaoping and Qian Qihu considered the impact of damage on a rock, described the damage degree of surrounding rocks in a deep circular tunnel according to the damage variables defined by the energy equivalence of damage mechanics, and analyzed the relationship between the damage of a deep rock and zonal fracture [14–16]. From the perspective of differential geometry, these existing models in geotechnical engineering all assumed that the undeformed body is in a three-dimensional Euclidean space, or the deformed body is a manifold. However, there are defects such as joints and cracks in the rock before excavation; therefore, the initial configuration should also be regarded as a manifold. In this paper, three configurations are constructed to describe the elastoplastic deformation of rocks in differential geometry, and the relationship between the strain tensor and interior metric tensor is established. Accordingly, an elastoplastic incompatible model is constructed. Finally, the elastoplastic finite element program is written in the FORTRAN language, and the stress and deformation of a circular tunnel under hydrostatic pressure is also analyzed.

2. The Incompatible Plastic Deformation Theory of Rocks

For rocks with elastoplastic deformation, we need to analyze three configurations. As shown in Figure 1 below, in the initial configuration Ω_0, the rock is undeformed, and X, the micro element at any point, is denoted as dX. The deformed rock is the current configuration Ω, which has undergone much plastic deformation. The micro element at any point, namely x of the deformed rock, is denoted as dx. In addition, the intermediate configuration Ω_* is also introduced. The intermediate configuration is assumed to be obtained from the initial configuration through pure plastic deformation or from the current configuration through the process of elastic unloading, where the micro element at any point ζ is denoted as $d\zeta$.

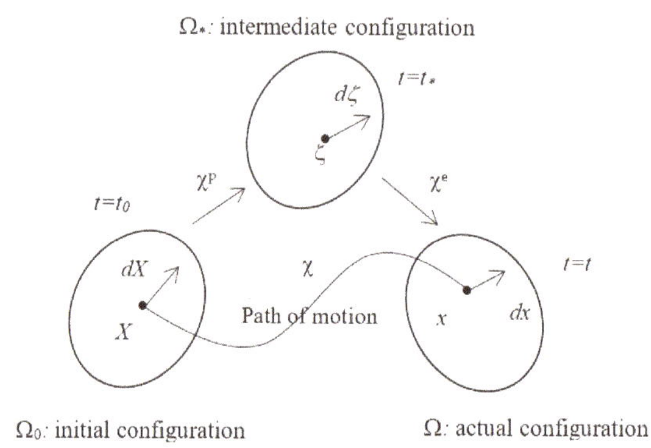

Figure 1. Three configurations of plasticity deformation.

A point X on Ω_0 is mapped to a point ζ on the intermediate configuration Ω_* under the mapping χ^p:

$$\chi^p : X \to \zeta \tag{1}$$

Due to the presence of these defects, the initial configuration is a material of manifold. In local coordinates, the neighborhood of point X has coordinates $(X^1, \ldots, X^3)$; the neighborhood of point ζ has coordinates $(\zeta^1, \ldots, \zeta^n)$; χ^p is a continuous mapping; and the transformation matrix between X and ζ is $\partial\zeta/\partial X$.

In differential geometry, χ^p is an immersion mapping of manifold M to manifold L, and the intermediate configuration is assumed to be a Riemannian manifold with a metric structure of a symmetric nonsingular covariant tensor field of the second order. In an analysis of local properties, it can be considered that M is embedded in L, inducing the push from the tangent space:

$$\chi_*^p : \eta^I(X)\partial_I \rightarrow \eta^I(X)A_I^\alpha\partial_\alpha = \overline{\eta}^\alpha\partial_\alpha \tag{2}$$

A metric is defined in a Riemannian manifold as

$$ds_*^2 = g_{\alpha\beta}d\zeta^\alpha d\zeta^\beta \tag{3}$$

The metric tensor field $g_{\alpha\beta}$ on L is dragged back to the metric field G_{IJ} on M by

$$ds_*^2 = g_{\alpha\beta}A_I^\alpha A_J^\beta dX^I dX^J = G_{IJ}dX^I dX^J \tag{4}$$

The metric tensor g of manifold L is determined by the distribution of rock defects and the degree of plastic deformation, while the metric tensor G of manifold M is determined by the distribution of these rock defects.

The element dX at a point X in the initial configuration can be related to the element $d\zeta$ at a point ζ in the intermediate configuration through nonholonomic transformation:

$$d\zeta = AdX \tag{5}$$

A is the nonholonomic coefficient. The components satisfy

$$d\zeta^\alpha = A_I^\alpha dX^I, e_\alpha = A_I^\alpha e_I \tag{6}$$

Point ζ on Ω_* is mapped to point x on Ω under mapping χ^e:

$$\chi^e : \zeta \rightarrow x \tag{7}$$

There is a tangent mapping at ζ:

$$\psi_* : T_\zeta(\Omega_*) \rightarrow T_x(\Omega) \tag{8}$$

The local coordinate of point ζ is $(\zeta^1, \ldots, \zeta^n)$, the corresponding coordinate of point x is $(x^1, \ldots, x^3)$, χ^e is a continuous mapping, and the transformation matrix between ζ and x is $\partial x/\partial\zeta$. The metric tensor of the current configuration N is determined by the metric tensor G of the intermediate configuration and the elastic deformation ε^e.

Because the elastic deformation is reversible, we have

$$dx = Fd\zeta \tag{9}$$

F is the deformation gradient, and its components satisfy

$$dx^i = F_\alpha^i d\zeta^\alpha, e_i = F_i^\alpha e_\alpha \tag{10}$$

The deformation of rocks can be considered as occurrences in a three-dimensional Euclidean space in a small range. The total deformation from the initial configuration to the current configuration is

$$\chi = \chi^e \cdot \chi^p \tag{11}$$

Since the rock is in a three-dimensional Euclidean space before and after the deformation, the total deformation is compatible, so χ is bijective, and according to the composite mapping theorem, if χ_p is injective, then χ_e is surjective.

The initial configuration is in a three-dimensional Euclidean space, and the length square of the element at point X is

$$ds_0^2 = \delta_{IJ} dX_I X_J \tag{12}$$

The intermediate configuration is a Riemannian manifold, whose element length squared is

$$ds_*^2 = g_{\alpha\beta} d\zeta^\alpha d\zeta^\beta = g_{\alpha\beta} A_I^\alpha A_J^\beta dX^I dX^J \tag{13}$$

The current configuration is also in a three-dimensional Euclidean space, and its element length squared is

$$ds^2 = \delta_{ij} dx^i dx^j = \delta_{ij} F_I^i F_J^j dX^I dX^J = F_I^K F_J^K dX^I dX^J \tag{14}$$

Therefore, the change in the squared length of the element from the intermediate configuration to the current configuration is

$$ds^2 - ds_*^2 = (F_I^K F_J^K - g_{IJ}) dX^I dX^J = 2\varepsilon_{IJ}^e dX^I dX^J \tag{15}$$

where ε_{IJ}^e is the elastic strain tensor. If we substitute the displacement equation into Equation (15), we will have

$$\begin{aligned} \varepsilon_{IJ}^e &= \tfrac{1}{2}(F_I^K F_J^K - g_{IJ}) = \tfrac{1}{2}[(\delta_{kI} + \tfrac{\partial u_k}{\partial x_I})(\delta_{kI} + \tfrac{\partial u_k}{\partial x_J}) - g_{IJ}] \\ &= \tfrac{1}{2}[\delta_{IJ} + \tfrac{\partial u_I}{\partial x_J} + \tfrac{\partial u_J}{\partial x_I} + \tfrac{\partial u_k}{\partial x_I}\tfrac{\partial u_k}{\partial x_J} - g_{IJ}] \\ &= \tfrac{1}{2}(\tfrac{\partial u_I}{\partial x_J} + \tfrac{\partial u_J}{\partial x_I} + \tfrac{\partial u_k}{\partial x_I}\tfrac{\partial u_k}{\partial x_J}) + \tfrac{1}{2}(\delta_{IJ} - g_{IJ}) \end{aligned} \tag{16}$$

where ε_{IJ} is the strain tensor in classic elasticity strain. When the deformation of the rock is small, the quadratic term in the above equation can be ignored, i.e.,

$$\varepsilon_{IJ}^e = \varepsilon_{IJ} + \frac{1}{2}(\delta_{IJ} - g_{IJ}) \tag{17}$$

The last two terms of Equation (17) are incompatible deformation terms—that is, the existence of plastic deformation makes the elastic deformation itself uncoordinated. Generally, the local coordinate basis $\{E_I\}$ of the initial configuration is different from the local coordinate basis $\{e_i\}$ of the current configuration. When the incremental deformation is small, the same coordinate basis vector $\{E_I\} = \{e_i\}$ can be selected. Then, the above equation can be written as

$$\varepsilon_{ij}^e = \varepsilon_{ij} + \frac{1}{2}\delta_{ij} - \frac{1}{2}g_{ij} \tag{18}$$

3. Stress-Strain Relationship of the Incompatible Plastic Deformation

In the present section, we consider a Stress-Strain relationship of the incompatible plastic deformation. A number of variables and symbols are listed as follows:

1. The total strain tensor ε_{ij};
2. The elastic strain tensor ε_{ij}^e;
3. The plastic strain tensor ε_{ij}^p;
4. The internal metric tensor g_{ij};
5. The elastic stiffness tensor C_{ijkl};
6. The elastic stress tensor σ_{ij};
7. The plastic multiplier $d\lambda$;
8. The plastic potential w;
9. The yield function F;
10. The flow vector a;
11. The hardening parameter κ;

12. The usual matrix of elastic constants D;
13. The incompatible parameter R;
14. The elastic-plastic stiffness matrix Dep;
15. The three principal stresses $\sigma_1, \sigma_2, \sigma_3$;
16. The three principal deviatoric stresses $\sigma_1', \sigma_2', \sigma_3'$;
17. The three stress invariants J_1, J_2, J_3;
18. The three invariants of deviatoric stress J_1', J_2', J_3';
19. The internal friction angle ϕ;
20. The cohesion c.

After the rock enters the state of plastic deformation, the strain at any point is composed of elastic strain and plastic strain. When the external load has a small increment, the total strain also has a small increment. The total strain increment consists of an elastic strain increment and a plastic strain increment, namely

$$d\varepsilon_{ij} = d\varepsilon_{ij}^{e} + d\varepsilon_{ij}^{p} \tag{19}$$

By substituting Equation (18) into Equation (19), we will obtain

$$d\varepsilon_{ij}^{e} = d\varepsilon_{ij} + \frac{1}{2}d\delta_{ij} - \frac{1}{2}dg_{ij} - d\varepsilon_{ij}^{p} = d\varepsilon_{ij} - \frac{1}{2}dg_{ij} - d\varepsilon_{ij}^{p} \tag{20}$$

The relationship between the elastic stress increment and strain increment can be determined by Hooke's law as

$$d\sigma_{ij} = C_{ijkl}d\varepsilon_{kl}^{e} \tag{21}$$

where C_{ijkl} is the elastic stiffness tensor. Since the plastic strain can be obtained through the flow law, the Stress-Strain relationship for ideal elastoplastic materials can be expressed as

$$d\sigma_{ij} = C_{ijkl}\left(d\varepsilon_{ij} - d\lambda\frac{\partial w}{\partial \sigma_{kl}} - \frac{1}{2}dg_{kl}\right) \tag{22}$$

where $d\lambda$ is an undetermined nonnegative scalar termed the plastic multiplier. During the plastic deformation, the stress point stays on the yield surface, and this supplementary condition is called the consistency condition, which is expressed by the following formula as

$$dF = \frac{\partial F}{\partial \sigma}d\sigma + \frac{\partial F}{\partial \kappa}d\kappa = a^{T}d\sigma - Ad\lambda = 0 \tag{23}$$

with the definitions of

$$a^{T} = \frac{\partial F}{\partial \sigma} = \left[\frac{\partial F}{\partial \sigma_x}, \frac{\partial F}{\partial \sigma_y}, \frac{\partial F}{\partial \sigma_z}, \frac{\partial F}{\partial \tau_{yz}}, \frac{\partial F}{\partial \tau_{zx}}, \frac{\partial F}{\partial \tau_{xy}}\right] \tag{24}$$

$$A = -\frac{1}{d\lambda}\frac{\partial F}{\partial \kappa}d\kappa \tag{25}$$

The vector a is termed the flow vector. Equation (22) can be immediately rewritten as

$$(1-R)d\varepsilon = [D]^{-1}d\sigma + d\lambda\frac{\partial F}{\partial \sigma} \tag{26}$$

where D is the usual matrix of elastic constants. Setting $dg_{kl}/d\varepsilon_{kl} = 2\,R$, according to Equation (18), we can obtain

$$R = \frac{dg_{kl}}{d\varepsilon_{kl}} = 1 - \frac{d\varepsilon_{kl}^{e}}{d\varepsilon_{kl}} = 1 - \frac{E}{E_e} \tag{27}$$

where E_e is the elastic modulus of an intact rock, and E is the elastic modulus of a rock with defects. The parameter R reflects the influence of microscopic defects on macroscopic deformation of the rock.

Premultiplying both sides of Equation (26) by using $d_D^T = a^T D$ and eliminating $a^T d\sigma$ by the use of Equation (23), we can obtain

$$d\lambda = \frac{(1-R)}{A + a^T D a} a^T d_D d\varepsilon \tag{28}$$

Substituting Equation (28) into Equation (26) we can obtain the complete incompatible elastic-plastic incremental Stress-Strain relation as

$$d\sigma = D_{ep} d\varepsilon \tag{29}$$

where

$$D_{ep} = (1-R)(D - \frac{d_D d_D^T}{A + d^T a}); d_D = Da \tag{30}$$

For numerical computation, it is convenient to write the yield function in terms of stress invariants. The advantage is that it permits the computer coding of yield function and the flow rule in a general form [17]. The principal deviatoric stresses σ_1', σ_2', σ_3' are given as the roots of the cubic equation [18]

$$t^3 - J_2' t - J_3' = 0 \tag{31}$$

Noting the trigonometric identity

$$\sin^3\theta - \frac{3}{4}\sin\theta + \frac{1}{4}\sin 3\theta = 0 \tag{32}$$

and setting $t = r\sin\theta$ and substituting into Equation (31), we can obtain

$$\sin^3\theta - \frac{J_2'}{r^2}\sin\theta - \frac{J_3'}{r^3} = 0 \tag{33}$$

Comparing Equation (32) and Equation (33), we can obtain

$$r = \frac{2}{\sqrt{3}}(J_2')^{\frac{1}{2}} \tag{34}$$

$$\sin 3\theta = -\frac{J_3'}{r^3} = -\frac{3\sqrt{3}}{2}\frac{J_3'}{(J_2')^{3/2}} \tag{35}$$

By noting the cyclic nature of sin (θ + 2 $n\pi$), we have immediately made the three possible values of sinθ which define the three principal stresses. The deviatoric principal stresses are given by $t = r\sin\theta$ upon substitution of the three values of sinθ in turn. Substituting for r from Equation (34) and adding the mean hydrostatic stress component give the total principal stresses as

$$\begin{Bmatrix} \sigma_1 \\ \sigma_2 \\ \sigma_3 \end{Bmatrix} = \frac{2}{\sqrt{3}}(J_2')^{\frac{1}{2}} \begin{Bmatrix} \sin(\theta + \frac{2\pi}{3}) \\ \sin\theta \\ \sin(\theta + \frac{4}{3}\pi) \end{Bmatrix} + \frac{J_1}{3}\begin{Bmatrix} 1 \\ 1 \\ 1 \end{Bmatrix} \tag{36}$$

with $\sigma_1 > \sigma_2 > \sigma_3$. For geomaterials, they typically have a frictional strength and different strengths in tension and in compression, so we can choose the Mohr–Coulomb yield criterion and write it in terms of J_1, J_2, and θ as

$$\frac{1}{3}J_1\sin\phi + (J_2')^{\frac{1}{2}}(\cos\theta - \frac{1}{\sqrt{3}}\sin\theta\sin\phi) = c\cos\phi \tag{37}$$

In order to calculate the D_{ep} matrix in (30), we need to express vector a in a form suitable for numerical computation as

$$a^T = \frac{\partial F}{\partial \sigma} = \frac{\partial F}{\partial J_2}\frac{\partial J_2}{\partial \sigma} + \frac{\partial F}{\partial (J_2')^{1/2}}\frac{\partial (J_2')^{1/2}}{\partial \sigma} + \frac{\partial F}{\partial \theta}\frac{\partial \theta}{\partial \sigma} \tag{38}$$

Differentiating Equation (35), we can obtain

$$\frac{\partial \theta}{\partial \sigma} = \frac{-\sqrt{3}}{2\cos 3\theta}\left[\frac{1}{(J_2')^{3/2}}\frac{\partial J_3}{\partial \sigma} - \frac{\partial J_3}{(J_2')^2}\frac{(J_2')^{1/2}}{\partial \sigma}\right] \tag{39}$$

Substituting Equation (38) in Equation (37) and using Equation (35), we can write a as

$$a = C_1 a_1 + C_2 a_2 + C_3 a_3 \tag{40}$$

where

$$a_1^T = \frac{\partial J_1}{\partial \sigma} = \{1,1,1,0,0,0\} \tag{41}$$

$$a_2^T = \frac{\partial (J_2')^{1/2}}{\partial \sigma} = \frac{1}{2(J_2')^{1/2}}\{s_x, s_y, s_z, 2\tau_{yz}, 2\tau_{zx}, 2\tau_{xy}\} \tag{42}$$

$$a_3^T = \frac{\partial J_3}{\partial \sigma} = \left\{\begin{array}{c} (s_y s_z - \tau_{yz}^2 + \frac{J_2'}{3}), (s_x s_z - \tau_{xz}^2 + \frac{J_2'}{3}), (s_x s_y - \tau_{xy}^2 + \frac{J_2'}{3}), \\ 2(\tau_{zx}\tau_{xy} - s_x\tau_{yz}), 2(\tau_{yz}\tau_{xy} - s_y\tau_{zx}), 2(\tau_{yz}\tau_{zx} - s_z\tau_{xy}) \end{array}\right\} \tag{43}$$

and

$$C_1 = \frac{\partial F}{\partial J_1} \tag{44}$$

$$C_2 = \frac{\partial F}{\partial (J_2')^{1/2}} - \frac{\tan 3\theta}{(J_2')^{1/2}}\frac{\partial F}{\partial \theta} \tag{45}$$

$$C_3 = \frac{-\sqrt{3}}{2\cos 3\theta}\frac{1}{(J_2')^{3/2}}\frac{\partial F}{\partial \theta} \tag{46}$$

The C_1, C_2, and C_3 are the constants related to the yield surface. The tunnel excavated in underground engineering can be regarded as a problem of two-dimensional plane strain. According to the Mohr-Coulomb yield criterion, the equation from Equation (40) to Equation (46) can be written as

$$a_1^T = \{1,1,0,1\} \tag{47}$$

$$a_2^T = \frac{1}{2(J_2')^{1/2}}\{s_x, s_y, 2\tau_{xy}, s_z\} \tag{48}$$

$$a_3^T = \frac{\partial J_3}{\partial \sigma} = \left\{(s_y s_z + \frac{J_2'}{3}), (s_x s_z + \frac{J_2'}{3}), 2s_z\tau_{xy}, (s_x s_y + \frac{J_2'}{3}),\right\} \tag{49}$$

and

$$C_1 = \frac{1}{3}\sin\phi \tag{50}$$

$$C_2 = \cos\theta\left[(1 + \tan\theta\tan 3\theta) + \sin\phi(\tan 3\theta - \tan\theta)/\sqrt{3}\right] \tag{51}$$

$$C_3 = \frac{\sqrt{3}\sin\theta + \cos\theta\sin\phi}{2J_2'\cos 3\theta} \tag{52}$$

4. Numerical Analysis

The FORTRAN language was used to write the elastic-plastic finite element program for the problem of two-dimensional plane strain. The parameters are listed as follows. It is assumed that the rock satisfies the associated flow rule, which is the Mohr–Coulomb yield criterion: the material constants include a Young's modulus $E = 21$ GPa, a Poisson's ratio $v = 0.3$, an internal friction angle $\phi = 30°$, and a cohesion $c = 5.85$ MPa. The problem chosen is a thick-walled cylinder subjected to the external pressure P, with plane strain conditions assumed in the axial direction. Axisymmetry normally only requires analyzing a segment of the cylinder. For example, a 90° segment (see Figure 2) is used here to take advantage of the symmetry conditions along the x and y axes.

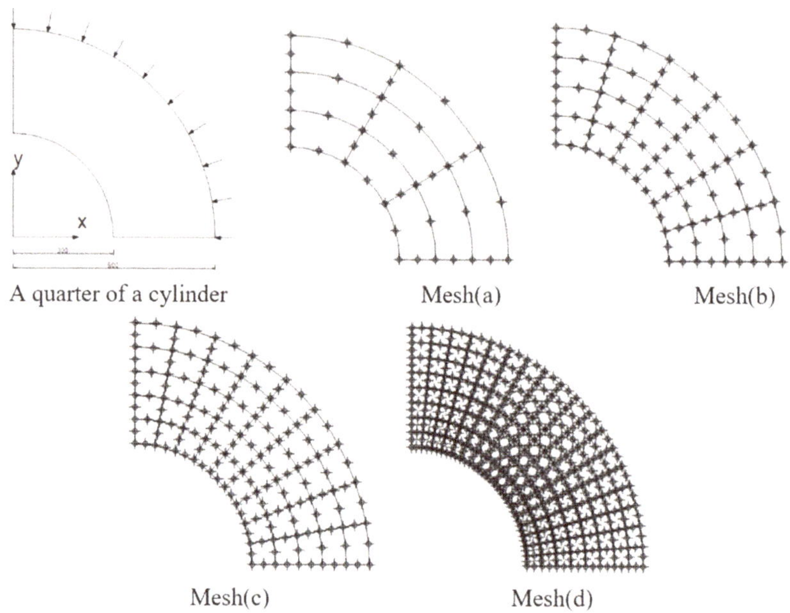

Figure 2. Different mesh divisions for the problem.

Before analyzing the problem, a sensitivity analysis of mesh is necessary. A number of different mesh divisions are shown in Figure 2, where an eight-node quadrangle element was chosen.

Referring to Figure 3, it can be seen that the results of meshes are quite reasonable since the results of radial stress and tangential stress are coincident with the results in the theoretical curve.

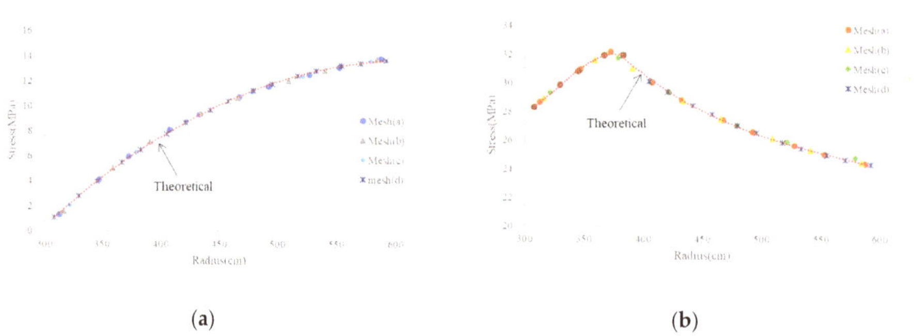

Figure 3. The distribution of radial stress and tangential stress in the thick cylinder; (**a**) Radial stress distribution under $P = 14$ MPa; (**b**) Tangential stress distribution under $P = 14$ MPa.

Thus, mesh(c) will be used because of the higher accuracy and relatively small amount of computation involved. After having decided on the element type and mesh division, it is now possible to prepare the data before putting them into the program. Such data are extracted directly from the drawing of the structure (see Figure 4) in which all information concerns nodes and element numbers, and load and boundary conditions are available.

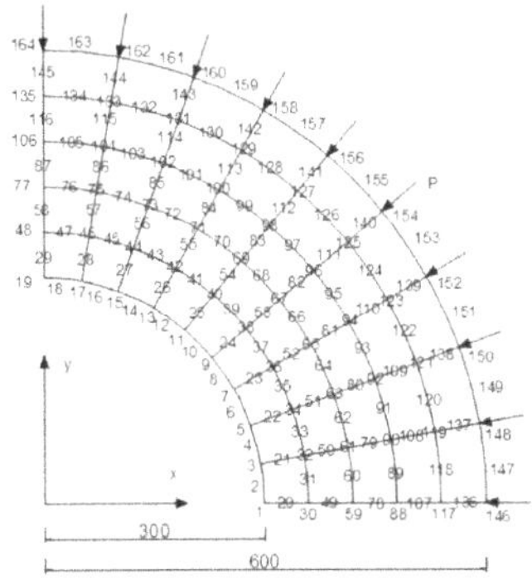

Figure 4. Details of the example problem.

When the surrounding rock is intact, or with no defects, the distribution of stress is shown in Figure 5 below. When the pressure level of the surrounding rock is at 10 MPa, the surrounding rock is in an elastic state, and the distribution of hoop stress and radial stress is completely consistent with the elastic solution. With the increase in the surrounding rock stress, when the pressure of the surrounding rock is at 12 MPa, the plastic state appears 0.5 m closer to the inside of the tunnel, and the hoop stress increases first and decreases with the radius. The rock to the left of the peak point is in a plastic state, while the rock to the right of the peak point is in an elastic state. As the pressure of the surrounding rock continues to increase (14 MPa, 16 MPa, 18 MPa), the plastic zone of the surrounding rock also expands, and the radius of the plastic zone can be 3.35 m, 4.25 m, and 5.25 m. When the pressure of the surrounding rock reaches 19.4 MPa, all rocks in the region come to a plastic state.

When the surrounding rock is a defect mass, the stiffness of the surrounding rock decreases due to the existence of the defect, per se, and the changes are shown in Figure 6 below. When $R = 0$, the rock has no defects, and the surrounding rock reaches a plastic state at 19.4 MPa. When $R = 0.2$, the surrounding rock reaches the plastic state at 19.2 MPa, indicating that the existence of defects reduces the stiffness of the surrounding rock, but the defects have little influence on the stress of the surrounding rock. When $R = 0.4$, the surrounding rock reaches a plastic state at 17.4 MPa, and the influence of the defects of the surrounding rock stress is further increased. When $R = 0.5$, the defects have a significant effect on the surrounding rock, and the surrounding rock reaches plasticity at 12.8 MPa.

Due to the existence of defects, the stiffness of the surrounding rock decreases, and the deformation increases. The variation rule of deformation with incompatible parameters when the stress of the surrounding rock is at 19 MPa is assessed, as shown in Figure 7 below. When R increases from 0.1 to 0.3, the displacement of the surrounding rock increases linearly; when R reaches 0.4, the displacement increases more greatly than what it was before.

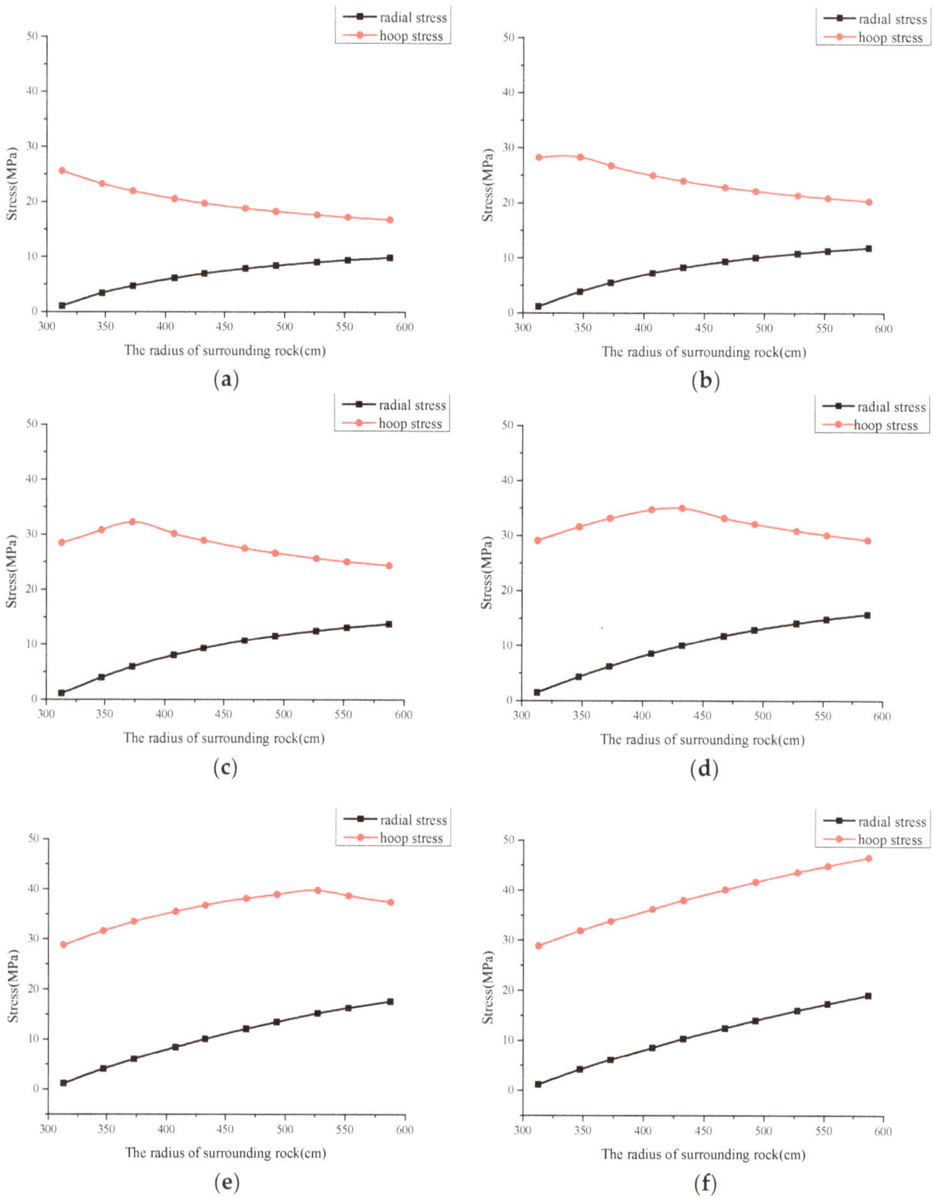

Figure 5. The distribution of radial stress and tangential stress varies with the confining pressure; (**a**) P = 10 MPa; (**b**) P = 12 MPa; (**c**) P = 14 MPa; (**d**) P = 16 MPa; (**e**) P = 18 MPa; (**f**) P = 19.4 MPa.

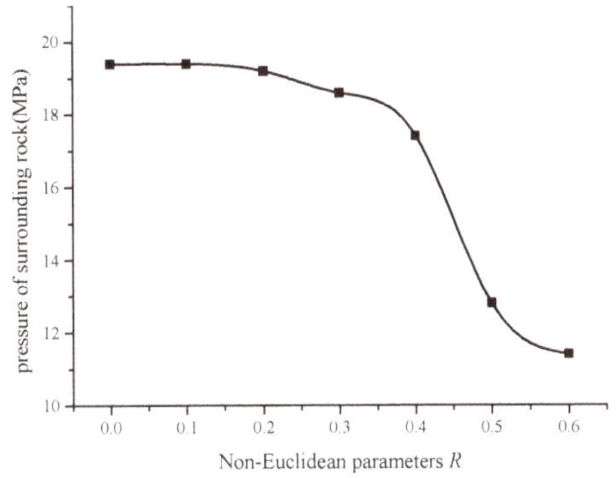

Figure 6. Variation in the confining pressure with incompatible parameter R when the rock reaches plasticity.

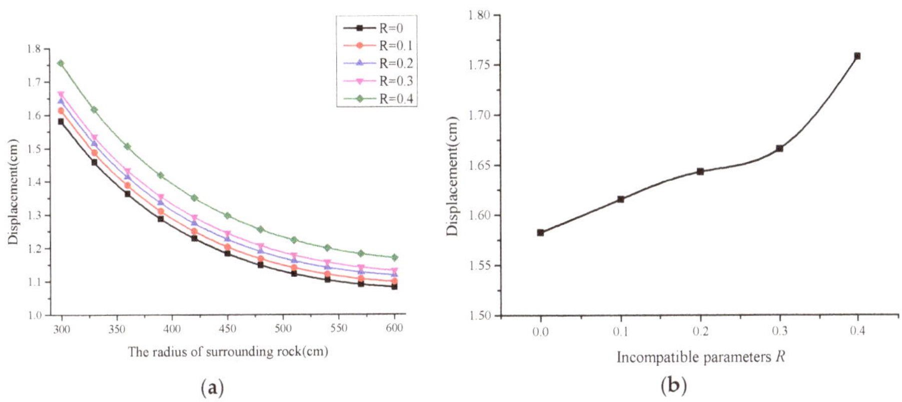

Figure 7. Variation in the deformation of the surrounding rock with incompatible parameters; (**a**) displacement of the surrounding rock; (**b**) displacement of the cylinder surface.

5. Conclusions

(1) The existence of defects is the cause of incompatible deformation, and plastic deformation is the result of incompatible deformation as well. Although the deformed object has undergone much plastic deformation, it still exists in a three-dimensional Euclidean space, so the rock before and after the deformation is in a Euler space. The incompatible plastic deformation leads to the incompatible elastic deformation, while the total deformation can be coordinated.

(2) When the deformation of the rock is small, the deformed and undeformed body can use the same coordinate base vector, and assuming the rock is an ideal plastic material, we can obtain the incremental plastic Stress-Strain relationship of the rock. The plastic strain increment is not only associated with the deviatoric stress state but also with the defect distribution. This distribution can be expressed as an incompatible parameter R. The larger the value of R is, the more defects there are in the rock and the smaller the elastic-plastic stiffness matrix will be.

(3) The excavation of a deep underground tunnel can be regarded as a problem of two-dimensional plane strain. An elastic-plastic finite element program is written in the FORTRAN language to analyze circular tunnels under hydrostatic pressure. As the existence of defects decreases the stiffness of the surrounding rock, the larger the

value of parameter R is, the faster the rock enters the plastic state and the larger the deformation of the surrounding rock will be.

Author Contributions: Conceptualization, Y.B. and B.J.; methodology, B.J.; software, Y.B.; validation, Y.B. and B.J.; formal analysis, Y.B.; investigation, Y.B.; resources, Y.B.; data curation, Y.B.; writing—original draft preparation, Y.B.; writing—review and editing, Y.B.; visualization, Y.B.; supervision, B.J.; project administration, B.J.; funding acquisition, B.J. All authors have read and agreed to the published version of the manuscript.

Funding: This work was financially supported by General Project of National Natural Science Foundation of China (Grant No. 51174196).

Institutional Review Board Statement: Not applicable.

Informed Consent Statement: Not applicable.

Data Availability Statement: The study did not report any data.

Acknowledgments: This work was financed by the General Project of National Natural Science Foundation of China (Grant No. 51174196).

Conflicts of Interest: The authors declare no conflict of interest.

References

1. Weingarten, G. *Sulle Superficie di Discontinuità nella Teoria della Elasticità dei Corpi Solidi*; Accademia dei Lincei: Rome, Italy, 1901.
2. Orowan, E. Plasticity of crystals. *Z. Phys.* **1934**, *89*, 605–659. [CrossRef]
3. Eshelby, J.D. The continuum theory of lattice defects. *Solid State Phys.* **1956**, *3*, 79–144.
4. Kondo, K. On the Geometrical and Physical Foundations of the Theory of Yielding. In *Proceedings of the Second Japan National Congress for Applied Mechanics*; Japan National Committee for Theoretical and Applied Mechanics: Tokyo, Japan, 1952; pp. 41–47.
5. Anthony, K. Nonmetric Connexions, Quasidislocations and Quasidisclinations. A Contribution to the Theory of Nonmechanical Stresses in Crystals (Nonmetric Connexions, Quasidislocations and Quasidisclinations Applied to Nonmechanical Stress Theory in Crystals). *Nat. Bur. Stand. Spec. Publ.* **1970**, *1*, 637–649.
6. Kröner, E. *Plastizität und Versetzungen, Mechanik der Deformierbaren Medien*; Akademische Arbeitsgemeinschaft: Leipzig, Germany, 1964; pp. 310–376.
7. De Wit, R. A view of the relation between the continuum theory of lattice defects and non-euclidean geometry in the linear approximation. *Int. J. Eng. Sci.* **1981**, *19*, 1475–1506. [CrossRef]
8. Myasnikov, V.; Guzev, M. Thermomechanical model of elastic–plastic materials with defect structures. *Theor. Appl. Fract. Mech.* **2000**, *33*, 165–171. [CrossRef]
9. Guzev, M.A.; Paroshin, A.A. Non–Euclidean Model of the Zonal Disintegration of Rocks around an Underground Working. *J. Appl. Mech. Tech. Phys.* **2001**, *42*, 131–139. [CrossRef]
10. Guzev, M.A. Structure of kinematic and force fields in the Riemannian continuum model. *J. Appl. Mech. Tech. Phys.* **2011**, *52*, 709–716. [CrossRef]
11. Makarov, V.V.; Guzev, M.A.; Odintsev, V.N.; Ksendzenko, L.S. Periodical zonal character of damage near the openings in highly-stressed rock conditions. *J. Rock Mech. Geotech. Eng.* **2016**, *8*, 164–169. [CrossRef]
12. Guzev, M.A.; Odintsev, V.N.; Makarov, V.V. Principals of geomechanics of highly stressed rock and rockifs. *Tunn. Undergr. Space Technol.* **2018**, *81*, 506–511. [CrossRef]
13. Guzev, M.A. Non-classical solutions of a continuum model for rock descriptions. *J. Rock Mech. Geotech. Eng.* **2014**, *6*, 180–185. [CrossRef]
14. Qian, Q.; Zhou, X. Non-euclidean continuum model of the zonal disintegration of surrounding rocks around a deep circular tunnel in a non-hydrostatic pressure state. *J. Min. Sci.* **2011**, *47*, 37. [CrossRef]
15. Zhou, X.; Song, H.; Qian, Q. Zonal disintegration of deep crack-weakened rock masses: A non-Euclidean model. *Theor. Appl. Fract. Mech.* **2011**, *55*, 227–236. [CrossRef]
16. Hou, Q.; Qian, Q.; Zhang, Y.; Zhou, X. The zonal disintegration mechanism of surrounding rock around deep spherical tunnels under hydrostatic pressure condition: A non-euclidean continuum damage model. *Acta Mech. Solida Sin.* **2013**, *26*, 373–3878.
17. Owen, D.R.J. *Finite Elements in Plasticity Theory and Practice*; Prineridge Press Limited: Swansea, Wales, 1980.
18. Frederick, D.; Chang, T.S. *Continuum Mechanics*; Allyn and Bacon: Boston, MA, USA, 1965.

processes MDPI

Article

A Parametric Numerical Study for Diagnosing the Failure of Large Diameter Bored Piles Using Supervised Machine Learning Approach

Mohamed E. Al-Atroush [1,*], Ashraf M. Hefny [2] and Tamer M. Sorour [3]

1 Engineering Management Department, College of Engineering, Prince Sultan University, Riyadh 11543, Saudi Arabia
2 Department of Civil & Environmental Engineering, United Arab Emirates University, Al-Ain 15258, United Arab Emirates; a.hefny@uaeu.ac.ae
3 Department of Civil Engineering, Faculty of Engineering, Ain Shams University, Cairo 11865, Egypt; tamer.sorour@eng.asu.edu.eg
* Correspondence: mezzat@psu.edu.sa; Tel.: +966-506-362-379

Abstract: The full-scale static pile loading test is without question the most reliable methodology for estimating the ultimate capacity of large diameter bored piles (LDBP). However, in most cases, the obtained load-settlement curves from LDBP loading tests tend to increase without reaching the failure point or an asymptote. Loading an LDBP until reaching apparent failure is seldom practical because of the significant amount of settlement usually required for the full shaft and base mobilizations. With that in mind, the supervised learning algorithm requires a huge labeled data set to train the machine properly, which makes it ideal for sensitivity analysis, forecasting, and predictions, among other unsupervised algorithms. However, providing such a huge dataset of LDBP loaded to failure tests might be very complicated. In this paper, a novel practice has been proposed to establish a labeled dataset needed to train supervised machine learning algorithms on accurately predicting the ultimate capacity of an LDBP. A comprehensive numerical parametric study was carried out to investigate the effect of both pile geometrical and soil geotechnical parameters on both the ultimate capacity and settlement of an LDBP. This study was based on field measurements of loaded to failure LDBP tests. Results of the 29 applied models were compared with the calibrated model results, and the variation in LDBP behavior due to change in any of the hyperparameters was discussed. Accordingly, three primary characteristics were identified to diagnose the failure of LDBPs. Those characteristics were utilized to establish a decision tree of a supervised machine learning algorithm that can be used to predict the ultimate capacity of an LDBP.

Keywords: large diameter bored pile; hyperparameters; supervised machine learning; finite element method; parametric study; load transfer; failure mechanism

Citation: Al-Atroush, M.E.; Hefny, A.M.; Sorour, T.M. A Parametric Numerical Study for Diagnosing the Failure of Large Diameter Bored Piles Using Supervised Machine Learning Approach. *Processes* **2021**, *9*, 1411. https://doi.org/10.3390/pr9081411

Academic Editor: Li Li

Received: 15 July 2021
Accepted: 13 August 2021
Published: 16 August 2021

1. Introduction

The design of foundations has to satisfy two principal necessities [1]. First, complete failure of the foundation must be prevented with a sufficient margin of safety. Usually, the safety factor is assumed in practice to obtain the maximum safe foundation load (load-based design approach). Second, the relative and total settlements of the foundation must be kept within limits that the superstructure can tolerate. Therefore, the settlement of the foundation under the working load has to be estimated to ascertain its effect on the superstructure (settlement-based design approach). Traditionally, both load and settlement control approaches are usually applied separately in the design process. However, Ref. [2] stated that the allowable load on the pile foundation should be obtained through a combined approach considering both soil resistance and its settlement inseparably acting together and each influencing the value of the other.

Significant contributions have been proposed to develop a settlement-based design approach for large diameter bored piles (LDBP) alongside the conventional capacity-based design approaches. Despite that, several studies [3–6] reported that most of the available methods for forecasting either the bearing capacity or the load-settlement behavior of LDBP invariably are associated with various degrees of uncertainty resulting from several factors. The mechanisms of pile foundations and pile–soil interaction are ambiguous, complex, and not yet entirely understood [7–10]. Additionally, several factors influence the pile settlement in a semi-infinite mass and a finite layer, as indicated by [11], i.e., the pile length to diameter ratio, Poisson's ratio, and others. Based on results of numerical analyses studies, it was argued by [12] that the skin friction and end bearing load-transfer responses of large diameter bored piles may invariably depend on the pile shaft length to diameter ratio (l/d), the relative strength and stiffness of the existing soils along the pile shaft and in the vicinity of the pile base, also the applied load level, and the accomplished amount of pile settlement. Those conclusions pinpoint that pile proportional and geometrical parameters (length and diameter) have a significant effect not only on pile settlement but also on the ultimate capacity of the pile, i.e., the portion of the load carried by end bearing resistance may be somewhat dominant in the cases of either an embedded short pile in a predominately homogeneous granular soil or an embedded long pile in a stronger/stiffer layer overlain by a weaker/soft layer. Incidentally, Ref. [12] proposed an l/d limit of "10" to distinguish between short and long piles.

Consequently, by necessity, many of the available methods have been mainly based on simplifications and assumptions [13]. This has led to limited success in terms of providing consistent and accurate predictions (Refs. [8,14–17]). Therefore, the in-situ pile loading test has been commonly accepted as the method providing accurate bearing capacity and settlement predictions. For design purposes, the full load–settlement response of the pile has to be well predicted and simulated; the designer can thus decide the ultimate load and comply with the serviceability requirement. Based on that, the loading test has been widely recommended by several international design standards (e.g., Refs. [18,19] and others).

Fellenius [20] defined the ultimate pile capacity at failure state, based on the pile load settlement performance as the load at which a considerable increase in the pile settlement occurs under a sustained or slight increase in the applied load. However, in most cases, load-settlement curves obtained from full-scale load tests conducted on large diameter bored piles tend to increase without reaching the failure point or an asymptote. Despite the reliability, loading of this class of piles (large diameter bored piles) until reaching apparent failure is seldom practical. An enormous amount of pile settlement is usually compulsory to achieve full friction and bearing mobilization [21–24]. Therefore, several extrapolation techniques, such as [25,26], among others, were developed to interpret the pile failure load (ultimate pile capacity) using the pile loading test data. Most of those methods are associated with several degrees of uncertainty as they were developed using the results of different cases of pile loading tests with a variety of pile geometries and geological conditions, and hence occasionally, any two give the same failure load (Refs. [5,15,27]). Despite that, these methods are widely accepted in many international design codes (e.g., Refs. [18,19]) to predict the ultimate capacity of large diameter bored piles if the in-situ loading test measurements do not indicate an apparent failure point.

Several international codes (e.g., Refs. [18,19,28]) suggest settlement-based failure methods for estimation of the ultimate capacity of large diameter bored piles in caseit is impossible to perform pile loading tests at the design phase. In addition, various correlations have been proposed in several pile design codes to predict the bearing capacity of large diameter bored piles based on the results of in-situ soil tests such as the Standard Penetration Test (SPT) and the Cone Penetration Test (CPT). However, most of the available forecasting methods either for the bearing capacity or the load-settlement behavior of large diameter bored piles in clay soils are deterministic in the sense that average/representative values of the soil properties incorporated along the affected zone of soil by the pile load are used. However, these pile design methods neglected the potentially existing variation of the in-situ soil properties along the pile shaft (inherent vertical soil variability) (Refs. [5,29–32]).

With that in mind, the load transfer method is generally a simple analytical procedure that can be applied in many complex situations, such as variation in the sections along a pile shaft and an inhomogeneous layered soil system. Therefore, there is still a need to develop a theoretically sound method that predicts a reliable value for the ultimate capacity of large diameter bored piles (LDBP). However, with the new technologies available recently, such as neural networks, artificial intelligence, and machine learning, the influence of both soil resistance and its settlement can be more accurately considered inseparably, acting together and each influencing the value of the other. Nevertheless, the machine should first learn the effect of each parameter influencing the pile behavior. Therefore, each influencing factor should be investigated individually to be used in the machine learning process [33–37]. The parameters whose values are used to control the learning process are called hyperparameters.

Supervised and unsupervised machine learning are the two basic approaches of artificial intelligence (AI) and machine learning. The main distinction between the two approaches is the use of labeled datasets [34,37]. Supervised learning uses labeled input and output data, while an unsupervised learning algorithm does not. These datasets are designed to train or supervise algorithms into classifying data or predicting outcomes accurately. Using labeled inputs and outputs, the model can measure its accuracy and learn over time. On the other hand, unsupervised learning uses machine learning algorithms to analyze and cluster unlabeled data sets. These algorithms discover hidden patterns in data without the need for human intervention. Thus, supervised learning models are ideal for sensitivity analysis, forecasting, and predictions, among others. Despite that, they require a huge data set to train the machine properly and obtain accurate results [34]. With that in mind, it is hard to have a huge data set for loading tests for the pile cases, especially for LDBP that rarely reach failure.

Certainly, the full-scale static pile loading test is the most reliable methodology for estimating the ultimate capacity of large diameter bored piles (LDBP). That is why several international geotechnical codes and foundation design standards (e.g., Refs. [18,19,28]) recommend this method to study and investigate the load transfer and failure mechanisms of this class of piles. However, in most cases, the obtained load-settlement curves from such tests conducted on LDBP tend to increase without reaching the failure point or an asymptote. Loading LDBP until they reach apparent failure is seldom practical because of the significant amount of pile settlement that is usually required to fully mobilize the pile shaft and reach the ultimate base resistances [21–27]. Huge test loads and hence high-capacity reaction systems should be used to accomplish the required enormous settlements. Thus, the targeted failure load may not always be practical to achieve, as reported in many case studies [21–27].

In this paper, a novel practice is proposed to establish a labeled dataset needed to train a supervised learning algorithm on accurately predicting the ultimate capacity of the LDBP. A comprehensive parametric study is carried out to investigate the effect of both pile geometrical and soil geotechnical parameters on both the ultimate capacity and settlement of LDBP installed in clayey soils. The procedure followed in this parametric study aims to explore the characteristic effect of each factor affecting the behavior of LDBP by comparing results of the numerical models with the measurements of a well-documented loaded to failure pile load test and assess the variation in both pile settlement and ultimate capacity due to change in either pile geometrical or soil geotechnical factors included in this study. Results of this parametric study are utilized to develop the decision tree needed to train supervised machine learning algorithms.

2. The Reference Case Study

A full-scale and well-instrumented load test was carried out by [38] at the location of Alzey bridge (Germany) to investigate the behavior of a large diameter bored pile (LDBP).

The instrumentation utilized in the LDBP loading test is described in Figure 1a. The length and diameter of the investigated pile were 9.50 m and 1.30 m, respectively. This LDBP was installed in overconsolidated stiff clay soil and subjected to axial loading cycles until achieving failure, as shown in Figure 1b. Test setup and the soil characteristics at the site location are also given in Figure 1b.

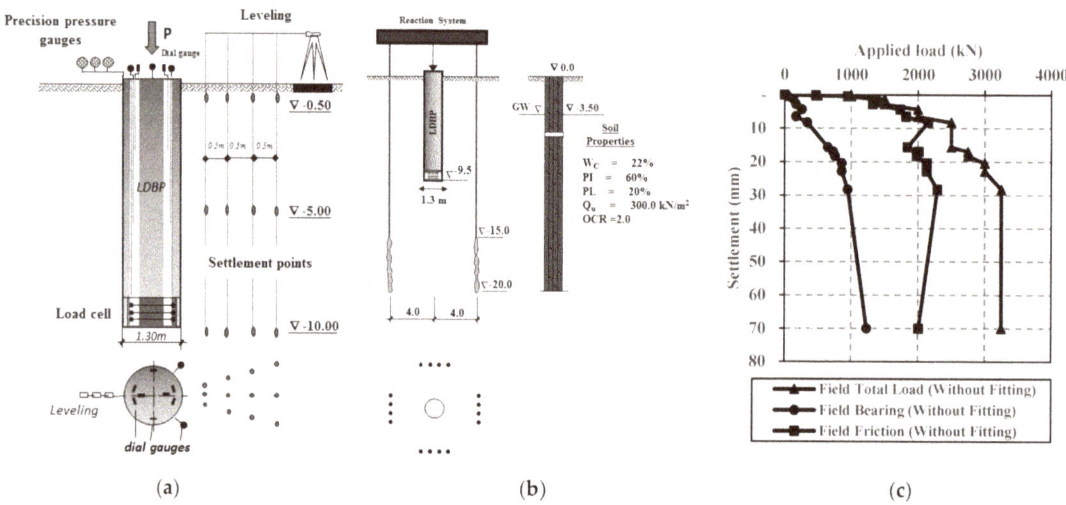

Figure 1. Large diameter bored pile loading to failure test (modified from [38]). (**a**) Measuring devices and instrumentation. (**b**) Test arrangement and typical soil profile with mechanical properties. (**c**) Field measurements of the loading test.

The main measurements of the well-instrumented load test are shown in Figure 1c. As shown, the significant increase in the measured settlement indicated apparent failure at the end of this test when the last load increment was applied on the pile head. More details of the loading test are available in [38].

3. The Reference Numerical Calibration Study

A numerical study has been carried out to simulate the response of the LDBP of the Alzey bridge case history [39]. Figure 2a shows the numerical model established to simulate the drained condition of overconsolidated (OC) stiff clay soil. The micro-fissures associated with the OC stiff clay are the main reason behind using the drained condition. These micro-fissures usually provide avenues for local drainage; soil along fissures has softened (increased water content) and is softer than intact material (more comprehensive discussion for using the drained condition with OC stiff clay is provided in [39]). Three constitutive soil models have been utilized to simulate the drained condition of the overconsolidated (OC) stiff clay soil. It was found that for this case history, the modified Mohr-Coulomb (MMC) constitutive model was superior to Mohr-Coulomb (MC) and the soft soil (SS) model in the simulation of the soil behavior [40,41]. The secant stiffness non-linear convergence method has been utilized to provide numerical stability required for the software solver to obtain convergence at substantial strain results (at failure). In addition, the sensitivity analyses performed also highlighted the significant effects of the mesh size and geometric dimensions on the analysis results. Results of this calibration study showed that excellent agreement was obtained between finite element results and the in-situ measurements of both the pile load settlement and load transfer relationships, as presented in Figure 2b,c.

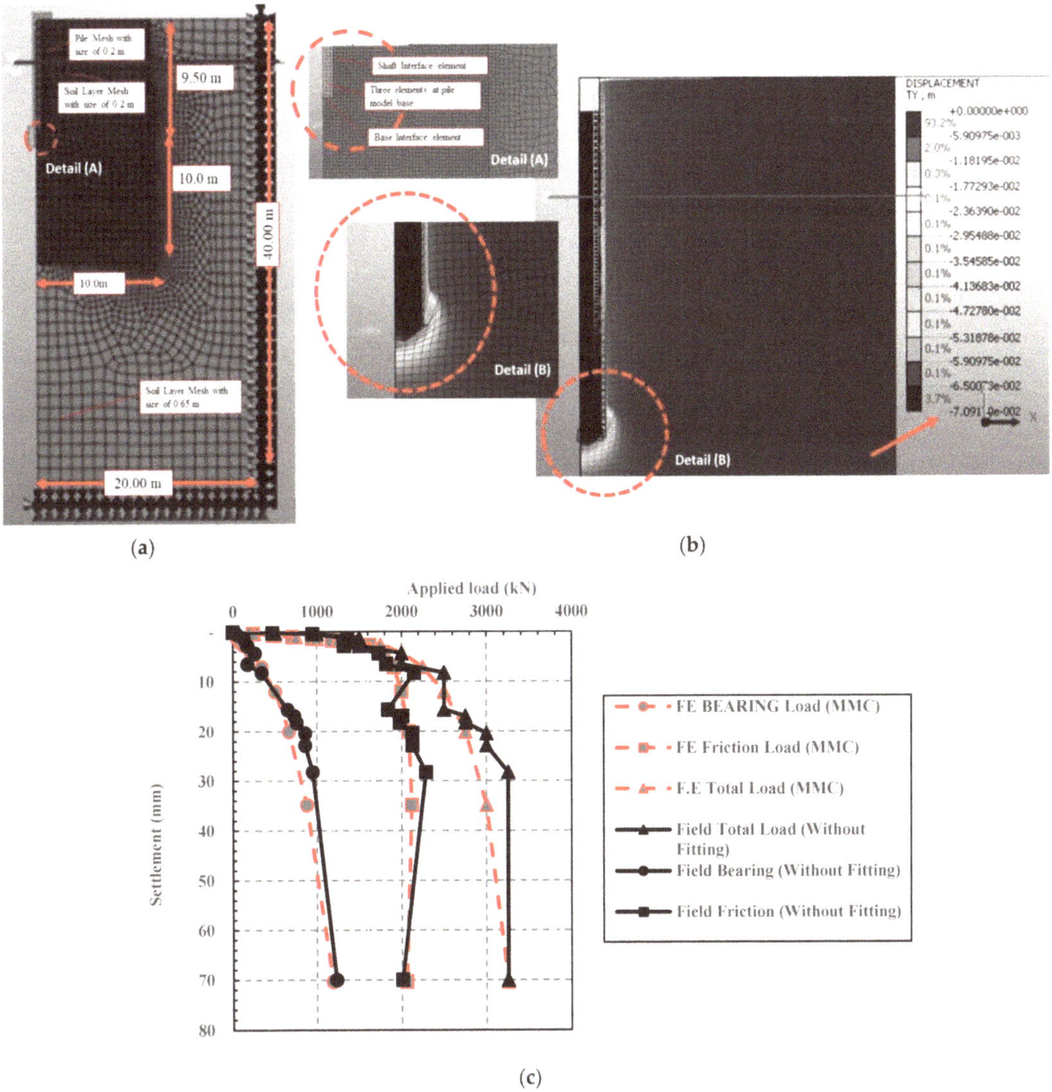

Figure 2. Numerical model established to simulate the response of the LDBP of the Alzey bridge case history (after [39]). (**a**) Details of boundary conditions. (**b**) Deformed shape of the finite element mesh (MMC) under the failure load. (**c**) Comparison between field measurements and the numerical results (after [39]).

Based on the numerical calibration analysis performed, the finite element method was capable of predicting not only the working capacity but also the ultimate capacity of the large diameter bored piles (LDBP). Moreover, the large induced pile settlement at the failure state could be determined. The procedures that followed to calibrate the numerical models are available in [39].

4. Methodology of the Parametric Study

Factors affecting the response of large diameter bored piles in clayey soils are classified in this study into two main categories (Figure 3). The procedure followed in this parametric study aims to explore the characteristic effect of each geometrical or geotechnical factor affecting the behavior of LDBP. Therefore, a particular sequence will be followed in this study; each factor is explored separately. Measurements of the reference case are used to assess the variation in both pile settlement and ultimate capacity due to change in the factor under investigation, and the influences of the other parameters are filtered out at this step. In this way, it is possible to examine the specific effect of each factor.

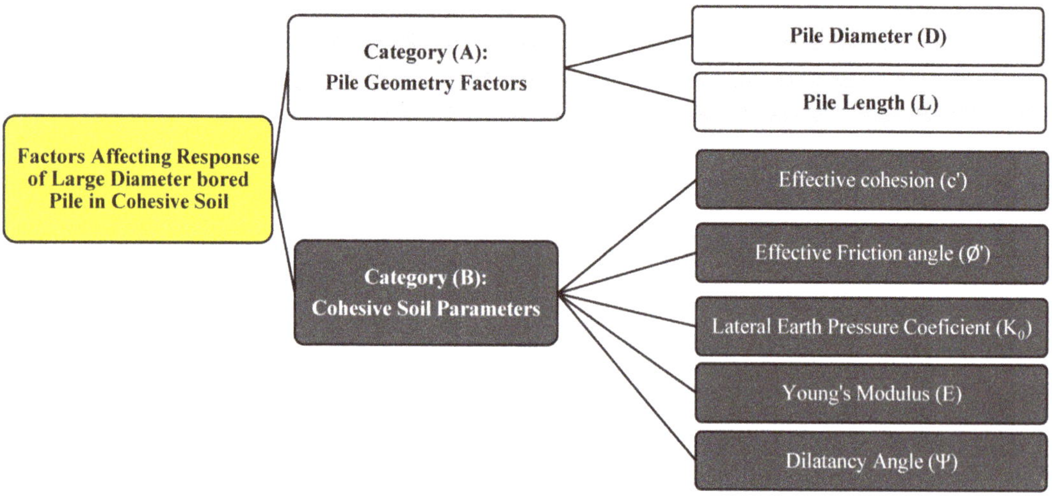

Figure 3. Factors affecting the behavior of large diameter bored pile in cohesive soil.

The first category (A) includes pile geometrical parameters (pile diameter D and pile length L). The effect of these parameters on the response of large diameter bored piles is assessed by implementing different values for pile diameter and length in the finite element analyses. At the same time, soil properties are kept with the same values adopted in the reference calibrated model [39] without any change to reveal the variance in pile behavior due to the change in either pile diameter or length. Because the pile's diameter is the pivotal parameter in this study, nine numerical models for piles with equal lengths and different diameters are established to reveal the change in ultimate bearing resistance under the pile base during the failure. In contrast, only three finite element pile models with equal diameters and various lengths ranging from double to triple the original pile case's length are established to investigate skin friction changes.

On the other hand, the second category (B) involves the cohesive soil geotechnical parameters: effective cohesion c′, effective friction angle Ø′, lateral earth pressure coefficient K_0, soil Young's modulus E, and dilatancy angle Ψ. The effect of each of these soil parameters on the response of LDBP in clayey soil will be individually evaluated by considering different values for each factor and assessing the change in the obtained ultimate pile capacity. Worth noting in this category, pile geometry parameters are kept with the same values adopted in the reference study [39]. Seventeen (17) new numerical analytical trials are performed for this purpose and compared with the calibrated model results.

4.1. Adopted Parameters in the Numerical Models

Modified Mohr-Coulomb (MMC) constitutive model has been used to define the drained condition of the homogenous overconsolidated stiff clay soil in all of the twenty-nine established numerical models for both categories A and B. The adopted parameters in each numerical model included in this study are summarized in Table 1. It is fundamental to note that the overconsolidation ratio (OCR) was set at a constant value of 2.0 for all cases [38], and its effect was not considered a variable in the parametric analyses.

Additionally, the adopted values for shear strength parameters were selected to cover the different OC stiff clay soil classes in various drained conditions.

Table 1. The adopted geometrical and geotechnical parameters in the parametric study.

	Pile Structural Parameters					Overconsolidated Stiff Clay Soil Parameters								
Model No	D	L	$E_{elastic}$	μ	G_c	$\gamma_{unsat}\backslash\gamma_{sat}$	v	C	Ø	$E_{50}^{ref} = E_{oed}^{ref}$	E_{ur}^{ref}	K_0	ψ	R
Units	m	m	kN/m^2	-	kN/m^3	kN/m^3	-	kN/m^2	o	kN/m^2	kN/m^2	-	o	-
1 (The Calibrated Model)	**1.3**	9.5	24,248,711	0.2	24	20	0.3	20	22.5	45,000	90,000	0.8	0.1	1
2	**0.4**	9.5	24,248,711	0.2	24	20	0.3	20	22.5	45,000	90,000	0.8	0.1	1
3	**0.5**	9.5	24,248,711	0.2	24	20	0.3	20	22.5	45,000	90,000	0.8	0.1	1
4	**0.6**	9.5	24,248,711	0.2	24	20	0.3	20	22.5	45,000	90,000	0.8	0.1	1
5	**0.7**	9.5	24,248,711	0.2	24	20	0.3	20	22.5	45,000	90,000	0.8	0.1	1
6	**0.8**	9.5	24,248,711	0.2	24	20	0.3	20	22.5	45,000	90,000	0.8	0.1	1
7	**1**	9.5	24,248,711	0.2	24	20	0.3	20	22.5	45,000	90,000	0.8	0.1	1
8	**1.2**	9.5	24,248,711	0.2	24	20	0.3	20	22.5	45,000	90,000	0.8	0.1	1
9	**1.5**	9.5	24,248,711	0.2	24	20	0.3	20	22.5	45,000	90,000	0.8	0.1	1
10	**2.0**	9.5	24,248,711	0.2	24	20	0.3	20	22.5	45,000	90,000	0.8	0.1	1
11	1.3	**13**	24,248,711	0.2	24	20	0.3	20	22.5	45,000	90,000	0.8	0.1	1
12	1.3	**19**	24,248,711	0.2	24	20	0.3	20	22.5	45,000	90,000	0.8	0.1	1
13	1.3	**26**	24,248,711	0.2	24	20	0.3	20	22.5	45,000	90,000	0.8	0.1	1
14	1.3	9.5	24,248,711	0.2	24	20	0.3	20	22.5	**20,000**	**40,000**	0.8	0.1	1
15	1.3	9.5	24,248,711	0.2	24	20	0.3	20	22.5	**30,000**	**60,000**	0.8	0.1	1
16	1.3	9.5	24,248,711	0.2	24	20	0.3	20	22.5	**60,000**	**120,000**	0.8	0.1	1
17	1.3	9.5	24,248,711	0.2	24	20	0.3	20	22.5	**80,000**	**160,000**	0.8	0.1	1
18	1.3	9.5	24,248,711	0.2	24	20	0.3	20	**10**	45,000	90,000	0.8	0.1	1
19	1.3	9.5	24,248,711	0.2	24	20	0.3	20	**15**	45,000	90,000	0.8	0.1	1
20	1.3	9.5	24,248,711	0.2	24	20	0.3	20	**30**	45,000	90,000	0.8	0.1	1
21	1.3	9.5	24,248,711	0.2	24	20	0.3	**5**	22.5	45,000	90,000	0.8	0.1	1
22	1.3	9.5	24,248,711	0.2	24	20	0.3	**10**	22.5	45,000	90,000	0.8	0.1	1
23	1.3	9.5	24,248,711	0.2	24	20	0.3	**30**	22.5	45,000	90,000	0.8	0.1	1
24	1.3	9.5	24,248,711	0.2	24	20	0.3	**50**	22.5	45000	90,000	0.8	0.1	1
25	1.3	9.5	24,248,711	0.2	24	20	0.3	**100**	22.5	45,000	90,000	0.8	0.1	1
26	1.3	9.5	24,248,711	0.2	24	20	0.3	20	22.5	45,000	90,000	**0.43**	0.1	1
27	1.3	9.5	24,248,711	0.2	24	20	0.3	20	22.5	45,000	90,000	**0.62**	0.1	1
28	1.3	9.5	24,248,711	0.2	24	20	0.3	20	22.5	45,000	90,000	0.8	**1.0**	1
29	1.3	9.5	24,248,711	0.2	24	20	0.3	20	22.5	45,000	90,000	0.8	**3.0**	1
30	1.3	9.5	24,248,711	0.2	24	20	0.3	20	22.5	45,000	90,000	0.8	**5.0**	1

D: pile diameter. L: pile length. $E_{elastic}$: Young's modulus of Pile material. μ: pile Poisson's ratio. Ø': soil effective friction angle. c': soil effective cohesion. v: soil Poisson's ratio. K_0: lateral earth pressure coefficient.t ψ: dilatancy angle. E_{50}^{ref}: secant stiffness in the standard drained triaxial test. E_{oed}^{ref}: tangential stiffness in oedometer test loading. E_{ur}^{ref} : elastic modulus at unloading. R: interface strength reduction factor.

4.2. Numerical Modeling and Sensitivity

The 29 numerical models of this study were established using MIDAS GTS NX finite element package. For category A, it was essential to perform sensitivity analysis to unify the effect of mesh size on the examined finite element models. This sensitivity study has been performed with the same procedure presented in the numerical reference study [39]. Results of these analytical trials showed that a fine mesh should be utilized to represent the pile element with a size that enables having at least two or three mesh elements in the vicinity of the pile base. This is important for bearing resistance results to eliminate the effect of mesh dependence at the level of the pile base [40]. Further, the boundary locations should be far away from the pile element for enough distance, not less than 15 times pile diameter (15D) in width, and at least 4 times the pile length (4L) in depth.

These dimensions should be considered to ensure that the locations of boundaries do not affect the analysis zone.

Based on the results of sensitivity analyses performed, 12 two-dimensional axisymmetric models are established to investigate the effect of the pile geometrical parameters Nine of them are prepared to simulate 10 separate single piles with equal lengths of 9.50 m and different diameters of; 0.40 m, 0.50 m, 0.60 m, 0.70 m, 0.80 m, 1.0 m, 1.20 m, 1.50 m, and 2.0 m. The other three numerical models are established to simulate three single piles with equal diameters of 1.30 m and different lengths of 13.0 m, 19.0 m, and 26.00 m. The 12 models' results were compared with the reference case study (1.3 m diameter and 9.50 m length) to assess the change in pile behavior due to change either in pile diameter or length.

In contrast, for category B, pile geometrical dimensions are taken with the same values of the reference case study, and the soil properties are changed according to the factor considered for evaluation in each section. The effect of the soil geotechnical parameters is investigated through 17 numerical models, and no change in the pile geometry is considered for this part of the study. Thus, the same mesh size and geometry dimensions adopted previously in the reference calibrated model have also been adopted to investigate the effect of category B parameters on pile response. According to [39], the positions of the model boundaries did not affect the analysis results of stresses and displacement around the pile when a numerical model height of 40 m and width of 20 m was adopted. Furthermore, the pile was represented by a very fine mesh with a mesh size of 0.216 m (quadratic 8-node elements). Additionally, with the same mesh size, a fine soil mesh media was established around and below the pile element. Gradually, soil mesh size was increased to be 1.0 m at the boundary locations.

According to the calibration study [39], very good agreements were obtained between field measurements of the Alzey bridge case study and the results of the numerical analyses, even at the failure state, when an interface strength reduction factor (R) of 1.0 was adopted in the analyses. Consequently, the same value of the reduction factor was utilized in the 29 parametric models.

4.3. Stages of Analysis

Similarly to the analytical procedure utilized in the numerical calibration study, the analysis sequence followed in this parametric study consists of three stages of analysis. The first stage represents the initial stresses of the soil due to the overburden effect. This stage is essential to simulate the initial condition of the soil before installing the pile. The second stage started by changing the pile volume to concrete material instead of soil material to represent the pile installation phase. It is worth noting that a rigid interface element has been used in the second stage to provide the required numerical stability at this stage, as it is rigidly connecting the paired nodes of soil and pile to avoid singularity. In the third stage, the rigid interface has been replaced with the frictional interface, and the uniform load is applied to the pile head using incremental load steps (250 kN per load increment). The applied load in each case is defined with a value greater than the estimated value of ultimate pile capacity (using [42]) for each of the 30 pile cases to allow the numerical solver to achieve the failure according to the adopted convergence criteria.

4.4. Load Transfer Mechanism and Failure Criteria

The maximum load where convergence can be achieved in the numerical analysis is considered as the failure load. Failure is also indicated by the apparent large settlement that is expected to be induced under the application of the ultimate load. The pile load transfer mechanism is investigated by determining the pile stress in the vicinity of the pile base and multiplying it by the cross-sectional area of each pile case to obtain the pile bearing resistance at each load increment. Thus, pile friction resistance can be calculated by deducting the obtained pile bearing load from the total applied load at each loading increment. Then, the relations between pile load settlement, pile friction, and bearing capacities at each loading increment can be plotted. In contrast, the formation sequence, shape, and size of the formed plastic bulb at the failure status are explored at the obtained ultimate load of each pile model.

5. Analysis Results and Discussion

5.1. Large Diameter Bored Pile Load-Settlement Relationship

For pile geometrical parameters (category A), load-settlement results of the nine numerical pile models with different diameters and equal length (9.50 m) are presented in Figure 4a. Additionally, the results of the three numerical equal-diameter (1.30 m) pile models with different lengths are shown in Figure 4b. Those results were compared with the load–settlement relationship of the calibrated model (Figure 1) to assess the change in pile load settlement performance due to variation in either pile diameter D or pile length L.

Figure 4a,b highlight the effect of pile diameter and length on the ultimate pile capacity and the induced settlement at failure. Hence, the ultimate capacity increases with increases in pile diameter and length. Consequently, the induced settlement at the failure state is also increased. A significant increase in pile settlement is also observed at the last load increment in the results of the 12 finite element models (category A), as the induced pile settlement at the last load increment (failure load) ranged from about one and a half to more than two times the obtained settlement value at the pre-last loading increment.

Similarly, Figure 4c–e highlights that ultimate pile capacity is increased, and consequently, the induced settlement at failure is also increased with increases of the following soil geotechnical parameters from category B: clay effective cohesion (c'), effective friction angle (Ø'), and soil dilatancy angle (Ψ). Conversely, the induced pile settlement at failure is decreased with increases in the lateral earth pressure coefficient (K_0), despite the observed increase in ultimate pile capacity, as shown in Figure 4f. This is attributed to the confinement of the surrounding soil due to stress increases according to lateral earth pressure coefficient increases. Conversely, the increase in soil Young's modulus (E) also caused a decrease in the induced pile settlement at failure, but without any observed change in ultimate pile capacity, as shown in Figure 4g.

5.2. Pile Load Transfer Mechanism

Based on the methodology explained before in Section 4, the ultimate total, bearing, and friction capacities are determined for each of the 29 numerical models performed in this study. The effect of each factor, either from category A or B, on the ultimate pile capacity is given in Figure 5a–g. Figure 5a–g highlights that the three ultimate capacities (total, bearing, and friction) are increasing with increases in pile diameter (D), length (L), effective clay cohesion (c'), effective friction angle (Ø'), and soil dilatancy angle (Ψ). However, Figure 5e reveals that the lateral earth pressure coefficient significantly affects the pile friction resistance; however, it has almost no effect on the ultimate bearing capacity. In contrast, Figure 5f indicates that soil Young's modulus has almost no effect on the ultimate friction, bearing, and total capacities of the large diameter bored pile. However, it has a major impact on the settlement of large diameter bored pile, especially at the failure state (see Figure 4g).

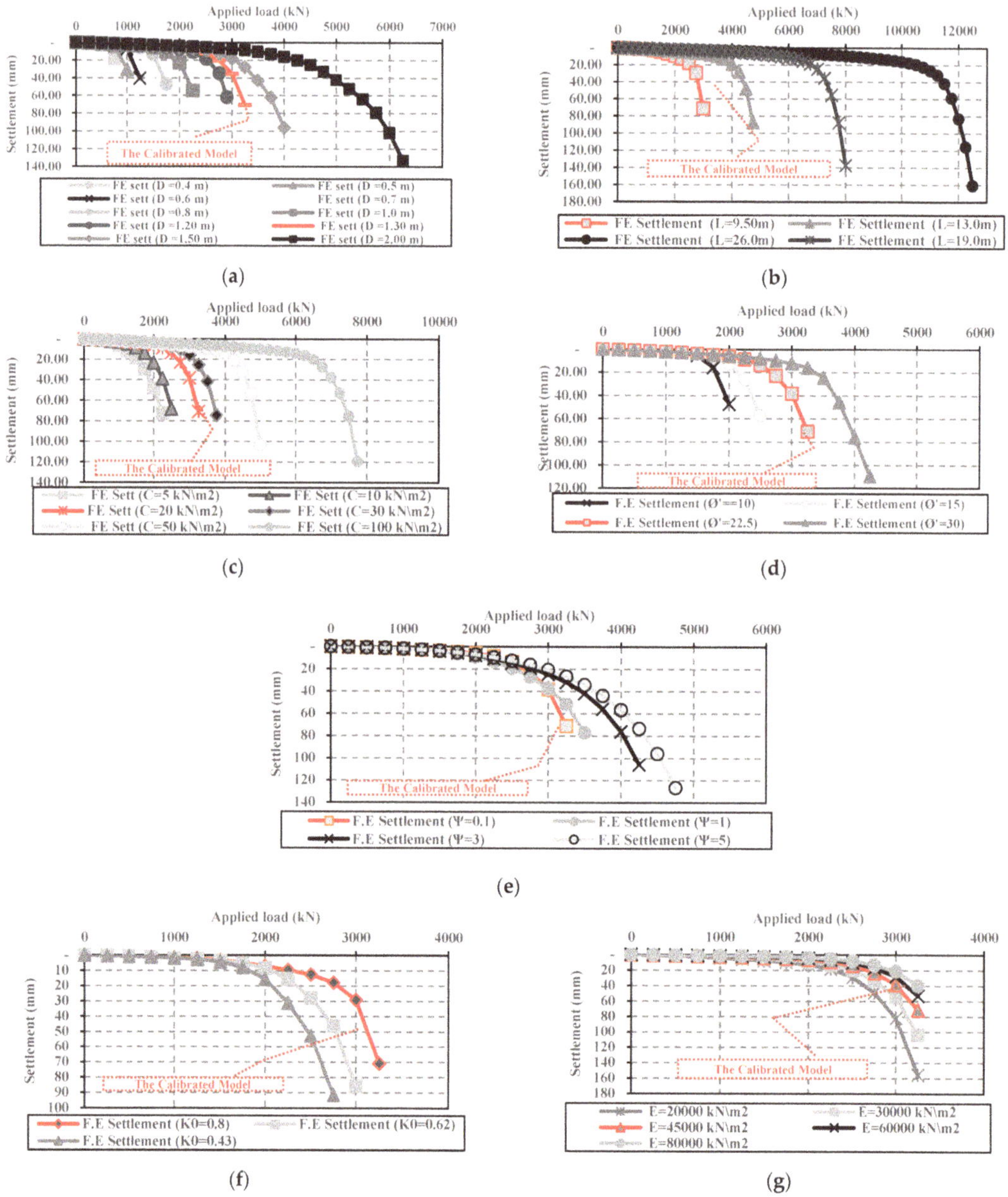

Figure 4. Effect of the pile geometrical and soil geotechnical parameters on the load settlement relationship. (**a**) Pile diameter (D). (**b**) Pile length (L). (**c**) Effective cohesion (c′). (**d**) Effective friction angle (Ø′). (**e**) Dilatancy angle (Ψ). (**f**) Lateral earth pressure coefficient (k_0). (**g**) Young's modulus (E_s).

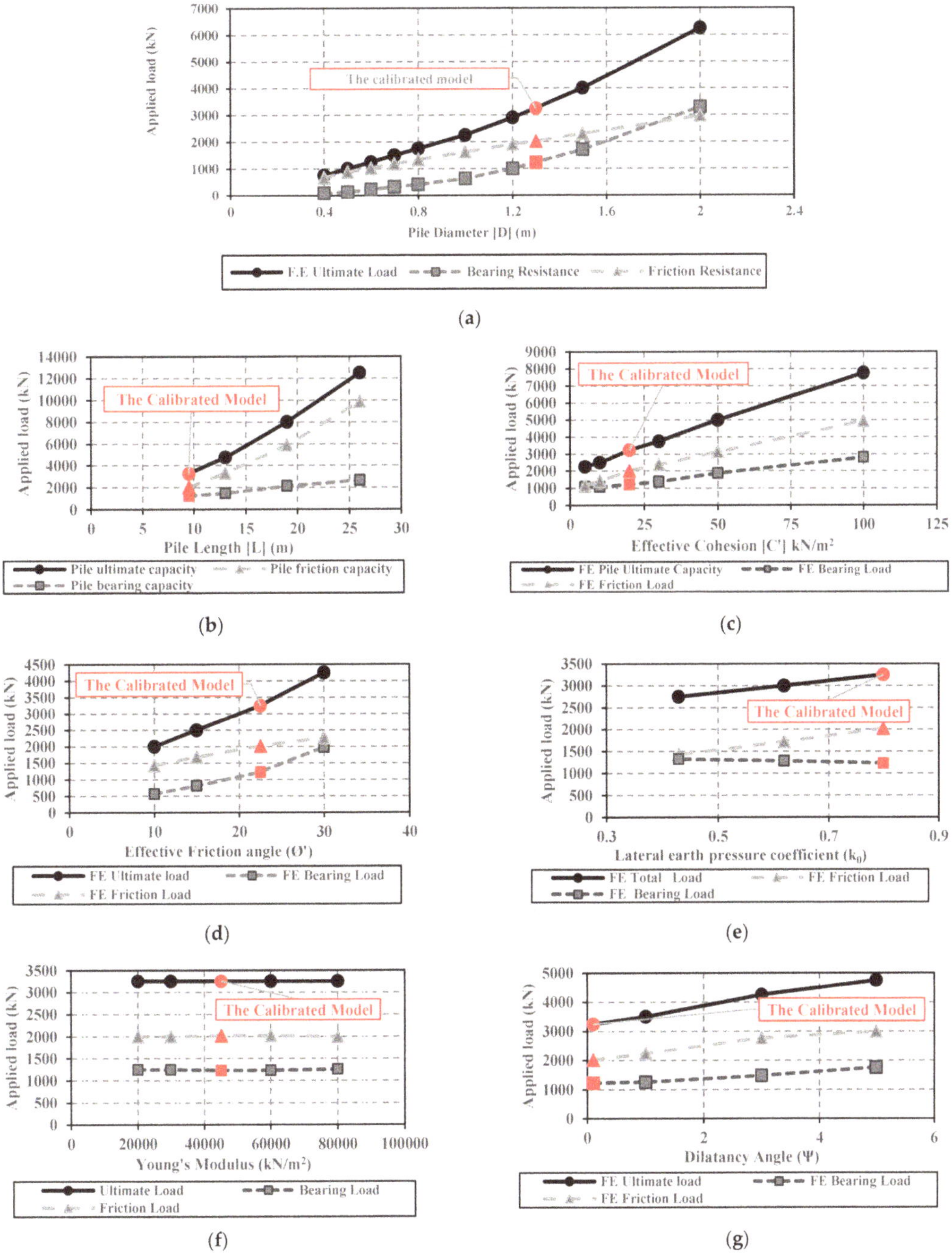

Figure 5. Variation of the ultimate bearing (Q_b), friction (Q_f), and total capacities (Q_t) with the different parameters. (**a**) Pile diameter (D). (**b**) Pile length (L). (**c**) Soil effective cohesion (c′). (**d**) Effective friction angle (Ø′). (**e**) Lateral earth pressure coefficient (K_0). (**f**) Young's modulus (E_s). (**g**) Soil dilatancy (Ψ).

In the same line, percentages of load transferred by bearing and by friction are calculated relative to total obtained ultimate capacity (Q_B/Q_T and Q_F/Q_T) for each of the 29 numerical models, and the variations of those percentages with each factor in both categories are given in Figure 6a–f.

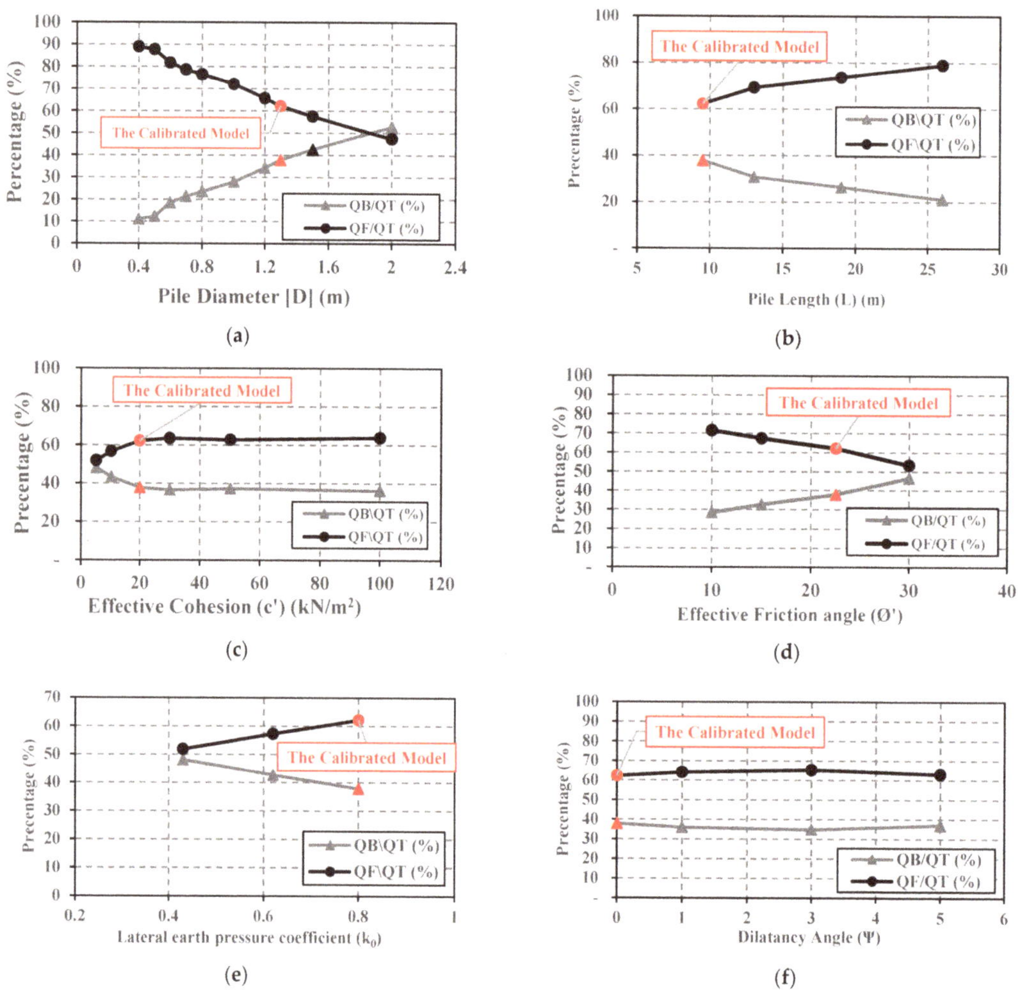

Figure 6. Variation of (Q_B/Q_T) and (Q_F/Q_T) with different parameters. (**a**) Pile diameter (D). (**b**) Pile length (L). (**c**) Effective cohesion (c′). (**d**) Effective friction angle (Ø′). (**e**) Lateral earth pressure coefficient (k_0) (**f**) Dilatancy angle (Ψ).

It can be seen from Figure 6a that the percentage of load transferred by bearing (Q_B/Q_T) increases from 10% to about 50%, with pile diameter increases from 0.40 m to 2.0 m. Conversely, the percentage of load transferred by friction decreased from about 90% for a pile with a diameter of 0.40 m to 47.3% for a pile with a diameter of 2.0 m.

It was also observed that the applied load was predominantly transferred by friction for the first nine cases (from 0.4 to 1.50 m in diameter). Especially in small diameter piles, the percentage of the load transferred by friction ranged from 90% (0.40 m small diameter pile) to 81% (0.60 m small diameter pile). An observed decrease in the percentage of the load transferred by friction was noted after the pile diameter increase from 0.5 m to 0.6 m, as the Q_f/Q_T decreased by about 10%.

For large-diameter piles (with diameters greater than 0.60 m), the load transferred by friction linearly decreased from 79% (a pile diameter of 0.70 m) to 58% (at a pile diameter of 1.50 m) with pile diameter increases. Thus, the total load friction and bearing percentages became nearly equal (52.6%:47.3%) in the last case (pile with 2.0 m diameter). These percentages indicate an apparent difference between the behavior of large and small diameter bored piles.

Figure 6b demonstrates that the applied load was predominantly transferred by friction (Q_f) for the four cases with different lengths. The percentage of load transferred by bearing decreases from about 40% to about 20% with pile length increases. Consequently, the percentage Q_f/Q_T is increased from about 60% (pile with length 9.50 m) to about 80% in the last case with the pile of 26 m length.

In summary, the percentages presented in Figure 6a,b indicate that the rate of load transferred by friction increases with increases in pile length. In contrast, this percentage (load transferred by friction) decreases with increases in pile diameter.

The effects from the category B parameters of effective clay cohesion (c'), effective friction angle (Ø'), lateral earth pressure coefficient (K), soil Young's modulus (E), and soil dilatancy angle (Ψ) on the percentage of load transferred by bearing or friction are given in Figure 6c–f.

It can be seen from Figure 6c that the percentage of load transferred by bearing (48%) nearly equals the percentage of load transferred by friction (52%) for a small value of soil cohesion of 5 kN/m^2. The percentage Q_F/Q_T increased from about 52% (5 kN/m^2) to near 63% for the case with an effective cohesion of 30 kN/m^2. For the last three cases with effective cohesion higher than 20 kN/m^2, the ratio between the load transferred by bearing and by friction of the total ultimate load tended to be constant (37% and 63%, respectively). Those percentages indicate that effective soil cohesion has the same impact on both bearing and friction pile resistances at the ultimate state.

Figure 6d demonstrates that the percentage of load transferred by friction decreases from about 70% (Ø' = 10°) to about 53% in the fourth case with an effective friction angle of 30°. In addition, the percentage Q_B/Q_T (47%) nearly equals the percentage of load transferred by friction (53%) in the last case with a friction angle of 30°. An observed increase is noticed in the percentage of load transferred by bearing with effective friction angle increases, as it increases from about 30% at an effective friction angle of 10° to about 47% at an effective friction angle of 30°. These percentages indicate that effective soil friction angle has a significant effect on the percentage of the load transferred by bearing, which is expected because of the related increase in soil bearing capacity with the increase in effective friction angle. Of note, in these trial analyses, the same value for the lateral earth pressure coefficient (K_0) was adopted for all four of the cases with different effective friction angles.

It can be seen from Figure 6e that the percentage of load transferred by friction increased from about 52% at k_o of 0.43 to about 62% in the third case with lateral earth pressure coefficient of 0.8; also, the percentage of load transferred by bearing (48%) nearly equals the percentage of load transferred by friction (52%) in the first case with k_o of 0.43. These percentages indicate that the soil lateral earth pressure coefficient has a significant effect on the percentage of the load transferred by friction, which is attributed to the related increase in soil shear strength with increases in k_0. On the other hand, Figure 6f highlights that the percentage of load transferred by friction and by bearing is almost unaffected by the change in dilatancy angle value, as about 63% of the applied load is carried by friction, and about 37% is transferred with bearing, in the four cases with different dilatancy angles.

In the next two sections, the results of the numerical models of this parametric study will be used to assess the calculated ultimate capacity of the LDBP using two different methods of both capacity-based and settlement-based design approaches. According to the available data in a field study [38], the Meyerhof 1976 capacity-based method [42] and

Egyptian code settlement-based design (ECP 202/4) [19] approaches were chosen for this evaluation because other methods, such as AASHTO LRFD [28], require soil testing results such as SPT or CPT test results that were not provided in the field study.

5.3. Average Ultimate Bearing Stress

Results of the numerical models are used to explore the effect of several geometrical and geotechnical parameters on the bearing stress below the pile base at the failure state. Pile diameter (D), length (L), clay effective cohesion (c′), and soil effective friction angle (Ø′) are the selected parameters to represent the two categories (A and B) in this evaluation. Numerical results are compared with estimated values of bearing stress using the chosen settlement-based design method of [19], as given in Figure 7a–d.

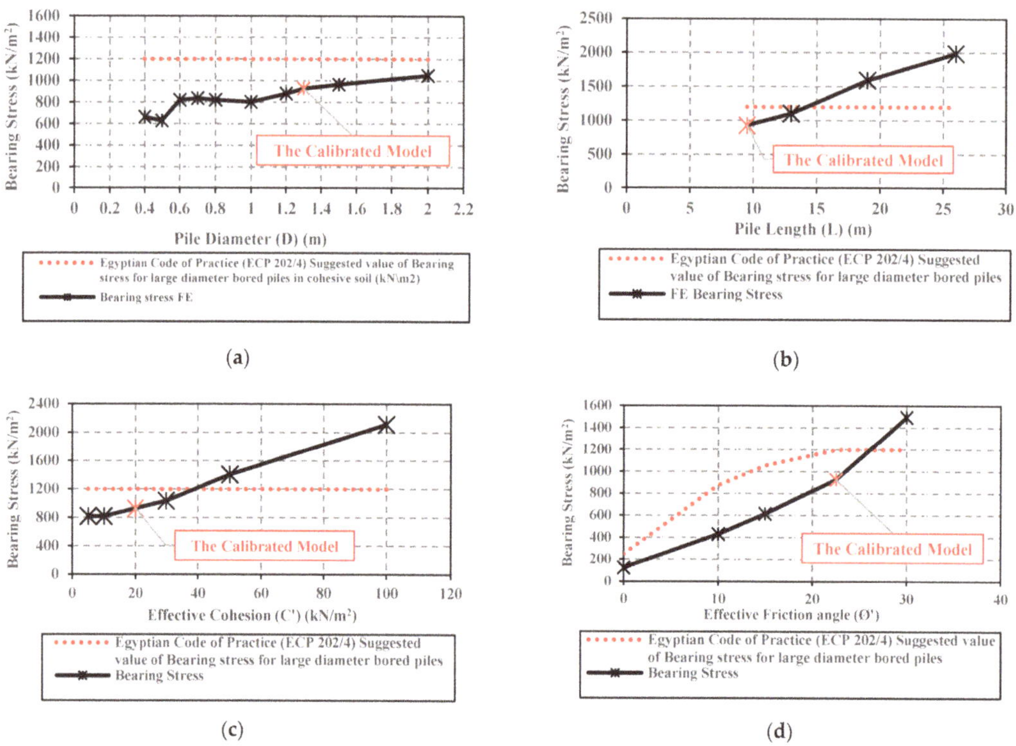

Figure 7. Influence in bearing stress with various parameters. (**a**) Pile diameter (D). (**b**) Pile length (L). (**c**) Effective cohesion (c′). (**d**) Effective friction angle (Ø′).

As shown in Figure 7a the almost equal value of bearing stress (about 600 kN/m^2) can be seen for piles with small diameters of 0.40 m and 0.50 m. A significant increase in bearing stress value was observed in the case of a pile with a diameter of 0.60 m, as the bearing stress increases to about 800 kN/m^2, which is about 33% greater than the obtained bearing stress at 0.4 and 0.5 m pile diameters. Further, almost the same value of bearing stress was obtained below the bases of piles with diameters of 0.6, 0.7, 0.8, and 1.0 m. Bearing stress increases again with pile diameter increases for piles with diameters greater than 1.0 m, to achieve its maximum value of 1050 kN/m^2 below the 2.0 m diameter pile base.

It can be seen from Figure 7b that the bearing stress increases with increases in pile length, which is expected at the drained condition because of the increase in vertical soil stress at the pile base level. The bearing stress increases from 920 kN/m^2 below the base of the pile with length 9.50 m, to nearly two times this value with a length of 26.0 m (1985 kN/m^2).

On the other hand, Figure 7c presents the obtained bearing stress values below the pile bases of the six finite element models with different soil effective cohesion values (category B). It can be seen from this Figure that the bearing stress increases with increases in soil effective cohesion. The bearing stress increases from about 800 kN/m^2 in the first case with an effective cohesion of 5 kN/m^2, to about three times that value (2113 kN/m^2) for the case with effective cohesion of 100 kN/m^2. Similarly, Figure 7d shows an increase in bearing stress due to the increase in soil effective friction angle.

Although the results of the numerical models showed that the ultimate bearing stress is influenced by the pile diameter, pile length, soil effective cohesion, and soil effective friction angle, the bearing stress estimated by the settlement-based method (ECP 202/4) is independent of pile diameter, pile length, and soil effective cohesion. Most of the available settlement-based methods depend on settlement-based criteria to estimate the ultimate bearing resistance of the pile at a particular settlement value. Those methods generally defined the nominal ultimate pile capacity (Q_{tu}) using different percentages of the pile head settlement to diameter ratio (s/d), such as 10%, according to ECP 202/4 [19].

5.4. Average Ultimate Unit Skin Friction

The effects of pile diameter (D), length (L), clay effective cohesion (c′), and soil effective friction angle (Ø′) on the soil unit skin friction are investigated using the results of the numerical models. The obtained numerical results are compared with the calculated values of the interface shear strength (according to Mohr-Coulomb failure criteria (σ_h tan $Ø_i$ + c_i)), to ensure that full mobilization has been achieved in each case.

As shown in Figure 8a–d, all numerical models reached the full friction mobilization. With that in mind, numerical results of the soil unit skin friction are also utilized to assess the ones determined using the chosen capacity-based method [42].

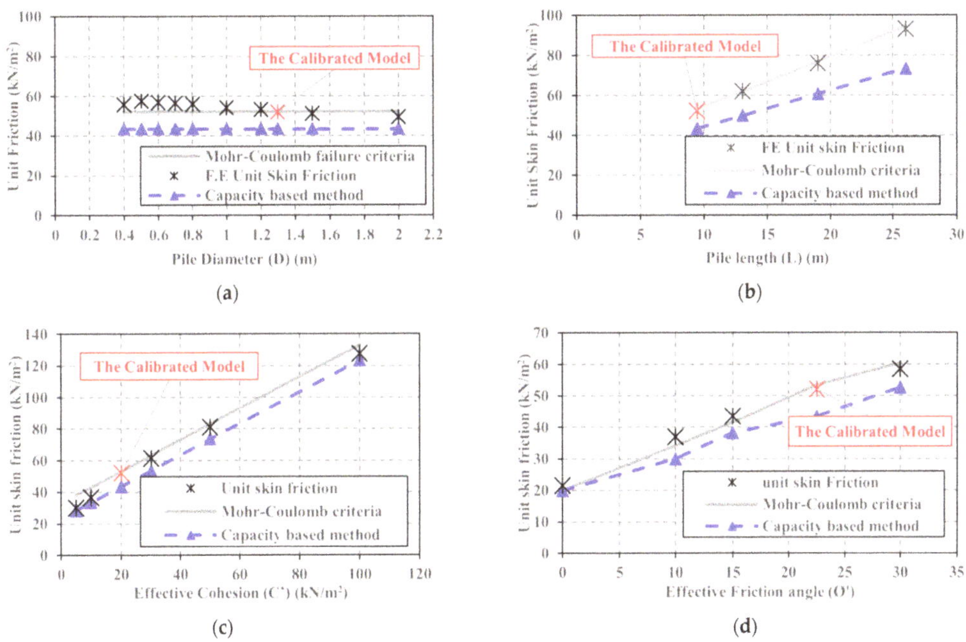

Figure 8. Variation of the average unit skin friction with various parameters. (**a**) Pile diameter (D). (**b**) Pile length (L). (**c**) Soil effective cohesion (c′). (**d**) Soil effective friction angle (Ø′).

As demonstrated in Figure 8a, the obtained results for the average unit skin friction of the 10 pile cases with different diameters are almost equal. This indicates that the unit skin friction is not affected by the pile diameter (D). In contrast, Figure 8b highlights that the unit skin friction of the four pile cases with different lengths increases with increases in pile length. This is attributed to the increase in vertical stress due to the increase in overburden pressure at the pile base level. Additionally, Figure 8c,d highlights that the obtained average unit skin friction results increase with the increase in effective soil cohesion and effective friction angle.

On the other hand, the calculated soil skin friction values using the capacity-based method [42] are close to the numerical results, except for the three cases with different lengths (Figure 8b) and the last two cases with greater friction angles (Figure 8d). The difference between the numerical results and those obtained using the Meyerhof formula ranges from 10% to 21%. However, the Meyerhof method underestimated the unit skin friction of all cases. Of note, the capacity-based method takes the effect of pile diameter (D), length (L), clay effective cohesion (c′), and soil effective friction angle (Ø′) into account, in contrast to the settlement-based methods that ignore the effect of several influencing factors, as presented in Section 4.3.

5.5. Size of Plastic Bulb below Pile Base

Figure 9a–d presents the shape of the formed plastic bulbs obtained from numerical analyses at the last load increment (failure load) for different pile model cases with various geometrical and geotechnical parameters. Plastic points are generally formed when the transferred stress achieves a value that almost equals the soil shear strength (according to Mohr-Coulomb's theory of rupture). Hence, the yielding occurs, and plastic strains are induced. The zone at which the plastic points are concentrated under the pile base is hereafter referred to as the plastic bulb. These plastic bulbs have been measured in diameter and length to assess the effect of each parameter considered in this study on the failure mechanism of large diameter bored piles.

Because the pile diameter is the main governing factor in this comparison, it was essential to investigate the differences between large and small diameter piles in terms of the size of the end bearing plastic bulb that formed under the pile base in the failure state. Figure 9a presents the change in width and length in the formed plastic bulb with pile diameter change. As shown, the plastic bulb is obviously increased in size and height with increases in pile diameter. The diameter of the plastic bulb (D_p) under the base of the first pile case with a diameter of 0.40 m is 2.14 m (~5D), and the greatest size of the plastic bulb is at 7.08 m (3.54D) when pile diameter increases to 2.00 m. The length of the plastic bulb below the pile bases (L_p) also increased with pile diameter increases, to about 5.14 m (~2.5D) below the 2.00 m diameter pile base. Of note, in the 10 cases with different diameters, the length of the plastic bulb below the pile base is nearly two and a half times the pile diameter.

In contrast, as shown in Figure 9b, plastic bulbs of almost identical diameter are obtained in all cases of piles with equal diameters and different lengths. A minor increase is observed with pile length increases. For the pile with a length of 9.50 m, the plastic bulb diameter measured at 3.68 times pile diameter and slightly increased to 4.05 D for the fourth case of the pile with a length of 26.0 m. Furthermore, the length of the plastic bulb slightly increased below the pile base from three times pile diameter at the pile with a length of 9.50 m to almost 3.10 of pile diameter at the pile with a length of 26.0 m.

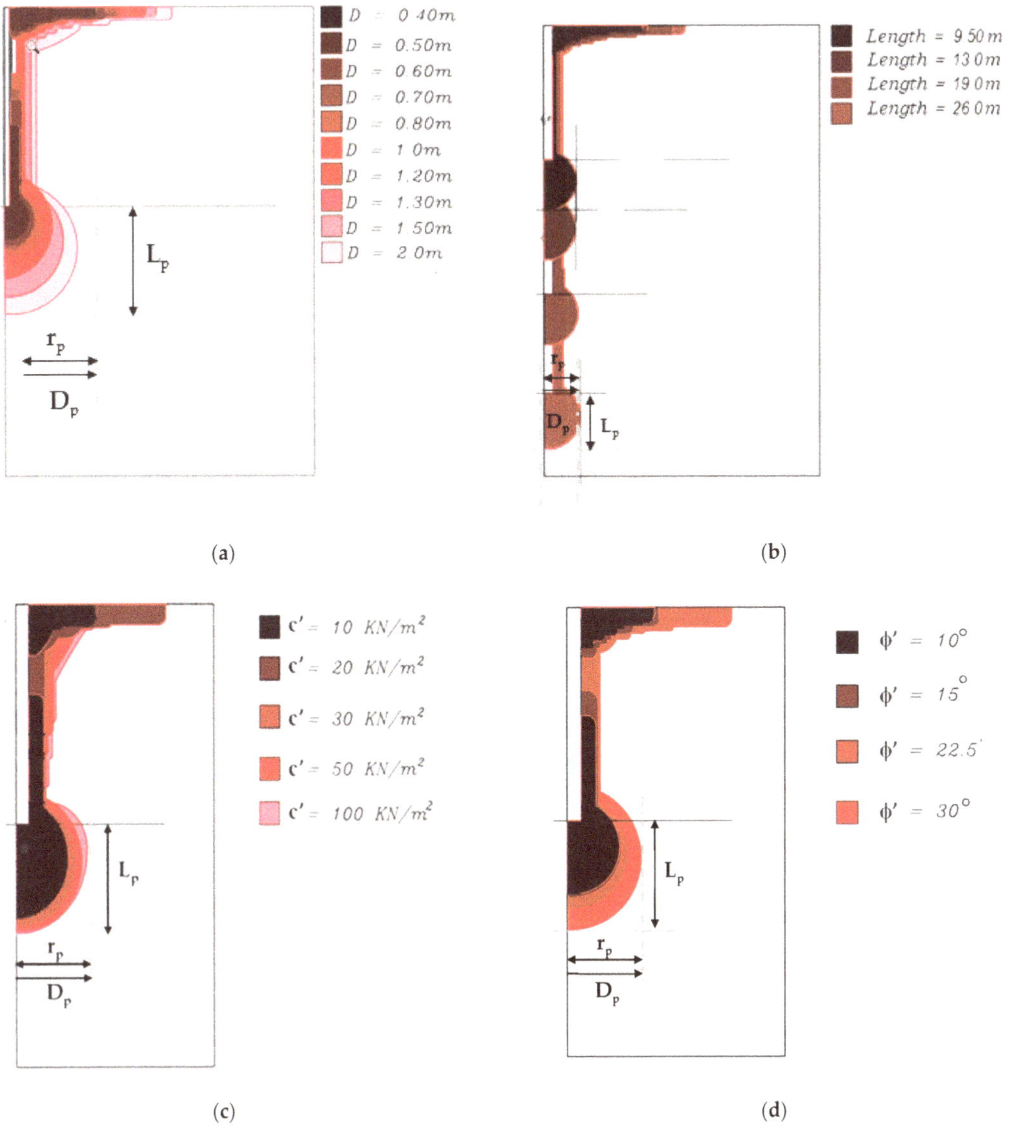

Figure 9. Influence in the formed plastic bulbs due to changes in different parameters. (**a**) Pile diameter (D). (**b**) Pile length (L). (**c**) Soil effective cohesion (c′). (**d**) Soil effective friction angle (Ø′).

Figure 9c indicates that the size of the formed plastic bulb is affected by effective soil cohesion. The formed plastic bulb increases in width and length as the soil effective cohesion increases, i.e., the plastic blub radius (r_p) is measured as 2.66 m (~4D) in the first case ($c' = 10$ kN/m^2) and increases to 3.6 m (5.54D) for the last case ($c' = 100$ kN/m^2). The height of the plastic bulb below the pile base ranges from 3.11 to 3.62 times the pile diameter. Similarly, as presented in Figure 9d, the plastic bulb also increased in size and length with soil effective friction angle increases.

6. Observations of the Ultimate Capacity of the Large Diameter Bored Piles

Based on the results of the numerical models investigated within this parametric study, it was noted that full friction mobilization occurs when the transferred shear stresses from the shaft interface achieve a value that almost equals the soil shear strength (according to Mohr-Coulomb theory of rupture). This may be described as the beginning of the failure stage, as in this phase, the pile friction resistance tends to be constant or slightly decreases (See Figure 1). The full friction mobilization was also indicated in the numerical models by the formed plastic points that extended above the pile base along an interface length of more than three times the pile diameter (3D). Additionally, in many cases, the plastic points extended above the base to fully cover the whole length of the pile shaft interface.

After full mobilization, the pile friction resistance tends to be constant (inactive constant friction situation). However, the pile is still able to sustain higher applied loads safely through its bearing resistance. The applied load is predominantly transferred by bearing within the failure stage. Consequently, the transferred high bearing stresses at the pile base level resulted in a significant increase in both the size and length of the formed plastic bulb below the pile base. Finally, the transferred bearing stress value will achieve the value of the soil's ultimate soil bearing capacity. In this situation, the pile will transition from the inactive constant friction situation into the sliding friction situation, and apparent failure will be observed through the large induced pile settlement at the end of this stage. Hence, the applied load at which the failure is achieved could be described as the ultimate load of the large diameter bored pile.

Results of the numerical model simulation indicate that the induced settlement at the failure state (S_f) ranges from 1.5 to more than 2 times the obtained settlement (S_{f-1}) at the pre-last load increment (90% of the ultimate load), as shown in the results summary presented in Figure 10.

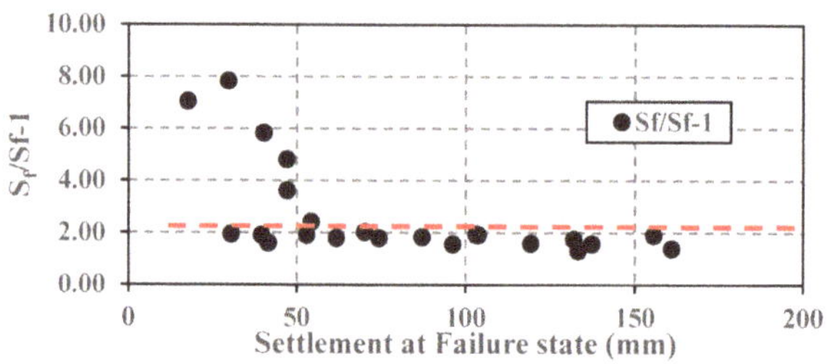

Figure 10. The ratio between the induced settlement at the failure load (S_f) and the settlement (S_{f-1}) under the load before the last load increment ($0.9Q_{ult}$).

To investigate the size and length of the end bearing plastic bulb at the failure stage, the ratio (D_p/D) between the plastic bulb diameter (D_p) and pile diameter (D) is calculated for (19) sample models with different pile proportions and soil properties as shown in Figure 11. In addition, the ratios (L_p/D) between the length of plastic bulb below the base level (L_p) to the pile diameter are obtained for the 19 models and shown in the same Figure. Results of the numerical models indicate that the plastic bulb diameter (Dp) ranges from 3 to 6 times the pile diameter (D). Moreover, the plastic bulb length (L_p) ranges from 2 to 4 times the pile diameter (D). These percentages are used to diagnose the expected plastic zone around the base of LDBP, as shown in Figure 12. Significant plastic deformations are highly expected to be induced in this area, which may cause the arching phenomenon that leads to the failure state and sudden excessive settlement of LDBP.

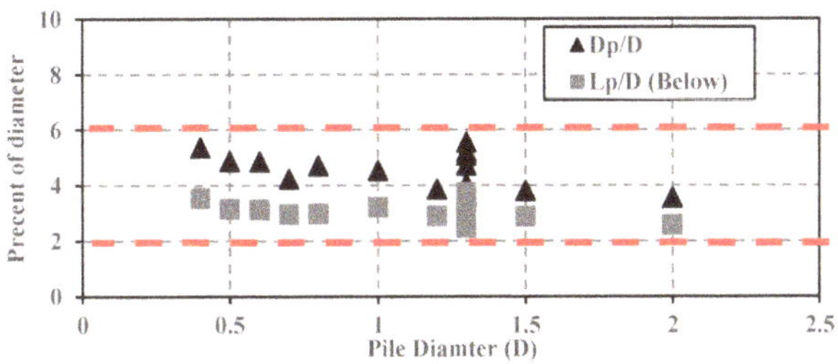

Figure 11. Relation between pile diameter and length and diameter of the formed plastic bulb below the pile base at the failure state.

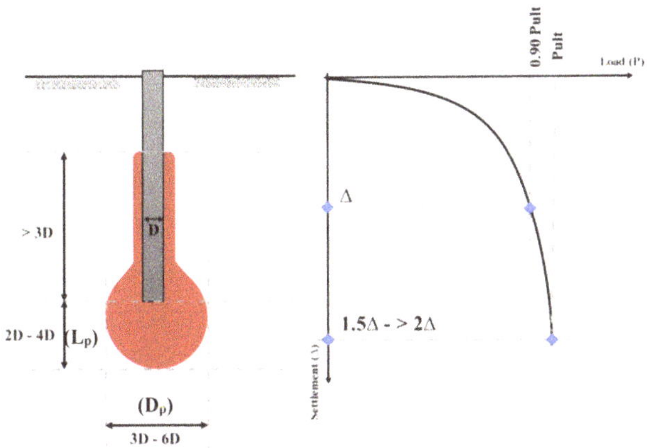

Figure 12. Schematic diagram showing the formed plastic bulb's predicted size and the induced settlement at the failure state of large diameter bored piles.

7. Application of the Study

Based on the results of the numerical parametric study performed, the failure of LDBP can be diagnosed by the ratio between the induced settlement at the failure load (S_f) and the settlement (S_{f-1}) at the pre-last load increment (90% of the ultimate load). Moreover, it can be identified through the size and height of the plastic bulb formed around and below the base of LDBP at the failure state, as explained in the previous section. In addition, the achievement of full friction mobilization can be ensured using Mohr-Coulomb failure criteria ($\sigma_h \tan Ø_i + c_i$). Thus, those three characteristics can be used to formulate a supervised machine learning algorithm. With that in mind, the parametric study results can extend the measurements of only one loaded to failure LDBP (Alzey case history) to provide the supervised algorithm with a data set for 29 cases with various pile geometrical and soil geotechnical parameters.

Figure 13 presents the decision tree that can be used to train the machine learning algorithm using the field measurements of the loading test and the parametric study results. The primary input in the tree will be the site investigation data needed to estimate the soil parameters, in addition to the pile geometry and field measurements of the available loading test. In the first layer of analysis, the algorithm should estimate both the ultimate pile capacity and settlement behavior. Secondly (layer 2), the estimated results shall be compared with the field measurements (layer 3) to assess the accuracy of the calculations. If the accuracy of the results exceeds the 95% benchmark (layer 4), the algorithm can open the data set of the parametric data set (Layer 5) and start performing the same procedure (forward process). In case the accuracy is below the benchmark, the algorithm shall re-analyze the data until hitting the benchmark (backward process).

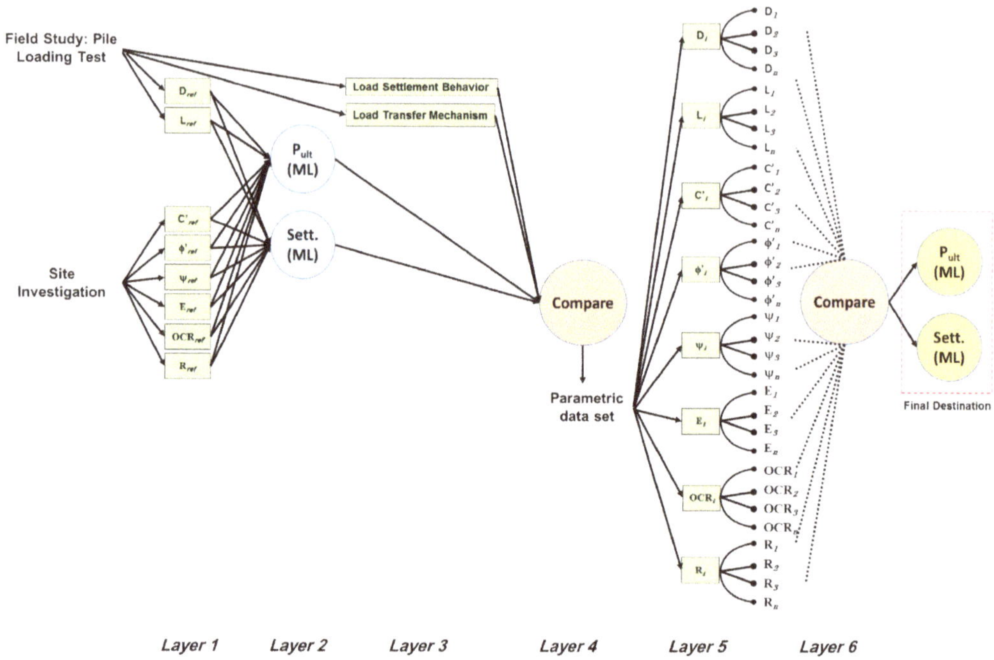

Figure 13. The suggested decision tree of a supervised machine learning algorithm for predicting the ultimate capacity of LDBP.

Consequently, the algorithm can be trained using the loading test measurements and the supervised data set of the parametric models. Thus, the algorithm will be ready to start the prediction process for different cases after achieving the targeted accuracy benchmark for all cases provided in the data set.

8. Conclusions

In this study, 29 numerical models were established to investigate the effect of different geometrical and geotechnical parameters on the response of large diameter bored piles in clayey soils by comparing their results with the results of the calibrated model of the Alzey Bridge case study. The following conclusions are drawn:

(1) The ultimate capacity of large diameter bored pile is increased with increases in each: pile diameter (D), pile length (L), effective soil cohesion (c′), and soil effective friction angle (Ø′). Consequently, the induced settlement at failure is also increased with the increase in any of these parameters.

(2) The initial stress coefficient has a minor effect on the results up to the working loads. However, the lateral earth pressure coefficient substantially affects both the ultimate capacity and the induced settlement at the failure state. Although the ultimate capacity increases with increases in the soil lateral earth pressure coefficient (k_0), the settlement of a large diameter bored pile at failure decreases with increases in the lateral earth pressure coefficient.

(3) Soil Young's modulus (E) does not affect the ultimate pile capacity. However, the settlement of a large diameter bored pile at failure is decreased with increases in this parameter.

(4) The average ultimate unit skin friction is not affected by pile diameter change. In contrast, it increases with each increase in pile length, effective soil cohesion, soil lateral earth pressure coefficient, or soil effective friction angle.

(5) The calculated average soil skin friction values using the Meyerhof (1976) capacity-based method are consistent with the numerical results of models with different geometrical and geotechnical parameters. However, Meyerhof's method underestimated the soil unit skin friction, with differences ranging from 10% to 21%.
(6) The ultimate bearing stress below the large diameter pile base is affected by pile diameter, pile length, soil effective cohesion, soil lateral earth pressure coefficient, and soil effective friction angle increases. However, several settlement-based methods proposed by different codes and design standards suggest constant bearing stress at a particular settlement value (i.e., 5% D) irrespective of pile geometry and without any discrimination for any class of the cohesive soils.
(7) The Meyerhof (1976) capacity-based method is generally a simple analytical procedure, taking the effect of pile diameter (D), length (L), clay effective cohesion (c'), soil lateral earth pressure coefficient (k_0), and soil effective friction angle (Ø') into account, so that it can be applied in many complex situations, such as variation in the sections along a pile shaft and an inhomogeneous layered soil system. Dissimilarly, settlement-based methods ignore the effect of several influencing factors.
(8) In the failure phase, the pile load transferred by friction tends to be constant or slightly decreased, and the applied load is predominantly transferred by bearing. Apparent failure is often observed through the large induced pile settlement at the end of this stage. The induced settlement at the failure state (S_f) is equal to or greater than 1.5 times the obtained settlement at 90% of the ultimate load.
(9) At the failure load, the formed plastic bulb vertically extends to distances of more than 3 times the pile diameter (3D) above the pile base and 2 to 4 times (2–4D) the pile diameter below the pile base. Additionally, the diameter of the formed plastic bulb ranges from 3 to 6 times the pile diameter (3–6D) at failure state.
(10) Results of the 29 numerical models applied in this parametric study can be used to develop a new theoretical method that predicts a more reliable value for the ultimate capacity of a large diameter bored pile installed in overconsolidated stiff clay soil. Moreover, the proposed decision tree can be typically utilized for several machine learning-supervised algorithms for different types and classes of soil.

Author Contributions: Conceptualization, M.E.A.-A. and A.M.H.; methodology, M.E.A.-A. and A.M.H.; experimental testing, validation M.E.A.-A., A.M.H. and T.M.S.; formal analysis, M.E.A.-A. and A.M.H.; investigation, M.E.A.-A., A.M.H. and T.M.S.; data curation, M.E.A.-A. and T.M.S.; writing—original draft preparation, M.E.A.-A.; writing—review and editing, M.E.A.-A., A.M.H. and T.M.S.; visualization, M.E.A.-A.; supervision A.M.H. All authors have read and agreed to the published version of the manuscript.

Funding: This research received no external funding.

Institutional Review Board Statement: Not applicable.

Informed Consent Statement: Not applicable.

Data Availability Statement: Not applicable.

Acknowledgments: The research is supported by the Structures and Materials (S&M) Research Lab of Prince Sultan University, Riyadh, Saudi Arabia.

Conflicts of Interest: The authors declare that they have no known competing financial interests or personal relationships that could have appeared to influence the work reported in this paper.

References

1. Meyerhof, G.G. The Ultimate Bearing Capacity of Foundations. Géotechnique. *ICE Virtual Libr.* **1951**, 2, 301–332. [CrossRef]
2. Fellenius, B.H. *Unified Design of Piles and Pile Groups. Transportation Research Record 1169*; Transportation Research Board: Washington, DC, USA, 1988; pp. 75–82.
3. Sawant, V.A.; Shukla, S.K.; Sivakugan, N.; Das, B.M. Insight into Pile Set-Up and Load Carrying Capacity of Driven Piles. *Int. J. Geotech. Eng.* **2013**, 7, 71–83. [CrossRef]
4. Wael, N. Evaluation of the Ultimate Capacity of Friction Piles. *Sci. Res. Eng. J.* **2012**, 4, 778–789. [CrossRef]

5. Tawfik, M.M. Assessment of Load Settlement Behavior of Large Diameter Bored Piles in Sand under Axial Compression Loads. Ph.D. Thesis, Faculty of Engineering, Ain Shams University, El-Khalyfa, Egypt, 2010; pp. 320–333.
6. El Gendy, M.; El-Arabi, I.A.; Kamal, M.A. Comparative analyses of large diameter bored piles using international codes. *J. Deep. Found. Inst.* **2014**, *8*, 15–26. [CrossRef]
7. Reese, L.C.; Isenhower, W.M.; Wang, S.T. *Analysis and Design of Shallow and Deep Foundations*; John Wiley: Hoboken, NJ, USA, 2006.
8. Omer, J.; Rokni, S. 3D Finite Element Assessment of the Effects of Energy Loops on the Load Capacity and Settlement of Bored Piles Installed in Sand. *Int. J. Geotech. Eng.* **2018**, *12*, 472–483. [CrossRef]
9. Shahin, M.A. Intelligent Computing for Modelling Axial Capacity of Pile Foundations. *Can. Geotech. J.* **2010**, *47*, 230–243. [CrossRef]
10. Rollins, K.M.; Clayton, R.J.; Mikesell, R.C.; Blaise, B.C. Drilled Shaft Side Friction in Gravelly Soils. *J. Geotech. Geoenviron. Eng. ASCE* **2005**, *131*, 987–1003. [CrossRef]
11. Poulos, H.G.; Davis, E.H. The Settlement Behavior of Single Axially Loaded Incompressible Piles and Piers. *Géotechnique ICE Virtual Libr.* **1968**, *18*, 351–371. [CrossRef]
12. Yang, J. Influence Zone for End Bearing of Piles in Sand. *Journal of Geotechnical and Geoenvironmental Engineering ASCE Libr.* **2006**, *132*, 1229–1237. [CrossRef]
13. Shahin, M.A. Artificial Intelligence in Geotechnical Engineering: Applications, Modeling Aspects, and Future Directions. In *Metaheuristics in Water, Geotechnical and Transport Engineering*; Elsevier: Amsterdam, The Netherlands, 2013; pp. 169–204. [CrossRef]
14. Abu-Kiefa, M.A. General regression neural networks for driven piles in cohesionless soils. *J. Geotech. Geoenviron. Eng. ASCE Libr.* **1998**, *124*, 1177–1185. [CrossRef]
15. Pal, M.; Deswal, S. Modelling pile capacity using Gaussian process regression. *Comput. Geotech.* **2010**, *37*, 942–947. [CrossRef]
16. Park, S.; Roberts, L.; Misra, A. Static load test interpretation using the *t-z* model and LRFD resistance factors for auger cast-in-place (AC IP) and drilled displacement (DD) piles. *Int. J. Geotech. Eng.* **2011**, *5*, 283–295. [CrossRef]
17. Boonyatee, T.; Lai, Q.V. A non-linear load transfer method for determining the settlement of piles under vertical loading. *Int. J. Geotech. Eng.* **2020**, *14*, 206–2017. [CrossRef]
18. DIN 4014. *German Association for Earthworks and Foundation Engineering*; Deutsches Institute fur Normung: Berlin, Germany, 1990.
19. ECP 202/4. *Egyptian Code for Soil Mechanics–Design and Construction of Foundations. Part 4, Deep Foundations*; The Housing and Building Research Center (HBRC): Cairo, Egypt, 2005.
20. Fellenius, B.H. *What Capacity to Choose from the Results of a Static Loading Test*; Deep Foundation Institute: Hawthorne, NJ, USA, 2001; pp. 19–22.
21. Meyerhof, G.G. Theory and Practice of Pile Foundations. In Proceedings of the first International Conference–Piling and Deep Foundations, Beijing, China; 1986; pp. 177–186.
22. Reese, L.C.; O'Neill, M.W. Field Load Tests of Drilled Shafts. In Proceedings of the International Seminar on Deep Foundations on Bored and Auger Piles, Rotterdam, The Netherlands, 7–10 June 1988; pp. 145–192.
23. Eid, M.; Hefny, A.; Sorour, T.; Zaghloul, Y.; Ezzat, M. Numerical Analysis of Large Diameter Bored Pile Installed in Multi Layered Soil: A Case Study of Damietta Port New Grain Silos Project. *Int. J. Curr. Eng. Technol. INPRESSCO* **2018**, *8*, 218–226. [CrossRef]
24. Mullins, G.; Dapp, S.D.; Lai, P. Pressure Grouting Drilled Shaft Tips in Sand. In Proceedings of the New technological and design developments in deep foundations, Geo-Denver, CO, USA, 5–8 August 2000; pp. 1–17. [CrossRef]
25. Hansen, J.B. Discussion on hyperbolic stress-strain response in cohesive soils. *J. Soil Mech. Found. Div. ASCE Libr.* **1963**, *89*, 241–242. [CrossRef]
26. Chin, F.K. Estimation of the ultimate load of piles from tests not carried to failure. In Proceedings of the 2nd. South-East Asian Conference on Soil Engineering, Singapore, 11–15 June 1970; pp. 81–90.
27. Eid, M.; Hefny, A.; Sorour, T.; Zaghloul, Y.; Ezzat, M. Full-Scale Well Instrumented Large Diameter Bored Pile Load Test in Multi-Layered Soil: A Case Study of Damietta Port New Grain Silos Project. *Int. J. Curr. Eng. Technol. Inpressco Int. Press* **2018**, *8*, 85–98. [CrossRef]
28. American Association of State Highway and Transportation Officials. *Bridge Design Specification, SI Units, Interim Revisions*, 3rd ed.; American Association of State Highway and Transportation Officials: Washington, DC, USA, 2005.
29. Alkroosh, I.S.; Bahadori, M.; Nikraz, H.; Bahadori, A. Regressive approach for predicting bearing capacity of bored piles from cone penetration test data. *J. Rock Mech. Geotech. Eng.* **2015**, *7*, 584–592. [CrossRef]
30. Nejad, F.P.; Jaksa, M.B. Load-settlement behavior modeling of single piles using artificial neural networks and CPT data. *Comput. Geotech.* **2017**, *89*, 9–21. [CrossRef]
31. Zhou, W.; Wang, L.; Guo, Z.; Liu, J.; Rui, S. A novel tz model to predict the pile responses under axial cyclic loadings. *Comput. Geotech.* **2019**, *112*, 120–134. [CrossRef]
32. Moshfeghi, S.; Eslami, A. Study on pile ultimate capacity criteria and CPT-based direct methods. *Int. J. Geotech. Eng.* **2018**, *12*, 28–39. [CrossRef]
33. Mohanty, R.; Suman, S.; Das, S.K. Prediction of vertical pile capacity of driven pile in cohesionless soil using artificial intelligence techniques. *Int. J. Geotech. Eng.* **2018**, *12*, 209–216. [CrossRef]

34. Delua, J. Supervised vs. Unsupervised Learning: What's the Difference? 2021. Available online: https://www.ibm.com/cloud/blog/supervised-vs-unsupervised-learning (accessed on 4 August 2021).
35. He, X.; Xu, H.; Sabetamal, H.; Sheng, D. Machine learning aided stochastic reliability analysis of spatially variable slopes. *Comput. Geotech.* **2020**, *126*, 103711. [CrossRef]
36. Samui, P. Support vector machine applied to settlement of shallow foundations on cohesionless soils. *Comput. Geotech.* **2008**, *35*, 419–427. [CrossRef]
37. Makasis, N.; Narsilio, G.A.; Bidarmaghz, A. A machine learning approach to energy pile design. *Comput. Geotech.* **2018**, *97*, 189–203. [CrossRef]
38. Sommer, H.; Hammbach, P. Großpfahlversuche im Ton für die Gründung der Talbrücke Alzey. *Der Bauingenieur* **1974**, *49*, 310–317.
39. Al-Atroush, M.E.; Hefny, A.; Zaghloul, Y.; Sorour, T. Behavior of a Large Diameter Bored Pile in Drained and Undrained Conditions: Comparative Analysis. *Geosciences* **2020**, *10*, 261. [CrossRef]
40. Wehnert, M.; Vermeer, P.A. *Numerical Analyses of Load Tests on Bored Piles, Numerical Models in Geomechanics*; NUMOG 9th: Ottawa, ON, Canada, 2004.
41. Ezzat, M.; Zaghloul, Y.; Sorour, T.; Hefny, A.; Eid, M. Numerical Simulation of Axially Loaded to Failure Large Diameter Bored Pile. *Int. J. Earth Energy Environ. Sci.* **2019**, *13*, 1–15. [CrossRef]
42. Meyerhof, G.G. Bearing Capacity and Settlement of Pile Foundations. *J. Geotech. Eng. Div. ASCE Libr.* **1976**, *102*, 197–228. [CrossRef]

processes

MDPI

Article

Precomputation of Critical State Soil Plastic Models

Vicente Navarro *, Virginia Cabrera, Gema De la Morena, Daniel González, Laura Asensio and Ángel Yustres

Geoenvironmental Group, University of Castilla-La Mancha, Avda. Camilo José Cela s/n, 13071 Ciudad Real, Spain; virginia.cabrera@uclm.es (V.C.); gema.delaMorena@uclm.es (G.D.l.M.); Daniel.Gonzalez@uclm.es (D.G.); Laura.Asensio@uclm.es (L.A.); Angel.Yustres@uclm.es (Á.Y.)
* Correspondence: vicente.navarro@uclm.es; Tel.: +34-926295300 (ext. 3264)

Abstract: In this paper, a simple precomputing procedure is proposed to improve the numerical performance of the technological application of critical state soil models. In these models, if associated plasticity is assumed, the normalization of the stress space allows both the yield surface and the plastic components of the elastoplastic matrix to be defined as a function of a single variable. This approach facilitates their parameterization and precomputation, preventing the repetition of calculations when the boundary value problems appear at the yield surface with the calculation of plastic strain. To illustrate the scope of the procedure, its application on a modified Cam Clay model is analysed, which shows that the method allows a significant reduction of about 50% (as compared with the conventional explicit integration algorithm) in the computational time without reducing the precision. Although it is intended for critical state models in soils, the approach can be applied to other materials and types of constitutive models provided that parameterization is possible. It is therefore a methodology of practical interest, especially when a large volume of calculations is required, for example when studying large-scale engineering systems, performing sensitivity analysis, or solving optimization problems.

Keywords: plasticity; critical state soil model; modified Cam Clay model; model normalization; precomputation

Citation: Navarro, V.; Cabrera, V.; De la Morena, G.; González, D.; Asensio, L.; Yustres, Á. Precomputation of Critical State Soil Plastic Models. *Processes* **2021**, *9*, 2142. https://doi.org/10.3390/pr9122142

Academic Editor: Avelino Núñez-Delgado

Received: 10 November 2021
Accepted: 25 November 2021
Published: 27 November 2021

1. Introduction

Simulation of soil yield processes is often computationally expensive, especially when analysing active clays such as MX-80 bentonite [1], FEBEX bentonite [2], or GMZ bentonite [3]. These materials, which usually have a high smectite content, when hydrated unconfined from partially saturated conditions, can undergo deformations of 500%. Under confined conditions, they can lead to swelling pressures approaching 10 MPa. This makes the simulation of their hydromechanical behaviour always computationally demanding, particularly when simulating complex domains or when simulations have to be repeated a large number of times. Even considering the development experienced in recent years by microelectronic technology, which has allowed the multi-core and many-core hybrid heterogeneous parallel computing platform to facilitate a very important advance in computing power, the efficiency of the calculation algorithm continues to be a key issue in the application of massive calculation processes. This is shown, for example, in [4] where the application of the Generalized Likelihood Uncertainty Estimation method in the probabilistic estimation of parameters in hydrology is analysed. In the field of soil mechanics, the improvement of computational performance is especially important in design or parameter identification processes, where the computational time can determine the viability of the study [5].

In some constitutive models, such as the critical state soil models, both the stress space [6] and the plastic terms $\mathbf{D}^{p}$ of the elastoplastic matrix $\mathbf{D}^{ep}$ [7–9] can be normalized. Thus, different stress states are associated with the same point of the normalized yield surface and with the same $\mathbf{D}^{p}$. The result could be that when solving boundary value problems, for a nonnegligible number of times, the calculations associated with the plastic behaviour of the same point (or in the vicinity of the same point) of the normalized stress space are repeated. It is therefore interesting to precompute $\mathbf{D}^{\mathrm{p}}$ for a grid of normalised stresses and to interpolate any value within the points of that grid when solving boundary value problems.

This approach is the resolution strategy proposed in the present work. In the following sections, the normalisation of the yield surface and $\mathbf{D}^{\mathrm{p}}$ is first described. Then, the CPU time savings that the precomputation entails are evaluated. Later, the criteria are defined to ensure that the proposed method does not compromise the quality of the simulation, and finally, an inspection exercise is performed to evaluate its performance in solving boundary problems.

2. Theoretical Background

In addition to saturated conditions, this paper also assumes axisymmetric conditions. These hypotheses do not reduce the scope of the proposed formulation and yet allow it to be explained more clearly. Furthermore, in line with this quest for clarity, the modified Cam Clay model (MCCm) [6] is adopted as the reference critical state soil model. In this way, the complexity of the constitutive formulation is reduced to a minimum, and the present work is focused on the description of the proposed calculation method. However, the widespread use of the MCCm in computational geotechnics [10–12] means that the methodology presented here is not trivial.

Usually, the yield surface f of the MCCm is formulated as

$$f \equiv \left(\frac{q}{M}\right)^2 - p(p_O - p) = 0, \tag{1}$$

where q is the von Mises deviatoric stress, M is the slope of the critical state line, and the isotropic preconsolidation pressure p_O is the hardening parameter. This parameter is a good normalisation parameter since by introducing the dimensionless stresses $P = p/p_O$ and $Q = q/p_O, f$ projects onto F (Figure 1), as defined by the expression

$$F \equiv \left(\frac{Q}{M}\right)^2 - P(1 - P) = 0. \tag{2}$$

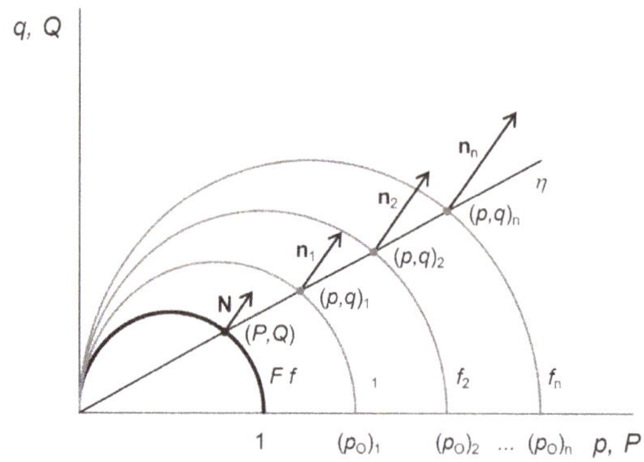

Figure 1. Projection of the different yield surfaces f_i on the normalized yield surface F.

All the points $(p,q)_i$ located on the same axis η, although they are associated with a different hardening parameters p_{Oi}, define the same dimensionless stress (P,Q) (Figure 1). Moreover, in all of them, the vector $\mathbf{n}$, normal to f, given by the expression

$$\mathbf{n} = \left(\frac{\partial f}{\partial p}, \frac{\partial f}{\partial q}\right) = \left(2p - p_O, \frac{2q}{M^2}\right) = p_O\left(2P - 1, \frac{2Q}{M^2}\right) = p_O\,\mathbf{N} \tag{3}$$

has the same direction of **N**, with its modulus varying according to the p_O factor. Therefore, assuming an associated plasticity model, the plastic terms of the elastoplastic matrix will be the same for all yield stresses which project onto the same dimensionless stress. Indeed, if the elastoplastic matrix $\mathbf{D}^{ep}$ defines the increase in the effective stress $d\sigma$ associated with an increase in the strain $d\varepsilon$ (the Voigt notation is used for both stress and strain tensors), then we have

$$d\sigma = \mathbf{D}^{ep} \cdot d\varepsilon \tag{4}$$

If the consistency equation is applied to the definition of the yield surface f and both the flow rule and the hardening law are considered, then $\mathbf{D}^{ep}$ can be expressed as (see, for example, [13,14])

$$\mathbf{D}^{ep} = \mathbf{D}^{e} \cdot \left(\mathbf{I} - \frac{\left(\frac{\partial f}{\partial \sigma}\right)^{t} \cdot \frac{\partial f}{\partial \sigma} \cdot \mathbf{D}^{e}}{\frac{\partial f}{\partial \sigma} \cdot \mathbf{D}^{e} \cdot \left(\frac{\partial f}{\partial \sigma}\right)^{t} - \frac{\partial f}{\partial \mathbf{x}} \cdot \frac{\partial \mathbf{x}}{\partial \varepsilon^{p}} \cdot \left(\frac{\partial f}{\partial \sigma}\right)^{t}} \right) = \mathbf{D}^{e} \cdot (\mathbf{I} - \mathbf{D}^{p}) \tag{5}$$

where $\mathbf{D}^{e}$ is the elastic strain matrix, $\mathbf{I}$ is the identity matrix with the dimension of $\mathbf{D}^{e}$, the symbol $(\cdot)^{t}$ indicates the transpose operator, and $\mathbf{x}$ defines the vector of hardening parameters. As noted, in the MCCm, $\mathbf{x}$ is equal to the scalar p_O. Its value is usually defined in critical state models by the hardening law

$$dp_O = \frac{1+e}{\lambda - \kappa}\, p_O\, d\varepsilon_V^{p} \tag{6}$$

where λ is the slope of the virgin compression line of the soil, and $d\varepsilon_V{}^{p}$ is the volumetric plastic strain. For the axisymmetric conditions adopted, where $d\varepsilon = (d\varepsilon_V, d\varepsilon_S)$ in which $d\varepsilon_S$ is the deviatoric strain, $\mathbf{D}^{p}$ can be calculated as

$$\mathbf{D}^{p} = \frac{\begin{pmatrix} (2p - p_O)^2 & (2p - p_O)\frac{2q}{M^2} f_v \\ (2p - p_O)\frac{2q}{M^2} & \left(\frac{2q}{M^2}\right)^2 f_v \end{pmatrix}}{(2p - p_O)^2 + f_v\left(\frac{2q}{M^2}\right)^2 + \frac{\kappa}{\lambda - \kappa} p_O (p - p_O)} \tag{7}$$

where f_v is a function of Poisson's ratio given by

$$f_v = \frac{9(1-2\nu)}{2(1+\nu)} \tag{8}$$

Dividing by $(p_O)^2$ results in the dimensionless expression

$$\mathbf{D}^{p} = \frac{\begin{pmatrix} (2P - 1)^2 & (2P - 1)\frac{2Q}{M^2} f_v \\ (2P - 1)\frac{2Q}{M^2} & \left(\frac{2Q}{M^2}\right)^2 f_v \end{pmatrix}}{(2P - 1)^2 + f_v\left(\frac{2Q}{M^2}\right)^2 + \frac{\kappa}{\lambda - \kappa}(2P - 1)} \tag{9}$$

Additionally, $\mathbf{D}^{e}$ is defined as

$$\mathbf{D}^{e} = \begin{pmatrix} K & 0 \\ 0 & 3G \end{pmatrix} \tag{10}$$

where K is the bulk modulus

$$K = \frac{1+e}{\kappa} p \tag{11}$$

and G is the shear modulus

$$G = \frac{3(1-2\nu)}{2(1+\nu)} K \tag{12}$$

These moduli depend on the void ratio and are updated as the soil is loaded and deformed. Therefore, the elastic stiffness matrix depends on the strain path followed, as well as the preconsolidation stress, which must be updated according to Equation (6) from the volumetric plastic strain. However, Equation (9) shows that $\mathbf{D}^p$ depends on P and Q only. Actually, it depends on only one parameter, since both P and Q are on the surface $F = 0$ (Figure 1), which, like $\mathbf{D}^p$, can be parametrised by the angle θ shown in Figure 2. Furthermore, in this figure, it is proposed to select N equispaced points (with an $\Delta\theta$ angle between them) at $F = 0$, calculating $\mathbf{D}^p{}_i$ for each of them. Subsequently, when solving a boundary value problem, the value of $\mathbf{D}^p$ is approximated by the interpolation $\mathbf{D}^{p^*}$ for angle values between two previously calculated values

$$\mathbf{D}^{p*}(\theta) = \frac{(\theta_{i+1} - \theta)}{(\theta_{i+1} - \theta_i)}\mathbf{D}^p{}_i + \frac{(\theta - \theta_i)}{(\theta_{i+1} - \theta_i)}\mathbf{D}^p{}_{i+1}, \theta_i \leq \theta < \theta_{i+1} \tag{13}$$

obtaining $\mathbf{D}^{ep^*}$ by

$$\mathbf{D}^{ep*}(\theta) = \mathbf{D}^e \cdot (\mathbf{I} - \mathbf{D}^{p*}(\theta)) \tag{14}$$

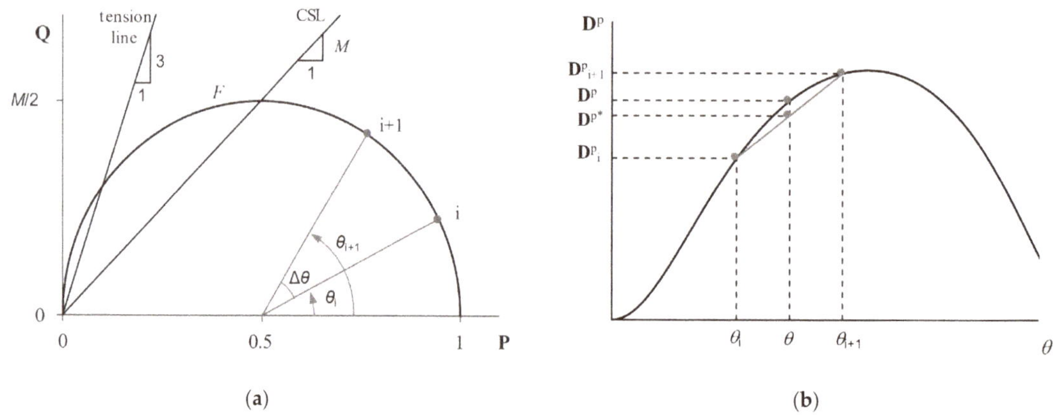

Figure 2. (**a**) Discretization of the dimensionless yield surface as a function of the angle θ, including the position of the tension line. (**b**) Diagram of the interpolation strategy. Note that for simplicity, the interpolation of the plastic matrices is represented in one-dimensional space.

The calculation procedure proposed in this work is based on precomputing the vector of matrices $\{\mathbf{D}^p{}_1, \mathbf{D}^p{}_2, \ldots, \mathbf{D}^p{}_N\}$, thus avoiding repeating the calculation of $\mathbf{D}^p$ during the resolution of the boundary value problems.

3. Evaluation of the Precomputation Efficiency

The efficiency of the proposed procedure depends on the cost of calculating $\mathbf{D}^p$. If this cost is not an important part of the total cost of the calculation process, the performance associated with using $\mathbf{D}^{p^*}$ might not be of interest. Figure 3a summarises the process followed to solve a computational step when integrating the constitutive model. Table 1 lists the number of operations associated with the calculation of $\mathbf{D}^e$ and $\mathbf{D}^p$. Although these numbers can vary depending on the way in which the algorithm is programmed, the changes are not significant in terms of the relative computational cost of each magnitude. Using the results of the benchmark defined in Appendix A (where it is indicated that the estimated cost of each multiplication is equivalent to 1.101 sums, with the cost of division being 3.698 sums), the costs shown in the last column of Table 1 were obtained. While it is advisable to be cautious about the reliability in absolute terms of the representativeness of these numbers, they do characterise the weight that each calculation carries in relative terms. Figure 3b shows the strategy followed when applying the precomputational procedure. As noted, $\mathbf{D}^e$ must be calculated as in the conventional method. In addition, interpolation of the $\mathbf{D}^p{}_i$ matrices must be made to obtain $\mathbf{D}^{p^*}$ by performing the operations indicated in Table 1. Using again the values of Appendix A, the computational cost of Table 1 is obtained, which is approximately 42% of the cost of calculating $\mathbf{D}^p$. The computational time savings are remarkable, and the application of the approach proposed here is therefore of interest.

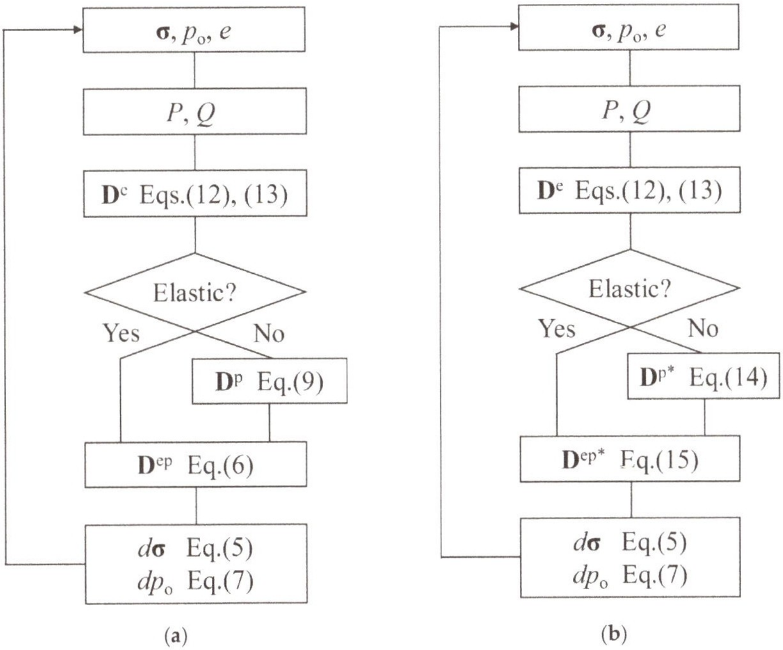

Figure 3. Integration of a computational step using (**a**) a conventional explicit algorithm and (**b**) the precomputational approach.

Table 1. Equivalent sum costs of the main stages that conform to a conventional explicit integration algorithm and an algorithm with the proposed procedure.

Stage	Sums	Multiplications	Divisions	Computational Cost
$\mathbf{D}^e$ matrix, C_{De}	3	6	2	17
$\mathbf{D}^p$ matrix, C_{Dp}	8	25	9	69
$\mathbf{D}^{p^*}$ matrix, C_{Dp^*}	4	16	2	29

4. Precomputation Density

To define the precomputation strategy, the minimum number of points N to be taken at $F = 0$ must be determined in such a way that the accuracy of the calculation when using $\mathbf{D}^{p^*}$ is comparable to the accuracy obtained when calculating $\mathbf{D}^p$. To accomplish this goal, a set of constant strain rate paths with a final strain of 20% and with different ratios between volumetric and shear strains, as in Figure 4, were inspected. In all of these inspection exercises, initial spherical conditions were assumed with an effective mean stress of $p = p_O/2 = 200$ kPa. The analysis was performed for the three soils, namely Weald clay, Klein Belt Ton, and kaolin, which are characterised in Table 2. The differences between their properties make the conclusions reached not limited to a particular material.

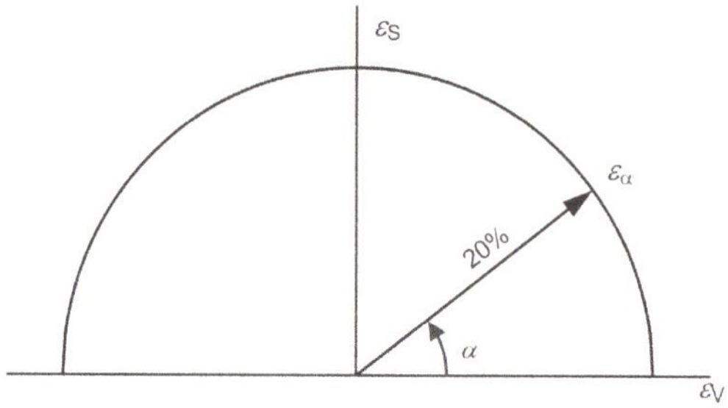

Figure 4. Strain paths ε_α as a function of the angle α considered to define the number of N subdivisions of the yield surface.

Table 2. Parameters used to characterise the behaviour of the clays assumed in the models (adapted from Schofield and Wroth [15]).

Parameter	Symbol	Weald Clay	Klein Belt Ton	Kaolin
Slope of the critical state line in the $p - q$ effective stress space	M	0.95	0.845	1.02
Slope of the virgin compression line in the $\ln p - e$ space	λ	0.093	0.356	0.260
Slope of the unloading–reloading line in the $\ln p - e$ space	κ	0.035	0.184	0.050
Void ratio at p = 1 kPa	e_1	1.088	3.677	2.764
Poisson's ratio	ν	0.3	0.3	0.3
Liquid limit (%)	LL	43	127	74
Plastic limit (%)	PL	18	36	42
Plasticity index (%)	PI	25	91	32

As a reference, all paths were simulated using an explicit Euler integration method. An intensive substepping process was applied, i.e., the number of calculation steps was divided into smaller substeps successively until the solution differed from the previous solution by only the sixth significant digit, which is a strategy similar to that used by Sołowski and Gallipoli [16]. These results were compared with the precomputed–interpolated solution using the same number of substeps during the explicit integration.

For each path, for each value of α in Figure 4, the number of points N to be computed on the yield surface was determined in such a way that the normalised root-mean-square deviation between the stress solution obtained by explicit Euler integration and the solution reached by calculation with the precomputation–interpolation procedure was less than 10^0, 10^{-1}, 10^{-2}, 10^{-3}, and 10^{-4}. Figure 5 shows the values of N obtained for each of these tolerances and for each value of α.

It is important to note that, as illustrated in Figure 6, when imposing a total strain of 20%, in all cases, a significant yield was attained. In other words, trajectories that were highly sensitive to the quality of the precomputation results were analysed.

In all paths, as the tolerance decreased, N increased. This trend can be clearly seen in Figure 7, where for each tolerance value, the maximum value of N obtained for any α was plotted. The behaviour of N was well defined and not erratic; additionally, Figure 7 can be used to define the number of points N to be taken to avoid exceeding a certain value of tolerance.

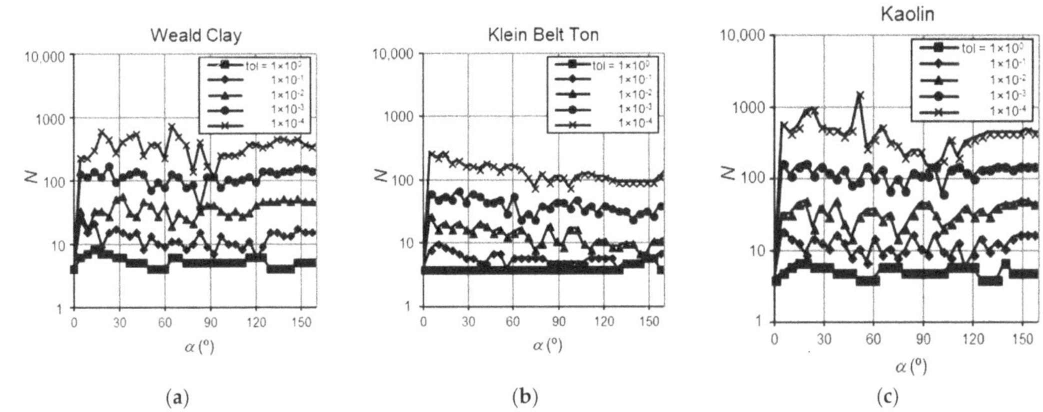

Figure 5. Minimum number N of subdivisions of the yield surface necessary to ensure a certain level of relative tolerance for each of the α strain paths defined in Figure 4 for (**a**) Weald clay, (**b**) Klein Belt Ton, and (**c**) kaolin.

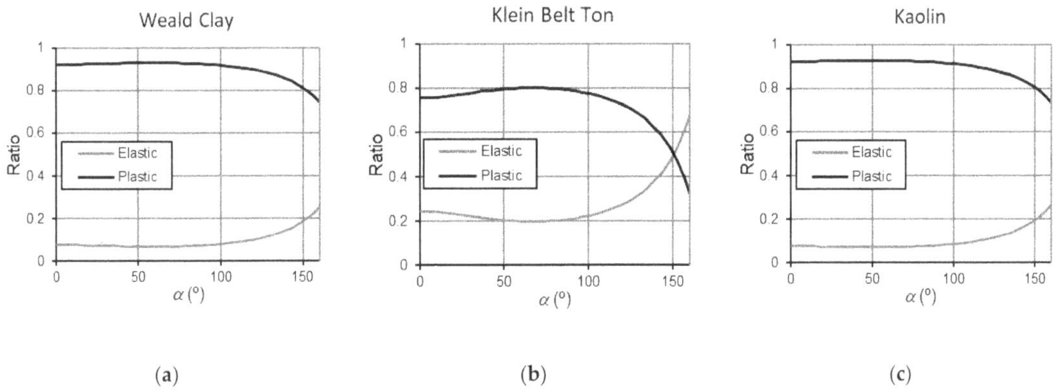

Figure 6. Ratio of elastic and plastic strain increments for the α paths defined in Figure 4 with respect to the number of computational steps for (**a**) Weald clay, (**b**) Klein Belt Ton, and (**c**) kaolin.

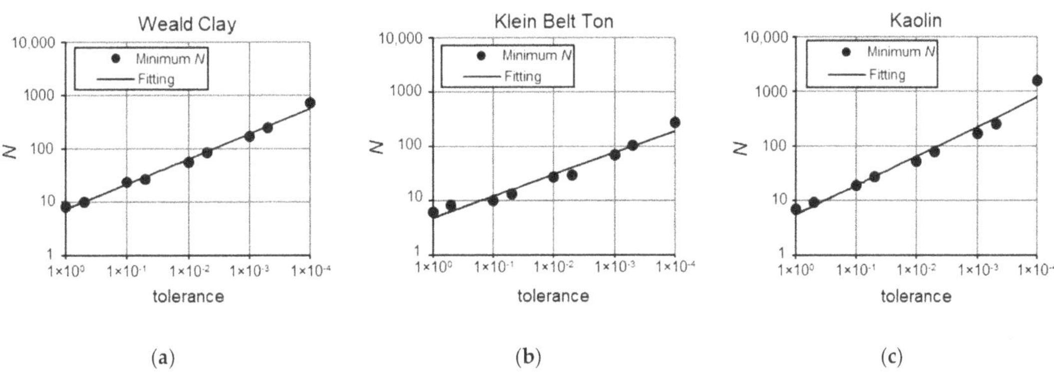

Figure 7. Fitting of the minimum number N of subdivisions of the yield surface is necessary to ensure a certain level of relative tolerance for any stress–strain path for (**a**) Weald Clay, (**b**) Klein Belt Ton and (**c**) Kaolin.

5. An Inspection on Solving Boundary Value Problems

Although the comparison in Table 1 highlights the interest of the precomputational approach, it does not illustrate its scope clearly enough, because it is associated with the simulation of individual stress–strain paths. Considering a finite element model in which conventionally the stresses are a state function, those paths would illustrate the behaviour of a single Gaussian point. It is reasonable to question what will happen when, in a real problem, the number of Gaussian points is in the hundreds or thousands.

It is not easy to answer this question. Savings will depend on the structure of the finite element code used and the calculation and information management strategies implemented. Each code will have a different response. However, in general, it is reasonable to assume that if N_{GP} is the number of Gaussian points considered in the mesh, the cost of a conventional explicit integration, χ_C, can be estimated as

$$\chi_C = N_{GP}\left(\eta_{\Delta\varepsilon e}\, C_{De} + \eta_{\Delta\varepsilon ep}\left(C_{De} + C_{Dp}\right)\right) \tag{15}$$

where the computational costs of $\mathbf{D}^e$ and $\mathbf{D}^p$, and C_{De} and C_{Dp} have been defined (as well as the computational cost of $\mathbf{D}^{p^*}$ and C_{Dp^*}) in Table 1, and for the total number of computational time steps needed to solve the boundary problem, $\eta_{\Delta\varepsilon e}$ defines the weighted value of the elastic increments, and $\eta_{\Delta\varepsilon ep}$ defines one of the elastoplastic increments. In turn, the cost of the precomputation–interpolation strategy, χ_{PI}, is estimated as

$$\chi_{PI} = N_{GP}\left(\eta_{\Delta\varepsilon e*}\, C_{De} + \eta_{\Delta\varepsilon ep*}\left(C_{De} + C_{Dp*}\right)\right) \tag{16}$$

where $\eta_{\Delta\varepsilon e^*}$ and $\eta_{\Delta\varepsilon ep^*}$ are the values analogous to $\eta_{\Delta\varepsilon e}$ and $\eta_{\Delta\varepsilon ep}$, respectively, when the integration of the constitutive model is done using the precomputation and interpolation procedure. Their values can, in principle, be different. However, since the discretization of the yield surface is proposed with a very demanding tolerance (10^{-4} in the analyses performed above), it is to be expected that $\eta_{\Delta\varepsilon e^*} \approx \eta_{\Delta\varepsilon e}$ and $\eta_{\Delta\varepsilon ep^*} \approx \eta_{\Delta\varepsilon ep}$. In this case, ρ, the ratio between the cost of a traditional explicit integration and the cost of the precomputation–interpolation strategy, is given by the expression

$$\rho = \frac{\chi_{PI}}{\chi_C} = \frac{\eta_{\Delta\varepsilon e}\, C_{De} + \eta_{\Delta\varepsilon ep}\left(C_{De} + C_{Dp*}\right)}{\eta_{\Delta\varepsilon e}\, C_{De} + \eta_{\Delta\varepsilon ep}\left(C_{De} + C_{Dp}\right)} = \frac{1 + \frac{\eta_{\Delta\varepsilon ep}}{\eta_{\Delta\varepsilon e}}\left(1 + \frac{C_{Dp*}}{C_{De}}\right)}{1 + \frac{\eta_{\Delta\varepsilon ep}}{\eta_{\Delta\varepsilon e}}\left(1 + \frac{C_{Dp}}{C_{De}}\right)} \tag{17}$$

The relative cost is independent of the mesh size. Furthermore, it is necessary to add to the computational costs considered in Equations (15) and (16) the costs associated with the generation of the database $\{\mathbf{D}^p{}_1, \mathbf{D}^p{}_2, \ldots, \mathbf{D}^p{}_N\}$. Consequently, it is to be expected that if the mesh is small, the relative importance of the precomputation cost is not negligible. However, the proposed methodology is oriented to large-scale computation, and thus it is a consistent estimator of the efficiency, and for a large computational volume, as shown in Equation (17), it is independent of the mesh size. The ratio $\eta = \eta_{\Delta\varepsilon ep}/\eta_{\Delta\varepsilon e}$ expresses the relative importance of the computation of plastic steps in a problem. As expected, given that the precomputation is aimed at optimizing the calculation of the plastic steps, its efficiency is conditioned by the relevance of the plastic strains over the elastic strains, in other words, by the value of η. Using the values of C_{De}, C_{Dp}, and C_{Dp^*} from Table 1, the change in ρ with respect to η was plotted and is shown in Figure 8. If $\eta = 0$, when the yield is not reached, precomputation is not applied. Only elastic steps are calculated, and both methods are equal ($\rho = 1$). However, as seen at the end of Section 3, as soon as some plastic steps are required, the precomputation is more efficient ($\rho < 1$). The larger the volume of the calculation of the plastic steps, the higher the efficiency. When the entire calculation is in the plastic regime (i.e., in normally consolidated conditions), the efficiency tends to a maximum value of 54%.

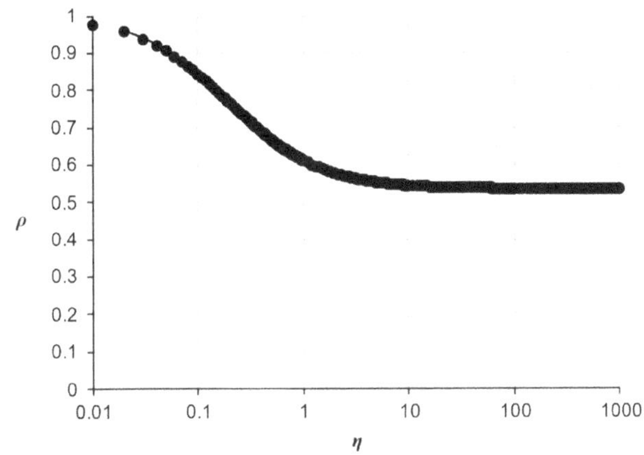

Figure 8. Dependence of ρ (ratio between the cost of a traditional explicit integration and the cost of the precomputation–interpolation strategy) on the value of η (relative importance of the computation of the plastic steps over the elastic steps).

6. Conclusions

A methodology was presented to parameterise, precompute, and approximate the plastic behaviour of a soil by interpolation. The procedure can be applied to materials whose constitutive models allow the parametrisation and subsequent normalisation of their yield surface as well as the plastic components of the elastoplastic matrix. In this work, the modified Cam Clay model was used as a reference model. After describing the precomputation and interpolation procedure, the efficiency of the method was evaluated by estimating the computational cost of one computational step. Then, a criterion was defined to determine the density of the precomputed database so that the quality of the approximations obtained with the proposed method is like that of solutions based on conventional explicit integration methods. Finally, an inspection was conducted to assess the efficiency of the proposed method in the solution of boundary value problems. It was concluded that the proposed method can reduce the computational burden to as much as 54% of the conventional cost, a significant savings when solving problems in technological applications. This finding is of particular interest in the case of modelling large-scale engineering systems, in sensitivity analysis, or in solving optimisation problems, where a large volume of calculations is required.

Author Contributions: Conceptualization, V.N.; methodology, V.N.; software, D.G.; validation, D.G. and G.D.l.M.; formal analysis, Á.Y.; investigation, L.A.; resources, L.A.; data curation, G.D.l.M. and V.N.; writing—original draft preparation, D.G. and G.D.l.M.; writing—review and editing, V.N., V.C., and Á.Y.; visualization, V.C.; supervision, V.N.; project administration, V.N.; funding acquisition, V.N. All authors have read and agreed to the published version of the manuscript.

Funding: This study was funded by Junta de Comunidades de Castilla-La Mancha and the European Regional Development Fund (European Union) through the project SBPLY/19/180501/000222.

Institutional Review Board Statement: Not applicable.

Informed Consent Statement: Not applicable.

Data Availability Statement: Not applicable.

Conflicts of Interest: The authors declare no conflict of interest.

Appendix A. Characteristics of Computers

To obtain a reasonable estimation of the computational cost of the operations (multiplications and divisions) indicated in Table 1, a benchmark was conducted for each type of operation with six computers (see Table A1) in such a way that the cost of the operation was converted into equivalent sums. Calculation loops were performed until the CPU time was stationary in relation to a similar addition loop for each computer and for both multiplication and division. Thus, calculating the average of the results obtained with the computers used, the cost of multiplication was estimated to be equivalent to 1.101 sums, and the cost of the division was equal to 3.698 sums.

Table A1. Characteristics of the computers used in the benchmarks to quantify the different computational costs.

Computer	Model	Processor	RAM	Architecture
1	DELL Inspiron 17 5000 Series	Intel® Core™ i7-4510U CPU @ 2.00 GHz	8.0 GB	64-bits
2	HP EliteBook 8560w	Intel® Core™ i5-2540 M CPU @ 2.60 GHz	16.0 GB	64-bits
3	DELL Optiplex 745	Intel® Core™ 2 CPU 6400 2.13 GHz	3.0 GB	32-bits
4	DELL Inspiron One	Pentium® Dual-Core CPU E5400 @ 2.70 GHz	4.0 GB	64-bits
5	HP EliteBook Folio 9470 m	Intel® Core™ i5-3427U CPU @ 1.80 GHz	12.0 GB	64-bits
6	HP Z840 Workstation	Intel® Xeon® CPUE5-2620 v3 @ 2.40 GHz	32.0 GB	64-bits

References

1. Cui, Y.-J. On the hydro-mechanical behaviour of MX80 bentonite-based materials. *J. Rock Mech. Geotech. Eng.* **2017**, *9*, 565–574. [CrossRef]
2. Lloret, A.; Villar, M.V. Advances on the knowledge of the thermo-hydro-mechanical behaviour of heavily compacted "FEBEX" bentonite. *Phys. Chem. Earth Parts A/B/C* **2007**, *32*, 701–715. [CrossRef]
3. Wang, Q.; Tang, A.M.; Cui, Y.-J.; Delage, P.; Gatmiri, B. Experimental study on the swelling behaviour of bentonite/claystone mixture. *Eng. Geol.* **2012**, *124*, 59–66. [CrossRef]
4. Zuo, D.; Kan, G.; Sun, H.; Zhang, H.; Liang, K. Improving computational efficiency of GLUE method for hydrological model uncertainty and parameter estimation using CPU-GPU hybrid high performance computer cluster. *Nat. Hazards Earth Syst. Sci. Discuss.* **2021**, *1*, 1–33. [CrossRef]
5. Khaledi, K.; Miro, S.; König, M.; Schanz, T. Robust and reliable metamodels for mechanized tunnel simulations. *Comput. Geotech.* **2014**, *61*, 1–12. [CrossRef]
6. Roscoe, K.H.; Burland, J.B. On the Generalized Stress-Strain Behavior of Wet Clay. In *Engineering Plasticity*; Heyman, J., Leckie, F., Eds.; Cambridge University Press: Cambridge, UK, 1968; pp. 535–609.
7. Potts, D.M.; Ganendra, D. An evaluation of substepping and implicit stress point algorithms. *Comput. Methods Appl. Mech. Eng.* **1994**, *119*, 341–354. [CrossRef]
8. Tamagnini, C.; Castellanza, R.; Nova, R. Implicit integration of constitutive equations in computational plasticity. *Rev. Française De Génie Civ.* **2002**, *6*, 1051–1067. [CrossRef]
9. Lloret-Cabot, M.; Sloan, S.C.; Sheng, D.; Abbo, A.J. Error behaviour in explicit integration algorithms withautomatic substepping. *Int. J. Numer. Methods Eng.* **2016**, *108*, 1030–1053. [CrossRef]
10. Sheng, D.; Sloan, S.W.; Yu, H.S. Aspects of finite element implementation of critical state models. *Comput. Mech.* **2000**, *26*, 185–196. [CrossRef]
11. Borja, R.I.; Lee, S.R. Cam-Clay plasticity, Part 1: Implicit integration of elasto-plastic constitutive relations. *Comput. Methods Appl. Mech. Eng.* **1990**, *78*, 49–72. [CrossRef]
12. Gens, A.; Potts, D.M. Critical state models in computational geomechanics. *Eng. Comput.* **1988**, *5*, 178–197. [CrossRef]
13. Potts, D.M.; Gens, A. A critical assessment of methods of correcting for drift from the yield surface in elasto-plastic finite element analysis. *Int. J. Numer. Anal. Methods Geomech.* **1985**, *9*, 149–159. [CrossRef]

14. Zienkiewicz, O.C.; Chan, A.H.C.; Pastor, M.; Schrefler, B.A.; Shiomi, T. *Computational Geomechanics with Special Reference to Earthquake Engineering*; John Wiley & Sons Inc.: New York, NY, USA, 1999.
15. Schofield, A.N.; Wroth, C.P. *Critical State Soil Mechanics*; McGraw-Hill: New York, NY, USA, 1968.
16. Sołowski, W.T.; Gallipoli, D. Explicit stress integration with error control for the Barcelona Basic Model.: Part II: Algorithms efficiency and accuracy. *Comput. Geotech.* **2010**, *37*, 68–81. [CrossRef]

processes

Article

Unsaturated Hydraulic Conductivity Estimation—A Case Study Modelling the Soil-Atmospheric Boundary Interaction

Md Rajibul Karim [1,*], David Hughes [2] and Md Mizanur Rahman [1]

[1] UniSA STEM, University of South Australia, Mawson Lakes Campus, Mawson Lakes, SA 5095, Australia; mizanur.rahman@unisa.edu.au
[2] Bothar Consulting Limited, Bryansford Village, Newcastle BT33 0AY, UK; davidabhughes@hotmail.com
* Correspondence: rajibul.karim@unisa.edu.au

Abstract: Pore water pressure changes due to soil-atmospheric boundary interaction can significantly influence soil behaviour and can negatively affect the safety and stability of geotechnical structures. For example, prolonged rainfall events can lead to increased pore water pressure and lower strength; repeated cycles of pore water pressure changes can lead to degradation of strength. These effects are likely to become more severe in the future due to climate change in many parts of the world. To analyse the behaviour of soil subjected to atmospheric boundary interactions, several parameters are needed, and hydraulic conductivity is one of the more important and is difficult to determine. Hydraulic conductivity deduced from laboratory tests are often different from those from the field tests, sometimes by orders of magnitude. The problem becomes even more complicated when the soil state is unsaturated, where the hydraulic conductivity varies with the soil's state of saturation. In this paper, a relatively simple alternative approach is presented for the estimation of the hydraulic conductivity of unsaturated soils. The method involved a systematic re-analysis of observed pore water pressure response in the field. Using a finite element software, the soil-atmospheric boundary interaction and related saturated/unsaturated seepage of an instrumented slope have been analysed, and results are compared with field measurements. The numerical model could capture the development of suction, positive pore water pressure and changes in water content with reasonable accuracy and demonstrated the usefulness of the hydraulic conductivity estimation method discussed in this paper.

Keywords: pore water pressure; hydraulic conductivity; alternative method; numerical analyses; unsaturated soil

Citation: Karim, M.R.; Hughes, D.; Rahman, M.M. Unsaturated Hydraulic Conductivity Estimation—A Case Study Modelling the Soil-Atmospheric Boundary Interaction. *Processes* **2022**, *10*, 1306. https://doi.org/10.3390/pr10071306

Academic Editor: Li Li

Received: 2 June 2022
Accepted: 29 June 2022
Published: 1 July 2022

1. Introduction

In many parts of the world, the frequency of extreme weather events has increased due to climate change. January 2019 was the hottest month on the record in Australia [1]; the period of May–July 2007 was the wettest in 250 years and caused extensive flooding in many parts of the UK [2]. Singapore recorded its wettest (since the record began in 1869) December in 2006 [3]. In 2011, Thailand was struck by monsoon and tropical cyclone rains from July to October, causing extreme flooding in the city of Bangkok. More extreme climate scenarios are predicted in Australia, the UK and many other parts of the world [4–9].

Climate change is having an impact on the transportation infrastructure in countries like the UK [10]. More than 160 slope failures were reported during the winter of 2000/2001 by the UK rail and road authority [11,12]. In many cases, meteorologically induced pore water pressure and strain softening resulting from pore water pressure cycling have been responsible for the failure of geotechnical structures [13–15].

The changes in pore water pressure in soils are often driven by meteorological parameters, vegetation and groundwater hydrology. For more economical and sustainable design and maintenance of geotechnical structures, it is important to understand the mechanics/processes that may affect their behaviour under the current and possible future climate scenarios.

Several past fields and laboratory studies have investigated the effect of atmospheric boundaries on the behaviour of infrastructure slopes [9,16–20]. Among the numerical studies on soil-slope systems, Briggs et al. [21], using a one-dimensional VADOSE/W [22] model, investigated the generation of pore water pressure in a railway embankment; Tsaparas et al. [23] modelled infiltration and compared with field observations of a sloping site in Singapore; Karthikeyan et al. [24] presented estimation of pore water pressure using software called SEEP/W [25] even though there was lack of agreement between the field measurements and model calculations. Rouainia et al. [26] investigated meteorologically induced pore water pressure generation and its effect on slope stability using a coupling of SHETRAN and FLAC-TP flow and showed the capabilities of sophisticated numerical modelling approaches. However, the complexity involved in their analyses was high and needed assumptions on several modelling parameters, which can be difficult to deduce using an objective approach.

One of the more important input parameters required for analysing the interaction between soil and the atmospheric boundary is the hydraulic conductivity (K) of soil. For a particular void ratio, K for saturated soils (K_{sat}) can be treated as a constant. However, the same does not apply to the case of K for unsaturated soils (K_{unsat}). It changes with soil suction even if the void ratio remains the same. The soil suction, on the other hand, is dependent on the soil's degree of saturation. Direct measurement of K will involve measurement of water flow. However, as soon as some flow has occurred, the water content will change, which will lead to a change in suction and eventually the K_{unsat}. Due to this, it is extremely difficult to measure K_{unsat} for soils. Several methods have been proposed in the literature to capture the suction-K_{unsat} correlation [27–29] and commonly, K_{unsat} is expressed as a function of K_{sat} and soil suction. For example, Van-Genuchten [27] proposed the following equation,

$$K_{unsat} = K_{sat}\frac{\left[1-(a'\varPsi)^{n-1}\left(1+(a'\varPsi)^{n}\right)^{-m}\right]^2}{\left(1+(a'\varPsi)^{n}\right)^{\frac{m}{2}}} \tag{1}$$

where a', n and m are curve fitting parameters and $\varPsi$ is the matric suction. In essence, with a reasonable estimation of the soil–water characteristics curve (SWCC) and K_{sat}, Equation (1) can be used to estimate the K_{unsat}.

A challenge in using Equation (1), often overlooked, is the estimation of K_{sat}. Several field tests (e.g., single or double ring infiltrometer, disc permeameter, Guelph permeameter, bailout test) and laboratory tests (e.g., Rowe cell, constant/falling head tests, Oedometer tests, triaxial hydraulic conductivity tests) have been developed and routinely conducted to deduce K_{sat}. However, K_{sat} deduced from a different field or laboratory tests can be very different, which is partly due to the underlying assumptions involved in those methods. Discrepancies of several orders of magnitude are not uncommon [9]. The field tests are often believed to yield more representative K_{sat} values compared to the laboratory tests. However, K_{sat} values deduced from field tests can also be affected by the presence of hidden cracks or fissures in the soil, local heterogeneity and the presence of higher hydraulic conductivity layers, which may not be detected during the field investigation. Thus, the conventional methods of deducing K_{sat} can have a low degree of reliability. For example, in the study site involved in this investigation (discussed in the next section), more than 100 times difference in K_{sat} values from different tests were observed. This makes it very difficult, and in some cases, impossible to objectively decide what value of K_{sat} to be used in a particular analysis [30]. The engineer is often forced to use either an average value or make a subjective judgement which can be influenced by the engineer's experience with similar sites or the lack of it. For a seepage analysis involving unsaturated soil, the problem becomes even more complex due to the dependency of K_{unsat} on the soil's degree of saturation.

This paper discusses a simple and objective approach for estimating K_{unsat} using a correlation similar to one presented in Eq. [1]. The method involves a systematic back-analysis of field observations to deduce K_{unsat} as a function of K_{sat}. The method of estimation of K_{sat} was originally proposed by Lo et al. [31] to estimate K_{sat} for prefabricated vertical drain improved soils. K_{sat}, in their case, was estimated by systematically analysing field settlement data. It allowed them to avoid explicit modelling of the smear zone and reduced K_{sat} due to smearing in that area which is often very difficult to determine due to limited time and budget associated with field investigation of a practical project. The effectiveness of the method in different problem domains has been demonstrated in the literature by Karim [32], Karim et al. [33], Manivannan et al. [34], Karim et al. [35] and Lo et al. [36]. The approach has been modified in this paper for analyses in the unsaturated domain, i.e., to estimate K_{unsat}. The effectiveness of the method has been put to the test by modelling a well-instrumented research site located in Southern England [20]. A selected period of monitoring data (from 1 January 2006 to 31 December 2008—a total of 3 years) have been used. A finite element software SEEP/W (version 2021.4) [37] has been used for this purpose.

In the next section, the research site is discussed in terms of its construction, geotechnical properties, instrumentation and monitoring details. Different aspects of the numerical analyses, including the mesh or geometry chosen, boundary conditions and other input parameters, are discussed in the next section. Subsequent sections discuss the results and conclusions drawn from this paper.

2. Newbury Cutting—The Research Site

The research site discussed in this paper, hereafter referred to as Newbury cutting, is located near A34 Newbury bypass in Southern England. The location of the site is presented in Figure 1. The site was extensively instrumented and monitored. Details of the site instrumentation, monitoring, geology and weather data can be found in Smethurst et al. [20] and Smethurst et al. [19]. For the sake of completeness, a brief description is presented here.

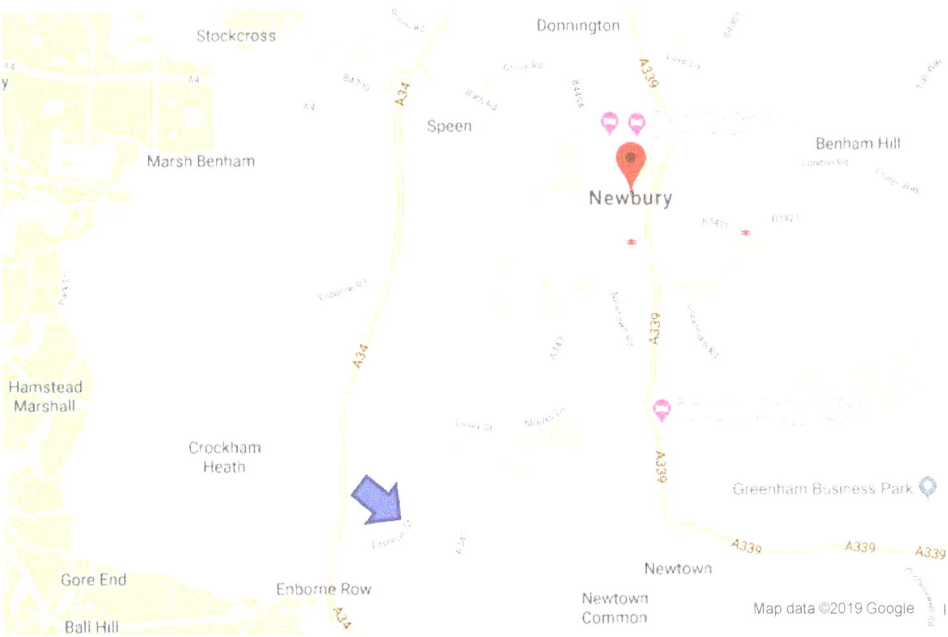

Figure 1. Location map of Newbury cutting with an arrow pointing to the site location [38].

The instrumented section of the cut slope was 8 m high and 28 m in length along the sloping direction, as presented in Figure 2. The slope was excavated in 1997. The site soil consists of stiff grey London clay of about 20 m thickness and its properties have been found to vary in the vertical and horizontal directions. The top 2.5 m of the slope, below the original ground level, was extensively weathered and hereafter referred to as weathered London clay. The presence of several bands of silty clay up to 50 mm thick and large flints was also detected during the site investigation. After the cutting was excavated, a 0.4 m layer of topsoil was placed over the cut surface to facilitate the planting of vegetation. To facilitate quick drainage, a fin drain was also installed near the toe of the slope.

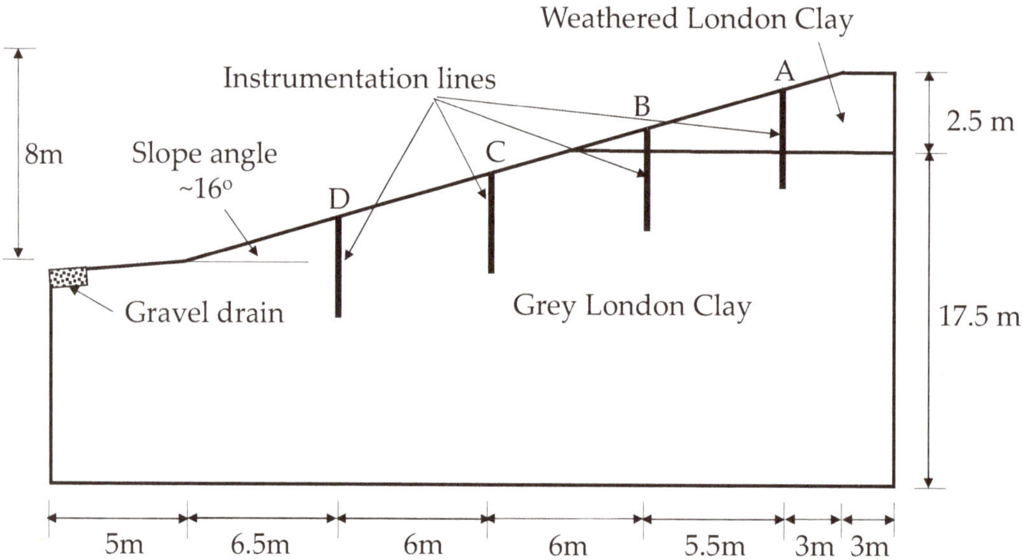

Figure 2. A cross-section of the Newbury cutting (dimensions after Smethurst et al. [20]).

A series of field and laboratory tests were conducted by Smethurst et al. [20] to deduce different soil parameters for the research site. The tests included Atterberg limit tests, triaxial (saturated) hydraulic conductivity test, dry unit weight test and field saturated hydraulic conductivity test using the bail-out method. The deduced K_{sat}, unit weight, and plasticity indices of the site soil are summarised in Table 1. It is to be noted that a large number of laboratory tests on undisturbed samples along with bailout tests (in the field from hand augured boreholes of up to 3 m deep) were used to deduce the K_{sat} values. 1 to 3 orders of magnitude differences between deduced K_{sat} values were reported by Smethurst et al. [20]. It was believed to be due to anisotropy and soil fabric, including silt partings and cracks and fissures.

The relationship between the soil water content and suction (also known as soil water characteristics curve—SWCC) for London clay was reported by Croney [39] and is presented in Figure 3 and can be assumed to be representative of the site soil [9]. It is to be noted that Croney [39] presented SWCC for both drying and wetting stages for London clay. As SEEP/W [37] does not allow the use of multiple SWCCs and also considering the difficulties associated with generating consistent wetting SWCC, the drying curve was used for this analysis. It is expected that if the wetting SWCC was used, the observed pore water pressure response would be different as K_{unsat} is a function of K_{sat} and SWCC. Change in one may require adjustment in the other parameter. However, its significance was outside the scope of this study.

Table 1. Hydraulic conductivity, unit weight and plasticity index of grey and weathered London Clay at the Newbury test site (after Smethurst et al. [20]).

Property	Grey London Clay		Weathered London Clay	
	Range	Average	Range	Average
K_{sat} from triaxial hydraulic conductivity test (m/day)	3.37×10^{-6}–5.7×10^{-5}	1.99×10^{-5}	4.32×10^{-5}–1.38×10^{-4}	7.52×10^{-5}
K_{sat} from borehole bailout tests (m/day)	1.99×10^{-4}–3.8×10^{-4}	3.2×10^{-4}	3.11×10^{-3}–4.32×10^{-3}	3.72×10^{-3}
Dry unit weight (kN/m^3)	13.2–15.2	14.6	13.2–16.2	16.0
Plasticity index (%)	32.5–36.4	34.8	31.7 *	31.7 *

* one out of five of weathered London Clay samples showed plasticity.

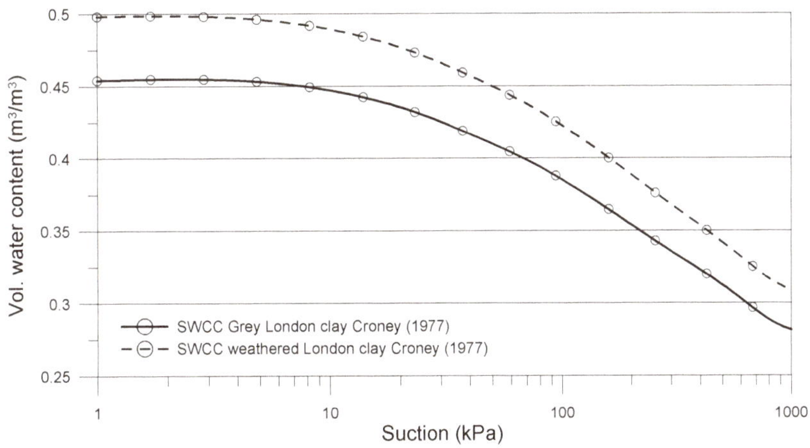

Figure 3. Transformed SWCCs for the Newbury cutting soil.

Moreover, it should be noted that SEEP/W [37] and most other numerical software use SWCC in terms of volumetric water content and the curves presented by Croney [39] were in terms of gravimetric water content. The gravimetric values were converted into volumetric values using the following relationship (Equation (2)) and the transformed SWCCs are presented in Figure 3.

$$w_{vol} = w_{gr} \times \frac{\gamma_{dry}}{\gamma_w} \quad (2)$$

where w_{vol} is the volumetric water content, w_{gr} is the gravimetric water content, γ_{dry} and γ_w are the dry unit weight of soil and unit weight of water, respectively. The average values for γ_{dry} for the two different soil layers were used for the calculations, as presented in Table 1.

The vegetation on the slope was predominantly rough grass and herbs with some small shrubs (0.5 m < tall). The grass and herbs were mowed periodically until October 2002 to help the development of shrubs planted on the slope. Several grown-up Beech, Oak and Silver birch trees were located near the top of the slope.

Instrumentation was done to monitor moisture content (TDR moisture probes), suction (standard water-filled tensiometers and equitensiometer), positive water pressure (piezometers), and water table location (dipping boreholes) at depths between 0.3 and 3.5 m. The instruments were installed in four groups, namely, cluster A (near the crest) to D (near the toe). A weather station was also installed at the site. A 350 mm deep interceptor drain was installed on the slope face (across the slope face) to capture surface runoff and interflow. The site was monitored from October 2002.

Figures 4 and 5 present average daily rainfall and runoff recorded (respectively) for the 3 years duration under consideration in this study. Other recorded meteorological parameters for the duration, i.e., average daily temperature, average daily wind speed (2 m above ground surface) and average daily humidity, are presented in Appendix A. These data will later be used in calculations.

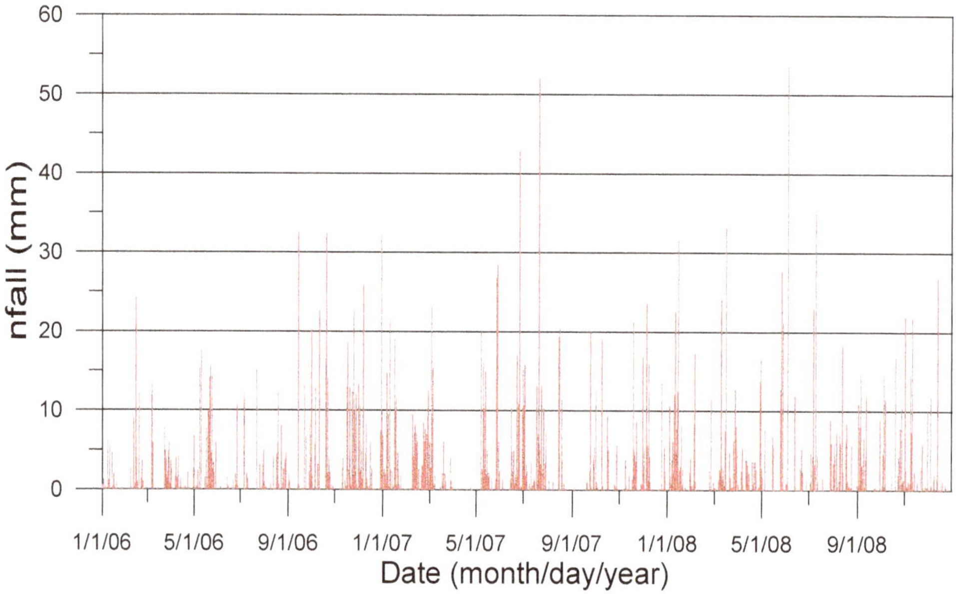

Figure 4. Measured daily rainfall at Newbury cutting.

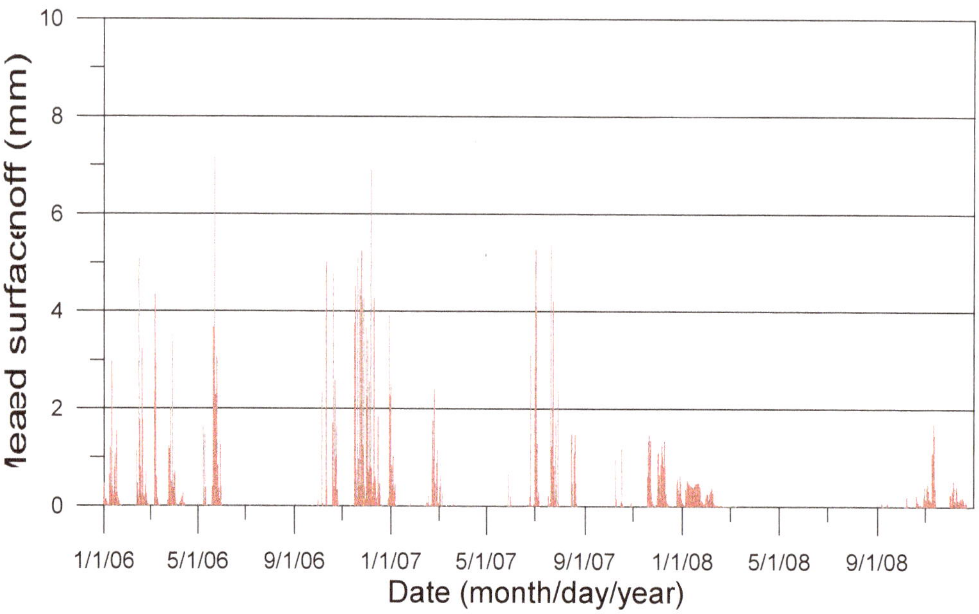

Figure 5. Measured daily runoff at Newbury cutting.

3. Numerical Analyses

SEEP/W [37] is a finite element software that allows analysis of groundwater seepage and pore water pressure distribution within porous media such as soil in both saturated and unsaturated states. The types of analyses can range from saturated steady-state problems to complex saturated/unsaturated time-dependent (transient) problems. The pore water pressure distribution from such analyses could be used as an input for other stress or stability analyses to assess the serviceability and ultimate limit state behaviour of the slopes.

The software requires SWCC and soil hydraulic conductivity function (i.e., variation of *K* with soil suction) as inputs for material definition. The effect of climate variables on the problem can be modelled using an equivalent hydrological boundary condition (explained later). Different modelling aspects of the problem are discussed in the next few sections.

3.1. Model Geometry

The discretised geometry used for the analysis is presented in Figure 6. To minimise the effect of imposed boundary conditions on the analysis results, the geometry was extended by 20 m towards the right and by 15 m towards the left. Several trials were run to ensure the effect of boundary and mesh size was minimum or insignificant to the observed results. Details of the trials are presented in Appendix B. A global mesh size of 0.8 m was used in the analysis. Except for the surface layer, the geometry was discretised using an automatically generated unstructured mesh using triangular and quadrilateral elements. The elements close to the boundary (to a depth greater than the maximum depth of measurement during field investigation) were then refined to half the global element size. The surface layer was discretised using quadrilateral elements. Quadrilateral elements have been shown to perform better in such scenarios as the gradients of primary unknowns are steeper in the direction normal to the surface [37]. The nodal convergence was checked. A total of 4200 nodes and 3965 elements (4-nodded quadrilateral and 3-nodded triangular) were used to discretise the geometry.

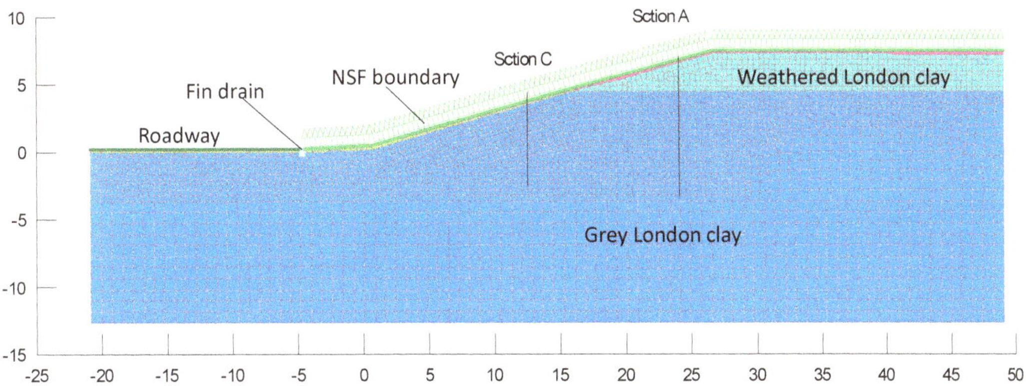

Figure 6. Discretised geometry was used for the SEEP/W analysis (dimensions are in m).

The soil in the slope had two distinct subdivisions, namely, weathered London clay and grey London clay layers. They were modelled using different materials. Because of the presence of plant roots and other organic matters, the *K* of the top surface of the soil slope was higher than the layers below. Separate surface layers (of 0.4 m thickness with element thickness of 0.1 m) using different material models were added to the geometry to account for these differences (i.e., surface layers corresponding to weathered London clay and grey London clay layers). The pore water pressure and other variables usually change rapidly close to the exposed soil surface, i.e., in the surface layer.

3.2. Material Models

SEEP/W [37] allows the soil to be modelled as saturated only or as a combination of saturated and unsaturated soil states. All four soil materials (i.e., grey and weathered London clay and their respective surface layers) were modelled using the saturated/unsaturated option. This way, the soil could stay in a saturated or unsaturated state during an analysis or change state depending on the analyses conditions. The SWCCs for the weathered and grey London clay have been discussed above. The SWCC for the surface layers was kept similar to the layers below. There were no specific K data available for the surface layers. The K_{sat} values for the soils were assumed as 10 times the values of the layers below [40].

3.3. Calculation for Input Hydrological Boundary Condition

The water balance equation [41] was used for the calculation of input hydrological boundary conditions to represent the effect of the atmospheric boundary (e.g., rainfall, solar radiation, humidity, wind speed) and vegetation as below,

$$\sum R - RO - ET - S + RE = 0 \tag{3}$$

where R is the rainfall, RO is the runoff, ET is the evapotranspiration, S is the change in stored water within the soil, and RE is the net recharge from surrounding soils. If we consider inflow and outflow for soil mass from the surrounding soil is equal, i.e., RE = 0, we can say S is equal to the magnitude of water percolating into the soil, net surface infiltration (NSF) through the surface boundary. Thus we can rewrite Equation (3) as below,

$$\sum R - RO - ET = NSF \tag{4}$$

The measured daily rainfall and runoff and calculated ET data were used for this purpose. To estimate ET, a reference ET was calculated first using the Penman-Monteith equation as below [42],

$$ET_r = \frac{0.408\Delta(R_n - G) + \gamma\frac{900}{T+273}u_2(e_s - e_a)}{\Delta + \gamma(1 + 0.34u_2)} \tag{5}$$

where ET_r is the reference evapotranspiration calculated in a 1-day time step, Δ is the slope of saturation vapour curve, R_n is the net radiation flux in MJ/m^2/day, G is the soil heat flux density in MJ/m^2/day, γ is the Psychrometric constant in kPa/°C, T is the mean temperature in °C 2 m above the ground level, u_2 is the wind speed in m/s at 2 m above the ground, and $(e_s - e_a)$ is the saturation vapour pressure deficit in kPa.

The Penman–Monteith equation [42] parameters presented in Table 2, along with the weather data presented in Appendix A have been used. Once the daily ET_r was calculated, ET was deduced using the following equation,

$$\begin{aligned} ET &= ET_r && \text{for } 0 \leq SMD \leq RAW \\ ET &= ET_r \times \tfrac{TAW-SMD}{TAW-RAW} && \text{for } SMD \geq RAW \end{aligned} \tag{6}$$

where SMD is the soil moisture deficit, RAW is readily available water, and TAW is the total available water. Calculated ET values are presented in Figure 7 and calculated daily NSF values are presented in Figure 8. NSF values represent the infiltration of water into the soil due to rainfall and also the removal of water due to evapotranspiration. NSF can be of either positive or negative magnitude. A positive value will indicate water is infiltrating into the ground and the rainfall is dominating the process. A negative value, on the other hand, will indicate water is leaving the soil due to evapotranspiration. The calculated NSF values were used as an input hydrological boundary condition for this analysis and represented the effect of climate and vegetation on the problem. It is to be noted that while using eq [3] to estimate NSF, the value of runoff needs to be known. However, it may not be a readily available measurement. In the absence of measured runoff values, a process outlined in Appendix C can be used to estimate runoff and thus NSF. Similar approaches have been used in the literature [9,43,44].

Table 2. Parameters used for the calculation of evapotranspiration.

Parameter	Value
Psychrometric constant (kPa/°C)	0.000665 × atm. pressure
Solar constant	0.082
Latitude (rad)	0.895877505
Albido or canopy reflection coefficient	0.23
Stefan–Boltzman constant (Mj/K^4/M^2/day)	4.903 × 10^{-9}
Elevation above sea level (m)	105

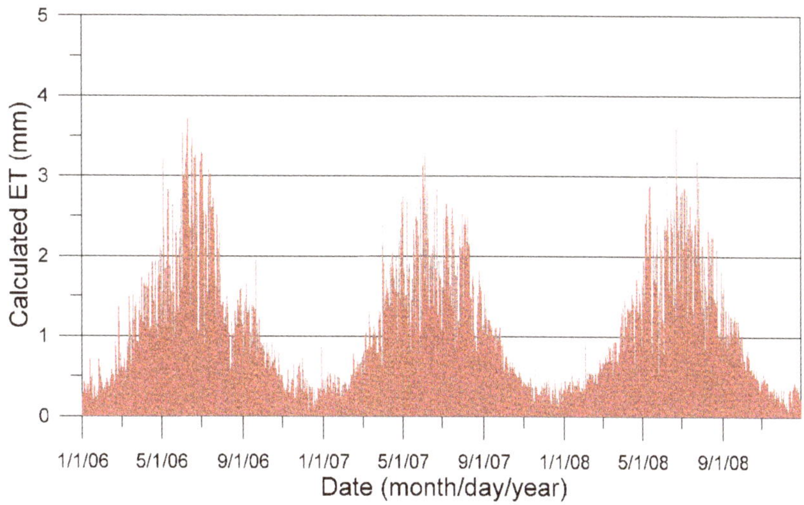

Figure 7. Calculated ET for Newbury cutting.

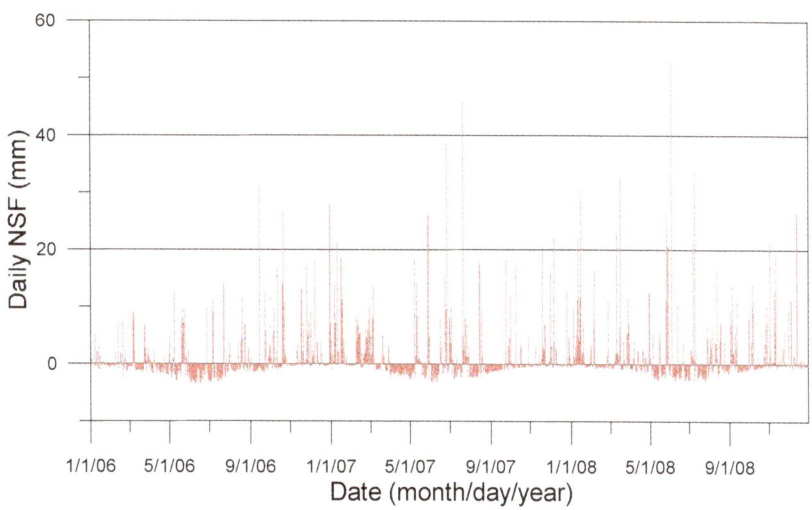

Figure 8. Net surface flux data for Newbury cutting.

It is to be noted that the use of a 1-day time step means everything occurring within a particular 24 h period is averaged out over that period and the effect of a shorter duration event may not be captured very accurately. It is possible to conduct a similar analysis with smaller time steps (e.g., hourly) if hourly measurements of all related parameters were available.

3.4. Other Boundary Conditions

The left, right and the bottom boundary of the problem was modelled as impermeable (no flow). Part of the ground surface with road surfacing was modelled as impermeable as well. The mesh and geometry sensitivity analyses show that the choice of these boundary conditions was appropriate. The fin drain installed near the toe of the slope was not modelled explicitly. Rather its effect was modelled using a zero pore water pressure boundary. A flux boundary condition was applied at the ground surface, as discussed earlier.

3.5. Initial Condition

The initial pore water pressure condition in the model was established using an initial water table. An educated guess was made by analysing the piezometer pore water pressure data on the date 1 January 2006. This might have some effect on the predicted pore water pressure profile, and to the author's understanding, a more complicated seepage analysis could be conducted to establish the initial pore water pressure profile. However, as will be shown later, the effect of assuming an initial water table gets diminished as the analysis progresses and somewhat becomes irrelevant beyond a few months of analysis, even when a very crude guess is made.

3.6. Estimation of K

For modelling unsaturated soils, SEEP/W [37] allows the choice of different functions that deduces K_{unsat} by relating soil suction to K_{sat}. A user can choose from available functions such as Fredlund et al. [29], Green and Corey [28] and Van-Genuchten [27]. In this investigation, the Van-Genuchten [27] function (Equation (1)) was used. Estimating an appropriate value for K_{sat} was a challenge as large (more than 100 times) differences were observed for values from different tests. As indicated earlier, in such scenarios, one has a choice of using an average value or using subjective judgment.

An alternative approach was used to estimate a representative K_{sat} value in this study. The first year of field-measured pore water pressure data was back analysed. Few trial analyses were conducted with K_{sat} being systematically varied until the best match was found. The field average K_{sat} as presented in Table 1, was used as the first trial value. SWCC and other parameters in the study remained unchanged (e.g., the ratio between K_{sat} in surface layers and the underlying layers). The best match K_{sat} values were then used to analyse the problem further. Piezometer readings from two different depths at two locations, A and C (see Figures 2 and 6), were used.

In principle, this process is not very different from conventional techniques (especially field tests) used to deduce K_{sat}. Interpretation of many of the field tests also involves several assumptions and back-fitting of field observations. For example, in a Guelph permeameter test, the flow of water into the soil is observed, and collected data are interpreted using saturated-unsaturated flow theories to estimate K_{sat}.

Figure 9 presents the effect of using different K values on the pore water pressure profile for location C at 2.5 m depth. Three different analyses are presented here, i.e., analysis using the field average K_{sat}, analysis using the best match K_{sat} and analysis using three times the best match K_{sat}. The figures show that K_{sat} may play a significant role in such pore water pressure analyses. The best match K_{sat} values for the weathered London clay and grey London clay were 0.015 m/d and 0.0025 m/d, respectively—approximately four and eight times their respective field average values. Plots at other locations are not presented here due to space limitations.

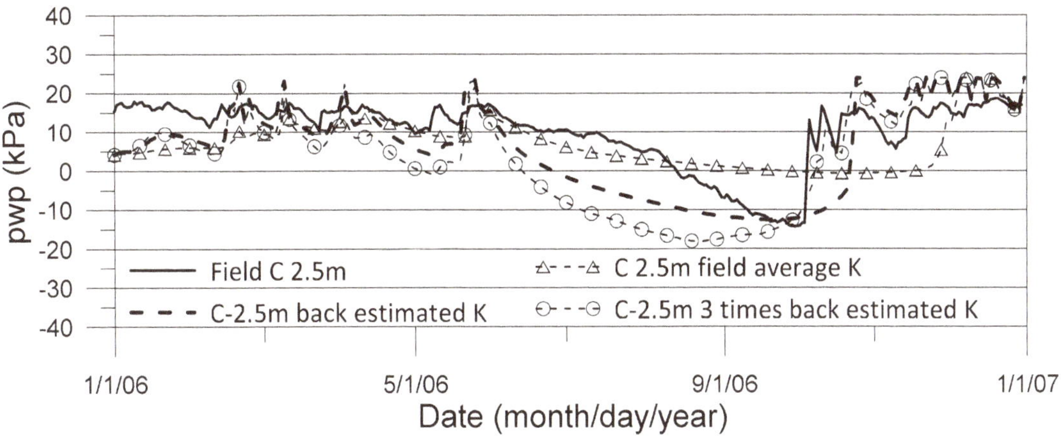

Figure 9. Different trial K_{sat} values and related pore water pressure at Section C 2.5 m depth from the surface.

4. Results

The analyses were run from 1 Jan 2006 to 31 December 2008, a total duration of 3 years, including the first year used for the estimation of K_{sat}. The results are discussed below.

Figure 10a,b show the field measured and calculated pore water pressure from two analyses (i.e., one using matched K and the other using average field value) at 1.5 and 2.5 m depths from the surface at instrumented section A and Figure 11a,b presents the same for location C. At location A, both analyses captured the negative pore-water pressure peaks with reasonable accuracy. Both analyses (using matched and field K) overestimated the magnitude of positive pore water pressure throughout the measurement period. At location C, the qualitative trend (peaks and troughs) was much better captured in the matched K analysis and the use of average field K in the analyses led to a significant underestimation of negative pore water pressure developed during the period of June-July of 2006. The number of peaks and troughs in the field measured values were better captured by using matched K in the analyses. The average K analysis trend was relatively smoother and often did not capture the smaller peaks and troughs in the trend (short duration events). Overall, the use of matched K improves the accuracy of the calculation.

The near-surface soil suctions were recorded by standard water-filled tensiometers at different depths (0.3, 0.6 and 0.9 m from the surface at locations A and C). Figure 12a,b present the measured suction and estimated responses from the numerical analyses (both using field average K and matched K) at 0.6 m depth at instrumented locations A and C, respectively. At location A the suction magnitude was underestimated between June and July 2007 and overestimated at other times by both analyses. The difference between the calculations from the two analyses was very small. At location C, the suction response was significantly overestimated by both analyses. The overestimation was slightly higher for the case of the field K analysis. The qualitative trend was reasonably well captured by both analyses with reasonable accuracy.

Figure 13 presents the measured and calculated volumetric water content variation with time. At location C, both analyses calculated volumetric water content with reasonable accuracy; however, at location A, both analyses significantly underestimated the changes in water content with time. This difference is likely to be due to local heterogeneity of the soil and the SWCC used in the analyses is unable to represent its behaviour. It is to be noted that in a saturated/unsaturated seepage analysis, the hydraulic conductivity and SWCC both affect the changes in water content (as K_{unsat} is a function of soil suction) and consequently developed pore water pressure. Estimating the K_{unsat} based on observation of pore water pressure response has allowed prediction of the same with greater accuracy. It is to be noted that the estimation of K_{unsat} could also be achieved by comparing the predicted and field observed water content data, and in such a case, it is expected that the prediction of water content could have improved accuracy. It is envisaged that with a better estimation of SWCC for different materials used in the simulation, a better estimation of water content could be achieved.

The interaction between soil, vegetation and the atmospheric boundary is complex and can be influenced by a number of variables. This paper presents a simplified process to model the interaction to capture the related changes in water pressure and soil water content and outlines an objective approach to estimate one of the most important parameters in seepage analyses (i.e., hydraulic conductivity). The simplification and assumptions include,

- the use of an SWCC from literature for a similar soil (not based on soil tested on this particular site)
- the use of the drying curve for representing both wetting and drying behaviour
- the assumption of uniform soil properties within different soil layers even though the site investigation revealed the grey London clay to be variable along with the vertical and horizontal directions.

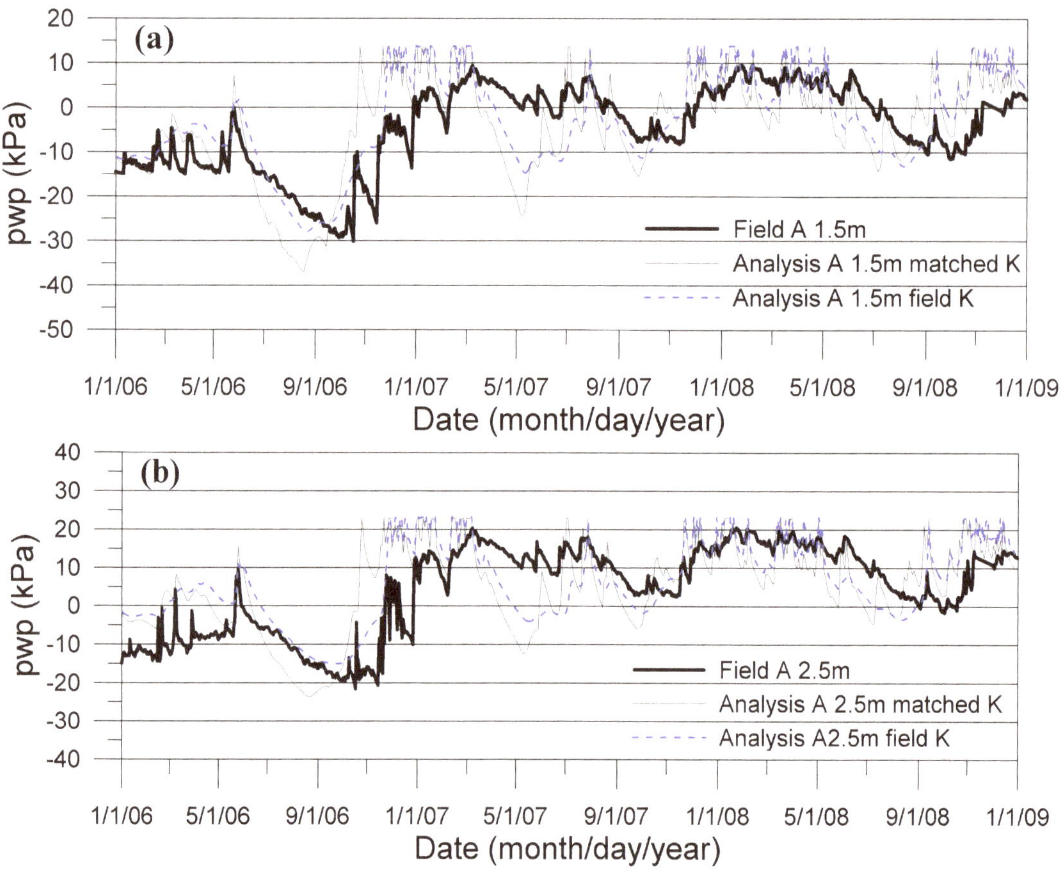

Figure 10. Field measured and calculated pore water pressure at (**a**) 1.5 m and (**b**) 2.5 m depths in section A.

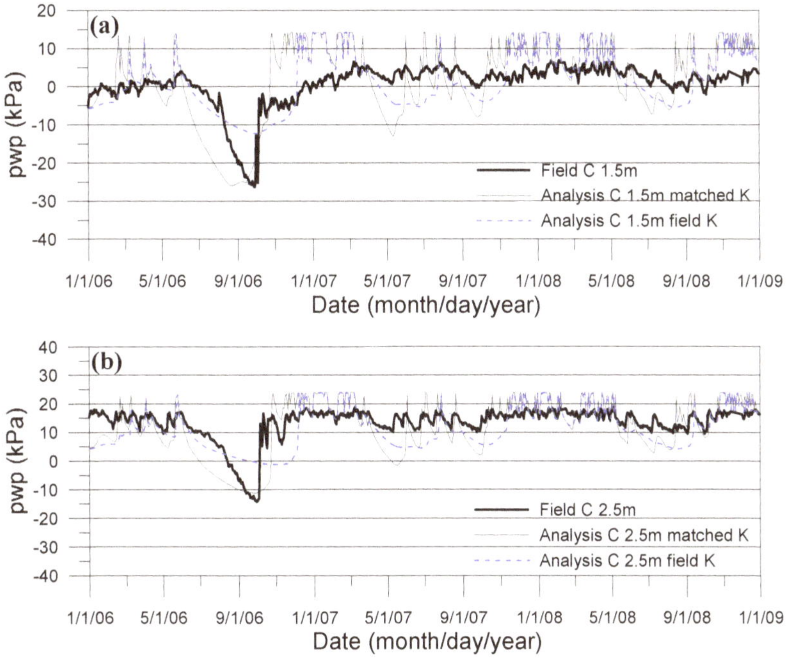

Figure 11. Field measured and calculated pore water pressure at (**a**) 1.5 m depth and (**b**) 2.5 m depth at section C.

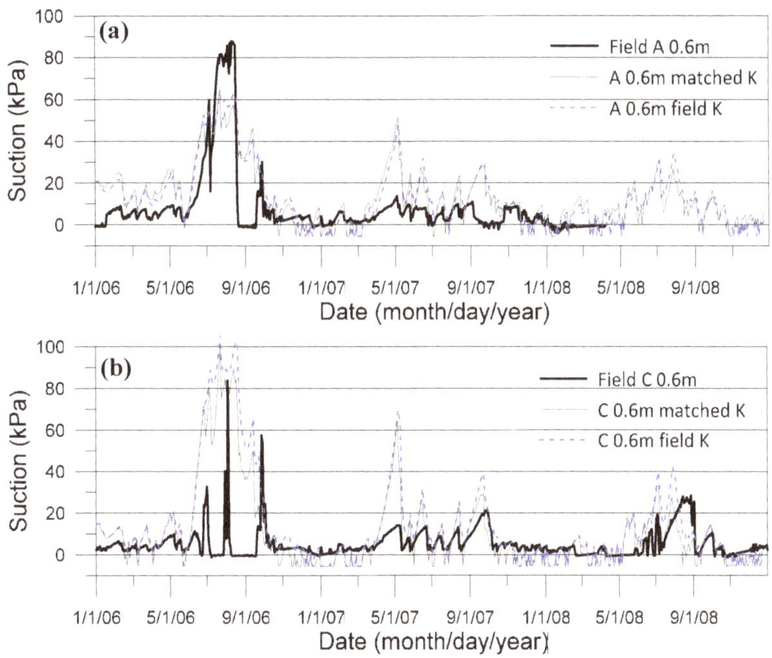

Figure 12. Measured and calculated soil suction at 0.3 m depth (**a**) at section A and (**b**) at section C.

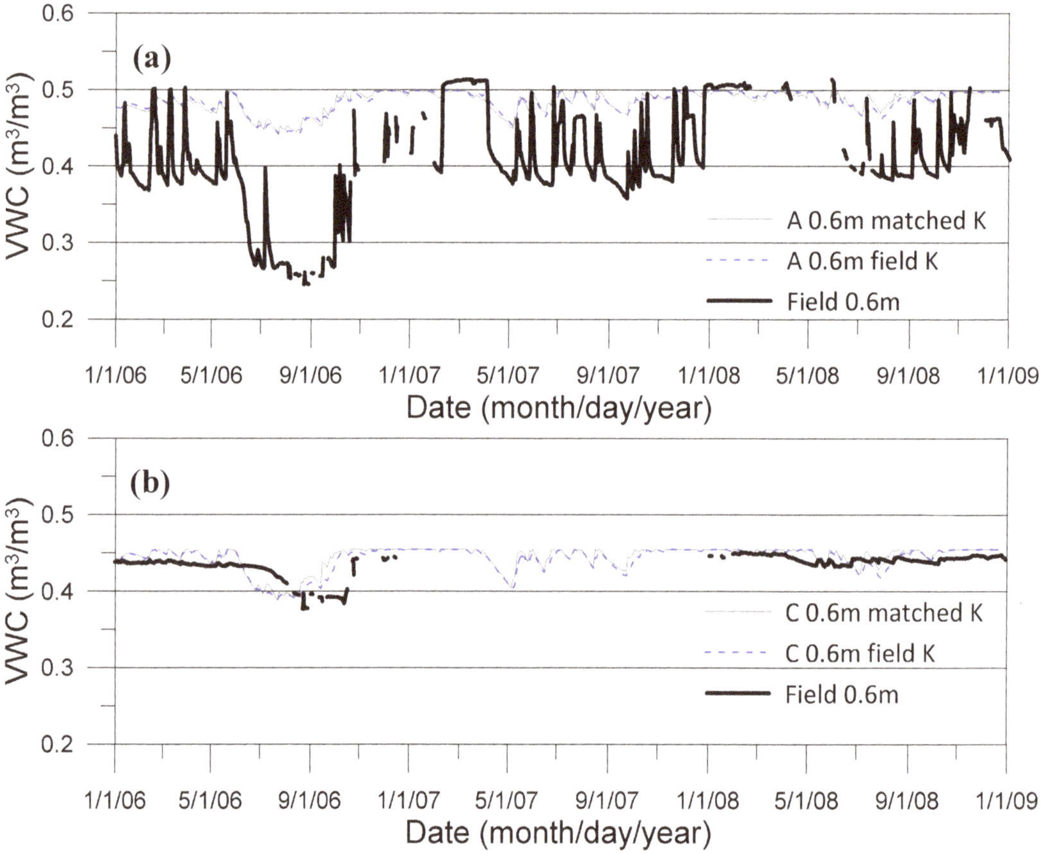

Figure 13. Measured and calculated volumetric water content 0.6 m depth (**a**) at location A and (**b**) at location C.

Despite these simplifications, the analyses presented here could capture the qualitative trend and quantitative changes in pore water pressure and water content with reasonable accuracy. The difference between the field averaged value of K_{sat} and matched K_{sat} was four to eight times. It is to be noted that standard laboratory or field tests for hydraulic conductivity often produce estimates that can be 100 or even 1000 times different from each other. Furthermore, the calculation here was found to be significantly less sensitive to K_{sat} compared to seepage or consolidation analyses in saturated soils. Due to this, the differences between the field average K analysis and matched K analysis were close to each other on many occasions, even though the matched K analysis showed a relatively better overall prediction for water pressure and water content. In other situations where more variability exists, the situation may be significantly different, and the use of field average K may produce far inferior predictions.

5. Discussion and Conclusions

A set of numerical analyses has been discussed in this paper. The analysis has been carried out with a finite element software SEEP/W [37]. All the inputs to the model have been determined from laboratory/field test results except for the hydraulic conductivity of the soil, which is likely to be affected by the presence of fissures or cracks in the soil and may not be captured by standard field or laboratory tests. An alternative approach is outlined here for the estimation of the hydraulic conductivity. Surface infiltration was used in the model as an input boundary condition and was calculated using the surface water balance equation. The results show that using the approach, it is possible to capture the seasonal, climate-induced pore water pressure variation in slopes, and the discussed method for estimation of hydraulic conductivity can be a useful tool for seepage analyses in unsaturated soils.

Author Contributions: Conceptualisation, M.R.K. and D.H.; methodology, M.R.K. and M.M.R.; software, M.R.K.; validation, M.R.K.; formal analysis, M.R.K.; investigation, D.H. and M.R.K.; resources, D.H.; data curation, D.H. and M.R.K.; writing—original draft preparation, M.R.K.; writing—review and editing, D.H. and M.M.R.; visualisation, M.R.K. and M.M.R.; supervision, D.H.; project administration, D.H.; funding acquisition, D.H. All authors have read and agreed to the published version of the manuscript.

Funding: Part of this research was carried out when the first author was working as a Research Fellow at the Queens University Belfast and was supported by an EPSRC-funded project, Infrastructure slopes Sustainable Management and Resilience Assessment—iSMART (http://www.ismartproject.org, accessed on 1 June 2022).

Institutional Review Board Statement: Not applicable.

Informed Consent Statement: Not applicable.

Data Availability Statement: Data related to publication can be accessed via contacting any of the authors in the paper.

Conflicts of Interest: The authors declare no conflict of interest

Appendix A

Figures A1–A3 showing recorded daily temperature, wind speed and relative humidity at Newbury cutting

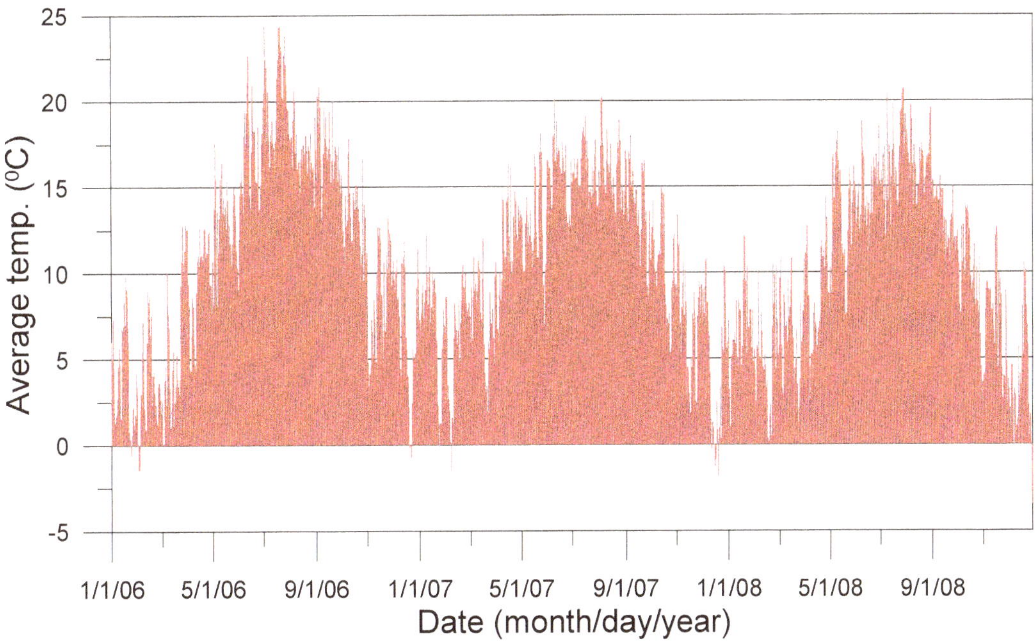

Figure A1. Average daily temperature at Newbury cutting.

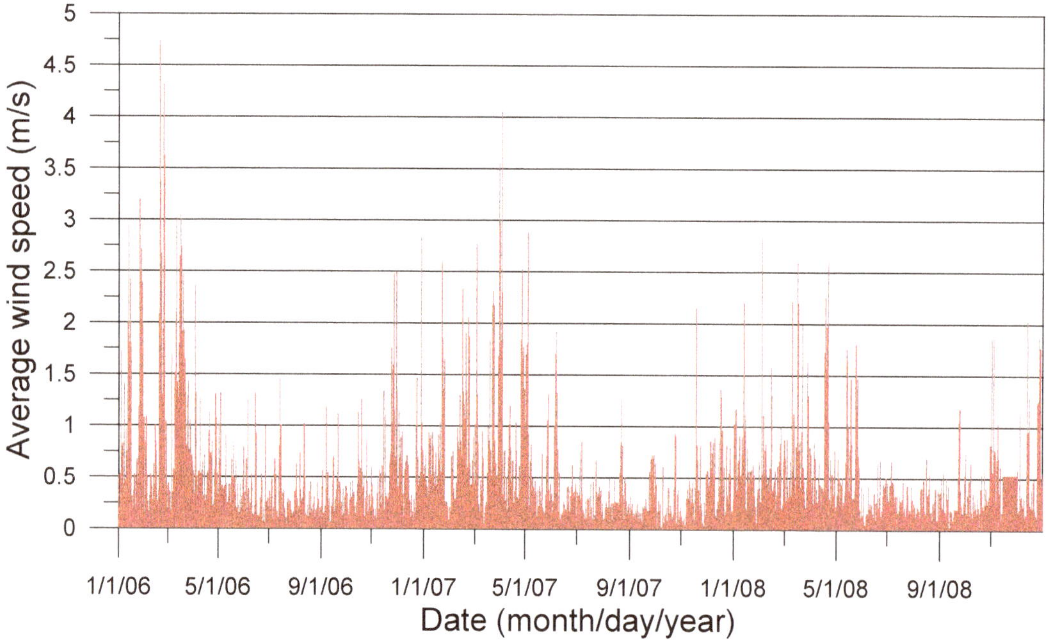

Figure A2. Average daily wind speed 2 m above the ground level at Newbury cutting.

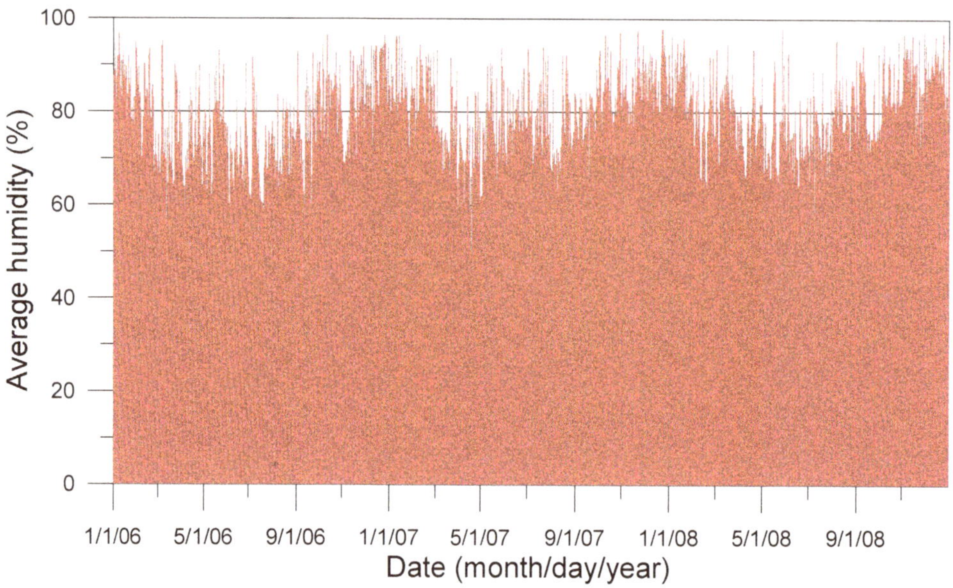

Figure A3. Average daily humidity at Newbury cutting.

Appendix B

To investigate the sensitivity of mesh size and the effect of imposed boundary conditions, additional analyses were conducted. Results from analysis based on mesh and geometry presented in Figure 6 were used as a benchmark and deviation from those results due to changes in the mesh and geometry condition was observed. For comparison, the analyses are named GM1—original analyses reported in this paper; GM2—analyses with the same geometry as GM1 but without mesh refinement (element size of 0.8 m except for the surface layers where the element thickness was 0.1 m) and GM3- with reduced depth and length of the geometry and 0.4 m global element size with a surface layer having 0.1 m element thicknesses). The mesh and geometries are shown in Figures A4 and A5, and calculated pore water pressure/suction are compared for location C at three different depths in Figure A6.

As can be seen in Figure A6, there were only very small differences between calculated responses from all the additional analyses. So the mesh size and the boundary could be treated as the reason for the problem being solved and the mesh and geometry combination GM1 was used for all the analyses to remove any concern of numerical ill-conditioning or boundary effect on results.

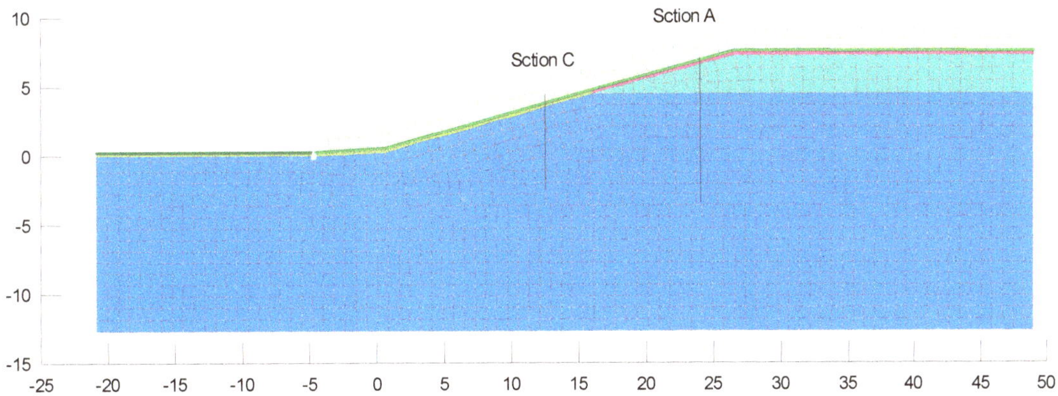

Figure A4. Geometry and mesh used for sensitivity analysis GM2.

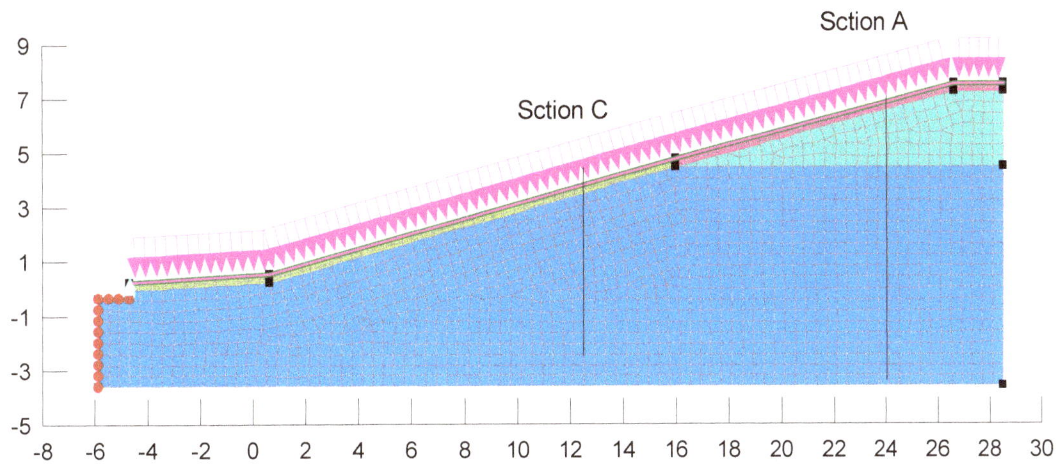

Figure A5. Geometry and mesh used for sensitivity analysis GM3.

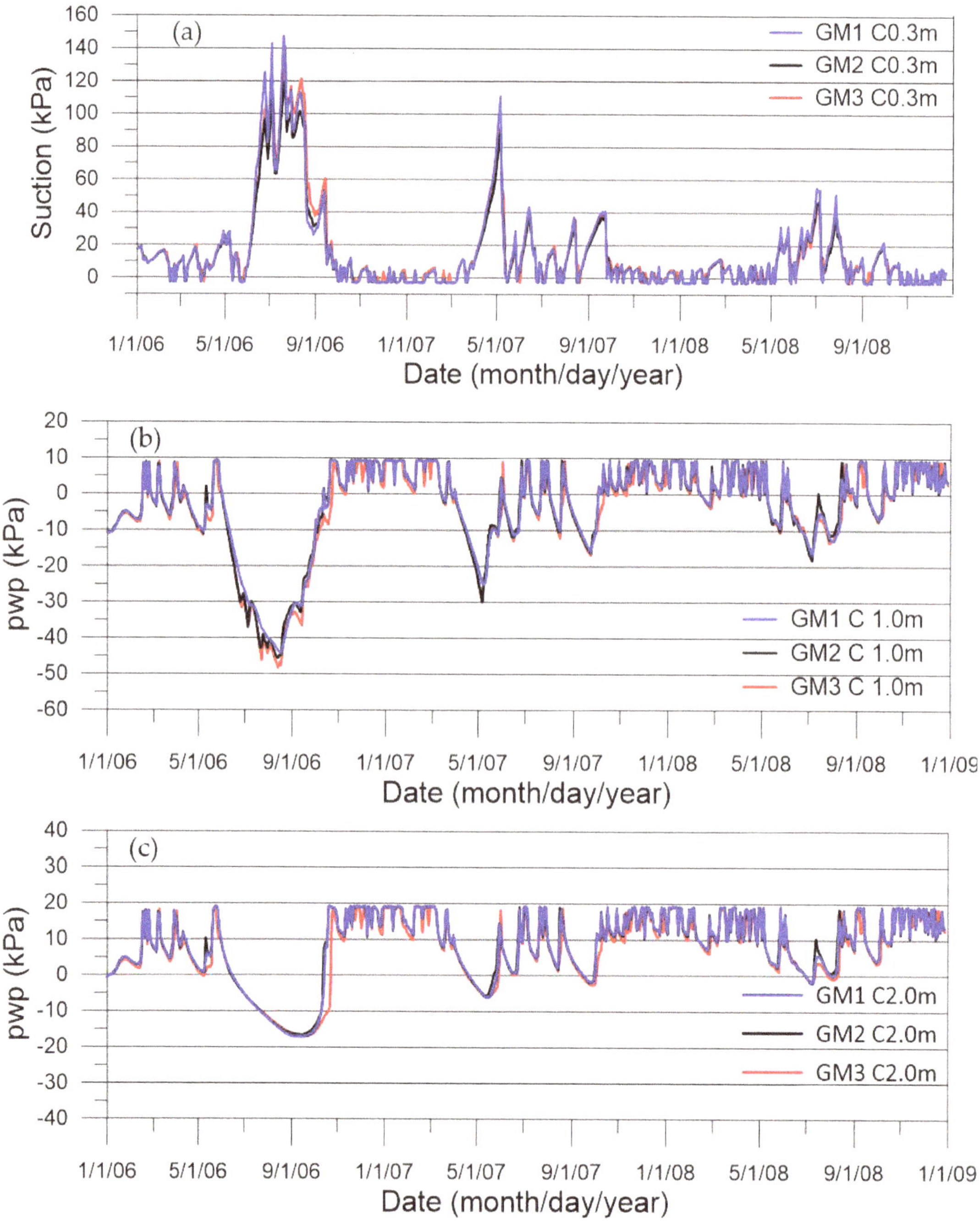

Figure A6. suction/pore water pressure response at instrumentation line C at (**a**) 0.3 m depth, (**b**) 1 m depth and (**c**) 2 m depth.

Appendix C

In many cases, the measured runoff data is not available. In such a case, a two-tank soil water storage model (SWSM) [9,43,45] can be used for the estimation of runoff. The model calculates how much water from a particular intensity of rainfall infiltrates the soil based on the available storage capacity and hydraulic conductivity of the soil. The amount of water that cannot infiltrate is treated as RO. That means, even if the soil has enough storage capacity, if the rainfall intensity is more than the soil's capacity to allow water in (hydraulic conductivity), the excess water will flow as runoff. In this case, both pore water pressure and runoff are a function of K and an iterative process as outlined below will be necessary to estimate K and thus *NSF*.

1. The first step in the process is to make an initial estimation of K_{sat}. It is difficult to objectively estimate a value for K_{sat} as the numbers can be scattered over a large range. A field average can be a good starting point.
2. Following this, RO can be estimated using the two-tank SWSM model and the first estimate of daily NSF can be calculated.
3. The deduced *NSF* can then be used as an input boundary condition in a first trial numerical analysis. Pore water pressure distribution from the simulation can then be compared with the field measured values.
4. In the next step, K_{sat} values can be systematically varied and step 1 to 3 repeated until a good match between the observed and calculated pore water pressures can be found. For every chosen K_{sat}, the calculation for surface RO and *NSF* needs repeating.

References

1. BOM. *Monthly Summary for Australia*; IDCKGC1A00; Bureau of Meteorology: Melbourne, Australia, 2019.
2. Toll, D.G.; Mendes, J.; Augarde, C.E.; Karthikeyan, M.; Phoon, K.K.; Gallipoli, D.; Lin, K.Q. Effects of climate change on slopes for transportation infrastructure. In Proceedings of the International Conference on Transportation Geotechnics, Nottingham, UK, 24–28 August 2008.
3. NEA. *Singapore Experienced Its Wettest December since 1968*; National Environment Agency: Singapore, 2006.
4. Murphy, J.M.; Sexton, D.M.H.; Jenkins, G.J.; Boorman, P.M.; Booth, B.B.B.; Brown, C.C.; Clark, R.T.; Collins, M.; Harris, G.R.; Kendon, E.J.; et al. *UK Climate Projections Science Report: Climate Change Projections*; Met Office Hadley Centre: Exeter, UK, 2009.
5. CSIRO. *Climate Change in Australia, Projections for Selected Australian Cities*; CSIRO: Canberra, Australia, 2015.
6. ClimateChangeAustralia. Climate Change in Australia. Available online: https://www.climatechangeinaustralia.gov.au/en/ (accessed on 2 November 2020).
7. Karim, M.R.; Rahman, M.M.; Nguyen, K.; Cameron, D.; Iqbal, A.; Ahenkorah, I. Changes in Thornthwaite Moisture Index and reactive soil movements under current and future climate scenarios-a case study. *Energies* **2021**, *14*, 6760. [CrossRef]
8. Karim, M.R.; Rahman, M.M.; Nguyen, H.B.K.; Newsome, P.L.; Cameron, D. TMI soil moisture index for South Australia under current and future climate scenarios. In Proceedings of the 20th International Conference on Soil Mechanics and Geotechnical Engineering, a Geotechnical Discovery Downunder, Sydney, Australia, 1–5 May 2022; Rahman, M.M., Jaksa, M., Eds.; Australian Geomechanics Society: Sydney, Australia, 2022; pp. 2233–2236.
9. Karim, M.R.; Hughes, D.; Kelly, R.; Lynch, K. A rational approach for modelling the meteorologically induced pore water pressure in infrastructure slopes AU—Karim, Md Rajibul. *Eur. J. Environ. Civ. Eng.* **2019**, *24*, 2361–2382. [CrossRef]
10. Loveridge, F.A.; Spink, T.W.; O'Brien, A.S.; Briggs, K.M.; Butcher, D.J.E. The impact of climate and climate change on infrastructure slopes, with particular reference to southern England. *Q. J. Eng. Geol. Hydrogeol.* **2010**, *43*, 461–472. [CrossRef]
11. Ridley, A.; McGinnity, B.; Vaughan, P.R. Role of pore water pressures in embankment stability. *Proc. Inst. Civ. Eng.* **2004**, *157*, 193–198. [CrossRef]
12. Turner, S. Climate Change Blamed as Landslip Incidents Treble. Available online: https://www.newcivilengineer.com/climate-change-blamed-as-landslip-incidents-treble/812153.article (accessed on 1 July 2015).
13. Sivakumar, V.; Hughes, D.; Clarke, G.; Glynn, D.T. A case study: Delayed failure of a deep cutting in lodgement till. *Proc. ICE Geotech. Eng.* **2007**, *160*, 193–202. [CrossRef]
14. Ng, P.B.; Poh, K.K.; Tenando, E. Slope failure after prolonged rainfall. In *BCA Seminar- Approach to Structural Inspection and Slope Stability*; Building and Construction Authority: Singapore, 2007.
15. Hughes, P.N.; Glendinning, S.; Mendes, J.; Parkin, G.; Toll, D.G.; Gallipoli, D.; Miller, P.E. Full-scale testing to assess climate effects on embankments. *Proc. Inst. Civ. Eng. Eng. Sustain.* **2009**, *162*, 67–69. [CrossRef]
16. Bolton, M.D.; Take, W.A. Seasonal ratcheting and softening in clay slopes, leading to first-time failure. *Géotechnique* **2011**, *61*, 757–769.

17. Briggs, K.M.; Smethurst, J.A.; Powrie, W.; O'Brien, A.S.; Butcher, D.J.E. Managing the extent of tree removal from railway earthwork slopes. *Ecol. Eng.* **2013**, *61*, 690–696. [CrossRef]
18. Clarke, D.; Smethurst, J.A. Effects of climate change on cycles of wetting and drying in engineered clay slopes in England. *Q. J. Eng. Geol. Hydrogeol.* **2010**, *43*, 473–486. [CrossRef]
19. Smethurst, J.A.; Clarke, D.; Powrie, W. Factors controlling the seasonal variation in soil water content and pore water pressures within a lightly vegetated clay slope. *Géotechnique* **2012**, *62*, 429–446. [CrossRef]
20. Smethurst, J.A.; Powrie, W.; Clarke, D. Seasonal changes in pore water pressure in a grass-covered cut slope in London Clay. *Géotechnique* **2006**, *56*, 523–537. [CrossRef]
21. Briggs, K.M.; Smethurst, J.A.; Powrie, W.; O'brien, A.S. Wet winter pore pressures in railway embankments. *Proc. Inst. Civ. Eng.* **2012**, *166*, 451–465. [CrossRef]
22. Geo-Slope. *Vadose Zone Modeling with VADOSE/W An Engineering Methodology*; Geo-Slope International Limited: Calgary, AB, Canada, 2013.
23. Tsaparas, I.; Toll, D.G. Numerical analysis of infiltration into unsaturated residual soil slopes. In Proceedings of the 3rd International Conference on Unsaturated Soils, Recife, Brazil, 10–13 March 2002.
24. Karthikeyan, M.; Toll, D.G.; Phoon, K.K. Prediction of changes in pore-water pressure response due to rainfall events. In *Unsaturated soils: Advances in Geo-Engineering, Toll, D.G., Ed.*; Taylor & Francis Group: London, UK, 2008; pp. 829–834.
25. Geo-Slope. *Seepage Modeling with SEEP/W an Engineering Methodology*; Geo-Slope International Limited: Calgary, AB, Canada, 2013.
26. Rouainia, M.; Davies, O.; O'Brien, T. Numerical modelling of climate effects on slope stability. *Proc. Inst. Civ. Eng. Eng. Sustain.* **2009**, *162*, 81–89. [CrossRef]
27. Van-Genuchten, M.T. A Closed-form Equation for Predicting the Hydraulic Conductivity of Unsaturated Soils. *Soil Sci. Soc. Am. J.* **1980**, *44*, 892–898. [CrossRef]
28. Green, R.E.; Corey, J.C. Calculation of hydraulic concuctivity: A further evaluation of some predictive methods. *Soil Sci. Soc. Am. J.* **1971**, *35*, 3–8. [CrossRef]
29. Fredlund, D.G.; Xing, A. Equations for the soil-water characteristic curve. *Can. Geotech. J.* **1994**, *31*, 521–532. [CrossRef]
30. Karim, M.R.; Lo, S.-C.R. Estimation of the hydraulic conductivity of soils improved with vertical drains. *Comput. Geotech.* **2015**, *63*, 299–305. [CrossRef]
31. Lo, S.R.; Mak, J.; Gnanendran, C.T.; Zhang, R.; Manivannan, G. Long-term performance of a wide embankment on soft clay improved with prefabricated vertical drains. *Can. Geotech. J.* **2008**, *45*, 1073–1091. [CrossRef]
32. Karim, M.R. Behaviour of piles subjected to passive subsoil movement due to embankment construction—A simplified 3D analysis. *Comput. Geotech.* **2013**, *53*, 1–8. [CrossRef]
33. Karim, M.R.; Manivannan, G.; Gnanendran, C.T.; Lo, S.-C.R. Predicting the long-term performance of a geogrid-reinforced embankment on soft soil using two-dimensional finite element analysis. *Can. Geotech. J.* **2011**, *48*, 741–753. [CrossRef]
34. Manivannan, G.; Karim, M.R.; Gnanendran, C.T.; Lo, S.-C.R. Calculated and observed long term performance of Leneghans embankment. *Geomech. Geoengin.* **2011**, *6*, 195–207. [CrossRef]
35. Karim, M.R.; Gnanendran, C.T.; Lo, S.-C.R.; Mak, J. Predicting the long-term performance of a wide embankment on soft soil using an elastic-visco-plastic model. *Can. Geotech. J.* **2010**, *47*, 244–257. [CrossRef]
36. Lo, S.R.; Karim, M.R.; Gnanendran, C.T. Consolidation and Creep Settlement of Embankment on Soft Clay: Prediction Versus Observation. In *Geotechnical Predictions and Practice in Dealing with Geohazards*; Chu, J., Wardani, S.P.R., Iizuka, A., Eds.; Springer: Dordrecht, The Netherlands, 2013; pp. 77–94.
37. Geo-Slope. *Heat and Mass Transfer Modelling with GeoStudio*; Seequent Limited (GEOSLOPE International, Ltd.): Calgary, AB, Canada, 2021.
38. GoogleMap. Available online: https://www.google.com/maps/place/Newbury,+UK/@51.3885288,-1.3609492,13z/data=!3m1!4b1!4m5!3m4!1s0x487402002f595ba9:0xc6646baff4a75c50!8m2!3d51.401409!4d-1.3231139 (accessed on 15 April 2019).
39. Croney, D. *The Design and Performance of Road Pavements*; Her Magistye's Stationery Office: London, UK, 1977.
40. Toll, D.G.; Rahim, M.S.; Karthikeyan, M.; Tsaparas, I. Soil atmosphere interactions for analysing slopes in tropical soils. In Proceedings of the 14th International conference of the International Association of Computer Methods and Advances in Geomechanics, Koyoto, Japan, 22–25 September 2014; pp. 1333–1338.
41. Blight, G. The vadose zone soil-water balance and transpiration rates of vegetation. *Geotechnique* **2003**, *53*, 55–64. [CrossRef]
42. Zotarelli, L.; Dukes, M.D.; Romero, C.C.; Migliaccio, K.W.; Kelly, T. Step by Step Calculation of the Penman-Monteith Evapotranspiration (FAO-56 Method). 2013. Available online: https://edis.ifas.ufl.edu/pdf/AE/AE45900.pdf (accessed on 1 June 2022).
43. McLernon, M. *Climate Driven Pore Awter Pressure Dynamics and Slope Stability within Glacial Till Drumlins in Northern Ireland*; Queen's Unviersity: Belfast, UK, 2014.
44. Hughes, D.; Karim, M.R.; Briggs, K.; Glendinning, S.; Toll, D.; Dijkstra, T.; Powrie, W.; Dixon, N. A comparison of numerical modelling techniques to predict the effect of climate on infrastructure slopes. In Proceedings of the Geotechnical Engineering for Infrastructure and Development—Proceedings of the XVI European Conference on Soil Mechanics and Geotechnical Engineering, ECSMGE 2015, Edinburgh, UK, 13–17 September 2015; pp. 3663–3668.
45. Karim, M.R.; Hughes, D.; Rahman, M.M. Estimating hydraulic conductivity assisted with numerical analysis for unsaturated soil—A case study. *Geotech. Eng.* **2021**, *52*, 12–19.

processes

MDPI

Article

Practice-Oriented Validation of Embedded Beam Formulations in Geotechnical Engineering

Andreas-Nizar Granitzer * and Franz Tschuchnigg

Institute of Soil Mechanics, Foundation Engineering and Computational Geotechnics, Graz University of Technology, 8010 Graz, Austria; franz.tschuchnigg@tugraz.at
* Correspondence: andreas-nizar.granitzer@tugraz.at

Abstract: The numerical analysis of many geotechnical problems involves a high number of structural elements, leading to extensive modelling and computational effort. Due to its exceptional ability to circumvent these obstacles, the embedded beam element (EB), though originally intended for the modelling of micropiles, has become increasingly popular in computational geotechnics. Recent research effort has paved the way to the embedded beam element with interaction surface (EB-I), an extension of the EB. The EB-I renders soil–structure interaction along an interaction surface rather than the centreline, making it theoretically applicable to any engineering application where beam-type elements interact with solid elements. At present, in-depth knowledge about relative merits, compared to the EB, is still in demand. Subsequently, numerical analysis are carried out using both embedded beam formulations to model deep foundation elements. The credibility of predicted results is assessed based on a comprehensive comparison with the well-established standard FE approach. In all cases considered, the EB-I proves clearly superior in terms of mesh sensitivity, mobilization of skin-resistance, and predicted soil displacements. Special care must be taken when using embedded beam formulations for the modelling of composite structures.

Keywords: embedded beam elements; finite element; mesh sensitivity; soil-structure interaction; deep foundation; 3D modelling

Citation: Granitzer, A.-N.; Tschuchnigg, F. Practice-Oriented Validation of Embedded Beam Formulations in Geotechnical Engineering. *Processes* **2021**, *9*, 1739. https://doi.org/10.3390/pr9101739

Academic Editor: Li Li

Received: 26 August 2021
Accepted: 26 September 2021
Published: 28 September 2021

1. Introduction

The dimensioning of deep foundation elements, such as single piles, pile groups, and piled rafts represents a standard task in geotechnical engineering. Broadly, the behaviour of these structural elements is characterized by two complex effects: (1) non-linear behaviour of the soil surrounding the piles and (2) separation between soil and pile as a result of large inertial forces, generated in the soil, around the pile heads [1]. In order to account for the extended-continuum nature of the embedding soil around piles, the finite element method (FEM), amongst other numerical techniques, has evolved as a widely accepted tool in academia and practice [2,3].

In this context, deep foundation elements are frequently considered as combinations of solid elements, conforming to linear elastic material of the pile, and zero-thickness interface elements allowing for realistic soil-structure interaction (SSI), including possible occurrence of detachment or interface sliding between pile and soil [4–7]. The so-called standard FE approach (SFEA) represents a very good approximation of related physical problems, especially regarding the simulation of ideal non-displacement piles experiencing subordinate installation effects (in the literature, they are also referred to as "wished-in-place" piles or drilled shafts [8,9]); thus, the SFEA is commonly used as a numerical benchmark solution to assess the credibility of novel pile formulations [10,11]. Nonetheless, the SFEA has deficiencies concerning the analysis of large-scale geotechnical entities comprising a high number of piles [12]. Most significantly, the SFEA requires the application of considerably small elements, implicitly induced by the slenderness inherent to piles; as a consequence, the number of degrees of freedom (DOFs) increases disproportionately, resulting in computational costs that must be regarded as unbearable for most practical purposes [13]. In addition, the SFEA requires an a priori consideration of the orientation and location of the piles during the geometrical setup-up of the model, which makes design optimization cumbersome in engineering practice [14].

To overcome these limitations, Sadek and Shahrour [15] proposed the so-called embedded beam (EB) formulation, which was further enhanced through the contributions of Engin et al. [16], as well as Tschuchnigg and Schweiger [12]; in more recent publications, the EB is also referred to as embedded pile [13,17]. In its actual configuration, EBs constitute line elements composed of beams and non-linear embedded interface elements, allowing for relative beam-soil displacements. Unlike the SFEA, the EB formulation defines piles independent from the solid mesh connectivity; consequently, EBs do not influence the spatial discretization of the soil domain, leading to a significant decrease in DOFs close to the thin geometrical entities, higher calculation efficiency, and lower modelling effort compared to the SFEA. In addition, the EB simplifies the structural analysis with respect to the evaluation of stress resultants [13].

Since the EB modelling technique considers SSI along the centreline only, it was originally intended for the analysis of axially-loaded small diameter piles installed in vertical direction [18]. Due to its deficiency to capture the effect of pile installation in the surrounding soil, it is further recommended to apply the EB framework for the modelling of piles undergoing inconsiderable installation effects, such as non-displacement piles or, more specifically, piles which are constructed using casings [12]. Applications of EBs are reported for single piles under compression [19] and tension [20], pile rows [21], pile groups [22,23], rigid inclusions [24], connected [25–29] and disconnected piled rafts [13]. Despite aforementioned recommendations regarding its field of application, and due to its exceptional ability to simplify the modelling process, EBs were found eligible for the modelling of almost any type of geotechnical reinforcement interacting with solid elements: Abbas et al. investigated both the lateral load behaviour [30] and the uplift load-carrying mechanism [31] of inclined micropiles modelled by means of EBs. El-Sherbiny et al. [32] applied EBs to study the ability of jet-grout columns to stabilize berms in soft soil formations. EBs found also use in the modelling of piles subjected to different modes of loading, namely lateral [33], passive [34,35], and dynamic loading [36]. More recently, use cases suggest the incorporation of EBs to capture the response of tension members, including tunnel front fiberglass reinforcements [37], rock bolts [38], ground anchors [39], and soil nails [40].

However, as relevant as it is poorly considered, the existing EB formulation suffers from severe limitations compared to the SFEA. Most significantly, this modelling approach introduces a mechanical incompatibility, resulting in stress singularities near the pile base [41]. This effect is mesh-dependent and becomes more determinant with increasing mesh refinement in the vicinity of an EB; accordingly, EBs require the employment of sufficiently coarse mesh discretizations in order to accurately predict their system response. Otherwise, increasing the mesh coarseness of the solid domain gives rise to inaccurate solutions and inconvenient convergence to the sought exact solution [13]; from a practical point of view, this may lead to an underestimation of foundation settlements. Hence, engineering judgment, in conjunction with mesh studies, are in demand to evaluate reliable mesh discretizations. Another important limiting factor concerns the SSI principle: as the latter is established along the centreline rather than the physical circumference, EBs cannot distinguish between lateral shear and normal stress components acting on the interface, which are therefore unlimited [18]. In this way, lateral plasticity is solely controlled by the evolution of failure points in the surrounding soil; this obstacle may give rise to inaccurate predictions in the case of complex loading situations [42].

To solve aforementioned problems, Turello et al. [41,43] developed the so-called embedded beam with interaction surface (EB-I). The novelty of this approach concerned the integration of an explicit interaction surface to the existing EB formulation, which allows the spread of non-linear SSI effects over the physical skin, instead of over the centreline. In this way, spurious stress singularities induced by the line element are markedly reduced. More recently, the improved embedded beam framework was further generalized in order to allow for inclined pile configurations as well as non-circular cross-section shapes [10,42]. Aforementioned publications found that the EB-I poses a significant enhancement, in terms of mesh sensitivity and global pile response, in comparison with the original EB. However, it should be noted that these papers were primarily devoted to the theoretical presentation of the EB-I concept, hence little attention was paid to its validation. For example, Turello et al. [44] investigated its ability to resemble pile group members considering linear elastic soil conditions, which must be regarded as considerable simplification with respect to the mechanical soil behaviour. Moreover, Smulders et al. [42] evaluated the EB-I performance solely based on the load-settlement behaviour of single piles, which is mainly controlled by the stiffness of the surrounding soil [11].

The present paper aims at assessing the relative merits of the improved embedded beam formulation, compared to the existing one, in more detail. To this end, the remainder of this paper is organized as follows: Section 2 reconsiders the theoretical framework of both embedded beam formulations and highlights important differences. Section 3 presents the numerical modelling approach applied in the numerical studies. Section 4 investigates the credibility of results computed with both embedded beam formulations based on a single-pile engineering case. In view of their intended use (i.e., to capture the behaviour of SFEA piles; see [11]), the performance of both embedded beam formulations is analysed based on a comparison with the results obtained using the SFEA, which serves as a benchmark model for numerical validation [45]. In this context, it is noted that the (numerical) validation approach applied differs from the traditional validation strategy [46], which generally uses experimental data as a benchmark. To the authors' knowledge, the underlying studies represent the first attempt to compare the structural performance of embedded beam formulations with respect to the mobilization of skin tractions, load separation between pile shaft and pile base, as well as displacements induced in the surrounding soil. In Section 5, both embedded beam models are utilized to approximate deep foundation elements of a recently constructed railway station, but emphasis is given to practical implications. Section 6 closes with the main conclusions of this work. In the course of this document, boldface uppercase and lowercase letters denote vectors or matrices, respectively. Italic element symbols are defined in the list of symbols.

2. Background

2.1. Embedded Beam Formulation

EBs combine the efficient use of defining beam elements independent of the solid mesh connectivity, with slip between the structural element and the solid mesh defined along the element axis. Over the last decade, many commercial software codes have incorporated an EB-type element; see Table 1. Though the latter differs in terms of element type name and recommended application, the working principle is broadly the same.

Table 1. Overview of EB-type elements employed in numerical software codes.

Numerical Code (Version) Simulation Method	Diana 3D (V10.4) [47] FEM	ZSoil 3D (V20.07) [48] FEM	RS3 (V4.014) [49] FEM	FLAC 3D (V7.0) [50] FDM	PLAXIS 3D (V21.00) [51] FEM
EB-type name	Pile foundations	Composite element	Embedded element	Pile structural element	Embedded beam element
Soil-structure interaction	Line interface along shaft, point interface at base				
Skin resistance in axial direction	Soil-dependent or pre-defined traction (kN/m)				
Base resistance in axial direction	Pre-defined force (kN)				
Application	Piles, ground anchors	Piles, soil nails, fixed end anchors	Piles, forepoles, beams	Structural support members	Piles, rockbolts, anchors

From a numerical point of view, EBs implemented in Plaxis 3D represent composite structural elements comprising three components:

1. Three-noded beam element with quadratic interpolation scheme and six DOFs (i.e., three translational DOFs, u_x, u_y, u_z; three rotational DOFs, φ_x, φ_y, φ_z). As Timoshenko's theory is adopted, shear deformations are explicitly taken into account. In order to improve the numerical stability in case of particularly fine mesh discretizations, an elastic zone is automatically generated in the surrounding soil. In this zone, all Gaussian points of the solid mesh are forced to remain elastic. As the zone size is controlled by the pile radius, the geometrical properties of the beam element influence the stress state in the surrounding soil [16].
2. Three-noded line interfaces with the quadratic interpolation scheme and three pairs of nodes instead of single nodes, one belonging to the beam and one to the solid element in which the beam is located. This component accounts for the SSI along the shaft based on a material law that links the skin traction vector $\boldsymbol{t}_{EB}^{skin}$ (kN/m) to the relative displacement vector $\boldsymbol{u}_{rel}$ (m):

$$\boldsymbol{u}_{rel} = \boldsymbol{u}^b - \boldsymbol{u}^s = \boldsymbol{N}^b \cdot \boldsymbol{a}^b - \boldsymbol{N}^s \cdot \boldsymbol{a}^s \tag{1}$$

$$\boldsymbol{t}_{EB}^{skin} = \boldsymbol{D}^e \cdot \boldsymbol{u}_{rel} \tag{2}$$

where $\boldsymbol{u}^b$ and $\boldsymbol{u}^s$ (m) denote the beam and solid displacement vector, while matrices $\boldsymbol{N}^b$ and $\boldsymbol{N}^s$ represent the interpolation matrix, including standard Lagrangian element functions of the corresponding beam and solid elements. The elastic constitutive matrix $\boldsymbol{D}^e$ (kN/m^2) is composed of the embedded interface stiffnesses in normal and tangential directions. Unlike the beam component, interface stiffness matrices are evaluated employing the Newton–Cotes integration scheme, hence element function values at the nodes are either one or zero. While skin traction components in radial direction are defined as unlimited, peak values in axial direction $t_{axial,max}$ (kN/m) may be linked to the actual soil stress state using a Coulomb criterion:

$$t_{axial,\ max} \leq \left(c' - \sigma_n^{avg\prime} \cdot \tan\varphi'\right) \cdot 2 \cdot \pi \cdot R \tag{3}$$

where c' (kPa) and φ' (°) represent the effective soil strength parameters, $\sigma_n^{avg\prime}$ (kPa) the averaged effective normal stress along the line interface and R (m) the radius.
3. Point interface with one integration point at the beam end. In its initial configuration, the latter coincides with a solid node. Similar to the line interface, SSI effects obey a material law defined in terms of $\boldsymbol{u}_{rel}$ (m) and the embedded foot interface stiffness K_{base} (kN/m) in an axial direction. On the contrary, the ultimate tip force is limited through a user-defined tip force F_{max} (kN), hence independent of the surrounding soil. Tensile stresses are internally suppressed through a tension cut-off criterion.

For details concerning virtual work considerations and the iteration procedure, the interested reader can refer to [47–51].

2.2. Improved Embedded Beam Formulation with Interaction Surface

To the best of found knowledge, the EB-I is yet to feature in commercial software codes. However, recent developments have been tested and incorporated in Plaxis 3D [18,42]. In order to get a better understanding of the enhancements compared to the EB, a few remarks on the implementation are noteworthy:

- Unlike EBs, which consider shaft SSI effects along the beam axis, the EB-I spreads the shaft resistance over a number of integration points located at the physical interaction surface; the same applies for SSI at the base, where the resistance is distributed over several points at the base, instead of a single point. In this way, singular stress concentrations, leading to local accumulations of plastic behaviour and large displacements along the axis, are avoided. In analogy to the EB, the skin traction vector t^{skin}_{EB-I} (kN/m^2) is expressed in terms of the relative displacement vector at the interaction surface $\boldsymbol{u}_{rel}$ (m) and the elastic stiffness matrix $\boldsymbol{D}^e$ (kN/m^3):

$$\boldsymbol{u}_{rel} = \boldsymbol{u}^b - \boldsymbol{u}^s = \boldsymbol{H}\cdot\boldsymbol{a}^b - \boldsymbol{N}^s\cdot\boldsymbol{a}^s \tag{4}$$

$$\boldsymbol{t}^{skin}_{EB-I} = \frac{\boldsymbol{D}^e}{2\pi R}\cdot\boldsymbol{u}_{rel} \tag{5}$$

  where $\boldsymbol{u}^b$ (m) denotes the mapped beam displacement at the interaction surface expressed in terms of the mapping matrix $\boldsymbol{H}$ and beam nodal DOFs $\boldsymbol{a}^b$. $\boldsymbol{u}^s$ (m) is the solid displacement vector at the interaction surface obtained by means of interpolation, within solid elements, located at the interaction surface. The latter is calculated based on vector $\boldsymbol{a}^s$ containing solid nodal DOFs and the interpolation matrix $\boldsymbol{N}^s$ including standard Lagrangian element functions of the corresponding solid elements. R (m) represents the beam radius and $\boldsymbol{D}^e$ (kN/m^2) represents the elastic stiffness matrix containing the interface stiffnesses. Since each interface element no longer poses a line, but a surface, the latter is divided by the shaft circumference. A similar approach is applied to resemble SSI at the base.
- In contrast to EBs, EB-Is are able to distinguish between two shear stress directions perpendicular to the beam axis. This allows them to enrich the slip criterion such that it accounts for any possible direction of slip failure occurring at the interaction surface:

$$\tau_{max} = \sqrt{\tau_s^2 + \tau_t^2} \leq c' - \sigma_n' \cdot \tan\varphi' \tag{6}$$

  where τ_{max} (kN/m^2) is the max. local shear stress, controlled by the (perpendicularly oriented) local shear stress components τ_s and τ_t (kN/m^2) acting on the interaction surface. σ_n' (kN/m^2) denotes the actual normal stress developing in the surrounding soil, c' (kN/m^2) and φ' (°) the effective soil strength parameters. Accordingly, the intrinsic slip criterion is not restricted to the axial direction, as is the case for the EB.
- Depending on the discretization and the number of integration points at the interaction surface, the EB-I connects multiple continuum elements to one beam element. From a numerical point of view, the embedded beam stiffness is spread over a higher number of nodes, leading to relative merits in terms of global stiffness matrix conditioning (i.e., smaller difference between max. and min. diagonal terms). As a consequence, the robustness of the numerical procedure is improved compared to the EB.

For the sake of clarity, Figure 1 compares the global response of a vertically loaded single pile, modelled by means of both embedded beam formulations, namely the existing embedded beam (EB) and the improved embedded beam with interaction surface (EB-I).

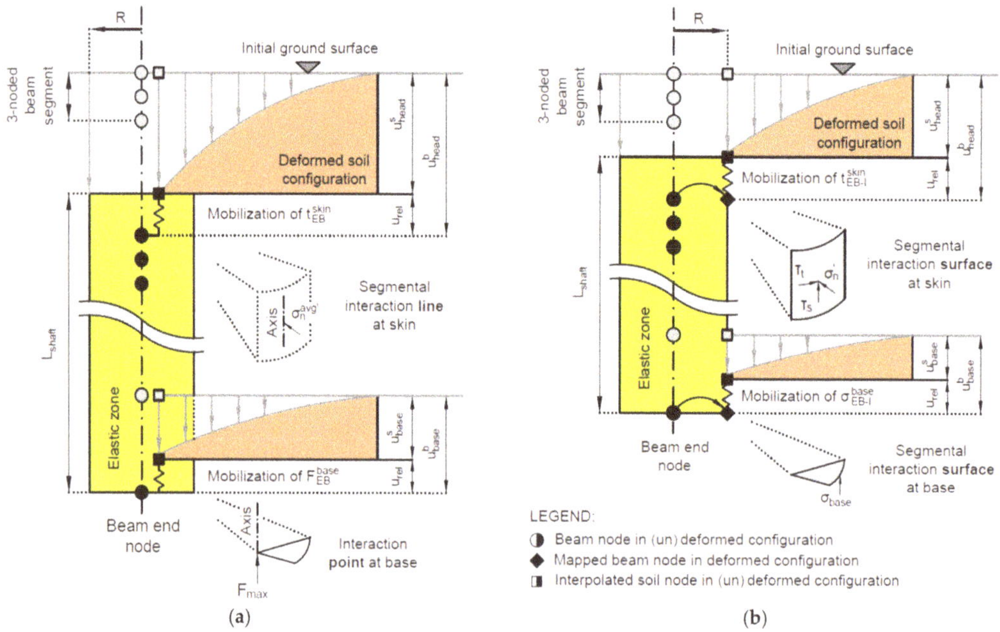

Figure 1. Global response of an axially loaded pile, considered by means of different modelling approaches: (**a**) embedded beam (EB); (**b**) embedded beam with interaction surface (EB-I). In both cases, the mobilization of skin tractions is primarily controlled by the relative displacement vector u_{rel} ("spring") and the slip criterion ("friction element").

3. Finite Element Modelling

3.1. Investigated Scenario and Model Description

Although single piles are rarely constructed in isolation, it is useful to consider their analysis for the numerical validation of new pile modelling methods [4]. Accordingly, the Alzey Bridge pile load test [52] serves as a reference scenario to compare the performance of the EB with the EB-I. The test program was originally intended to optimize the foundation design of a highway bridge in slightly overconsolidated Frankfurt clay. Recently, it has gained increasing popularity for the performance assessment of novel pile formulations; for example, see [16,42].

The numerical simulations, carried out in the present study, consider different pile modelling approaches and mesh discretizations. All simulations are carried out with Plaxis 3D [51]. For the sake of numerical consistency, in terms of solid element type and domain dimensions, all results are obtained using full 3D models, instead of axisymmetric 2D models or reduced 3D models for the SFEA. In view of the number of elements and DOFs required to discretize the domain, this allows for a direct comparison between SFEA and EB/EB-I models. Following the results of trial simulations, the model domain has been defined such that boundary effects are reduced to an acceptable limit; see Appendix A. Consequently, the model dimensions $B_m = L_m/D_m$ are set to 26/19 m, which is equal to 20 and 2 times the pile diameter and length, respectively; see Figure 2a. With regard to the boundary conditions, conventional kinematic conditions are considered: a fully fixed support, with blocking of the displacements in all directions, at the lower horizontal model boundary and a horizontally fixed support with freedom of vertical displacements along lateral model boundaries. Moreover, mesh sensitivity analyses, concerning the (numerical) benchmark solution, ensure that the reference model is stable and reliable; see Appendix B.

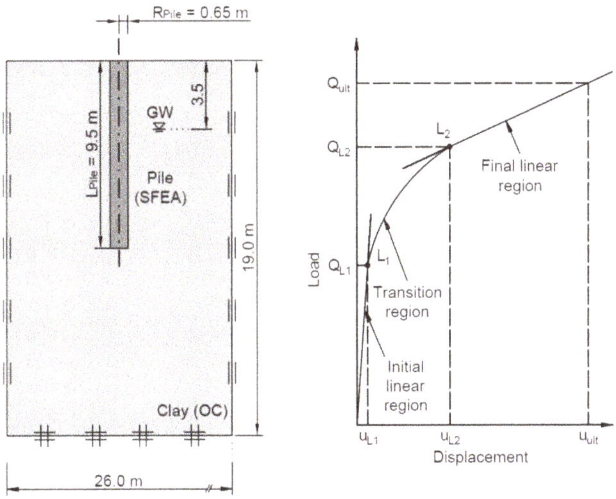

(a) (b)

Figure 2. (**a**) Model geometry, boundary conditions, and (**b**) characteristic load-displacement regions after [53].

Regardless of the pile modelling approach, the majority of the performed finite element analyses (FEA) consist of five phases. All simulations start with the definition of the initial conditions, where the initial stress field is generated by means of the K0 procedure; this is reasonable given a horizontal ground surface and homogeneous soil. The next phase concerns the pile installation; depending on the considered pile configuration, this includes the assignment of hardened concrete properties and the activation of interface elements between the pile and the surrounding soil (SFEA), or the activation of the respective embedded beam element (EB, EB-I). Hence, the piles are assumed to be homogenous in strength, stiffness, and weight, thereby ignoring effects of improper concreting [6]. According to the reference scenario, the soil around the pile is assumed to be in its initial state, which may be regarded as reasonable for non-displacement piles, where installation effects are of subordinate importance with respect to the pile behaviour [8]. Unless otherwise stated, the pile is finally subjected to displacement-driven head-down loading.

In the course of the evaluation of results, the mesh size effect on derived quantities (for example, the mobilization of skin and base resistance) is compared at different displacement levels. To satisfy practical relations on the one hand, and reduce bias on the other hand, the prescribed displacements are specified beforehand, following the L1-L2 method [53]; see Figure 2b.

Assuming drained conditions, the following prescribed displacements are determined as a function of the pile diameter $\mathrm{D_{Pile}}$:

- $u_{L1} = \mathrm{u_{Pile}}/\mathrm{D_{Pile}} = 0.34\%$, approximating the end of the initial linear region [54].
- $u_{L2} = \mathrm{u_{Pile}}/\mathrm{D_{Pile}} = 4\%$, indicating the initiation of the final linear region [55].
- $u_{ult} = \mathrm{u_{Pile}}/\mathrm{D_{Pile}} = 10\%$, complying with the ultimate pile resistance defined in [56].

As shown in Figure 2b, u_{L1} and u_{L2} correspond to the pile head displacement $\mathrm{u_{Pile}}$ at the end of the initial linear region, as well as at the initiation of the final linear region. u_{ult} marks the pile head displacement at the end of loading.

3.2. Pile Modelling Approach

As suggested by [10,38,41], the numerical behaviour of embedded beam formulations is numerically validated against SFEA results (i.e., full 3D FE model). In the latter case, the swept meshing technique [57] is applied to discretize the domain with 84,148 10-noded tetrahedral elements. In this way, the solid elements representing the SFEA pile are discretized with a structured swept mesh, allowing for both a faster mesh generation process, as well as a considerable mesh size reduction compared to the default solid mesh.

In order to minimize the interpolation error associated with the mesh discretization of the model, the mesh is refined inside the pile structure, and within a zone of $3 \cdot D_{Pile}$ below the tip and beside the shaft, respectively; see Figure 3. To address possible occurrence of relative displacements and plastic slip parallel to the soil-structure contacts, standard zero-thickness interface elements are considered at soil-structure contacts [58,59]. The interface elements are extended beyond the physical pile boundary in order to reduce the effect of singular plasticity points developing close to the pile edge [11,60]. The interface strength is specified with a Coulomb criterion that limits the max. shear stress τ_{max} (kN/m^2) by

$$\tau_{max} = \left(c' - \sigma'_n \cdot \tan\varphi'\right) \cdot R_{inter} \tag{7}$$

where σ'_n (kN/m^2), φ' (°), and c' (kN/m^2) are the effective interface normal stress, interface friction angle, and interface cohesion. In accordance with recommendations given in [61], an interface reduction factor R_{inter} = 0.9 is adopted in the analyses, which represents a typical value to account for SSI between concrete piles and fine grained soils.

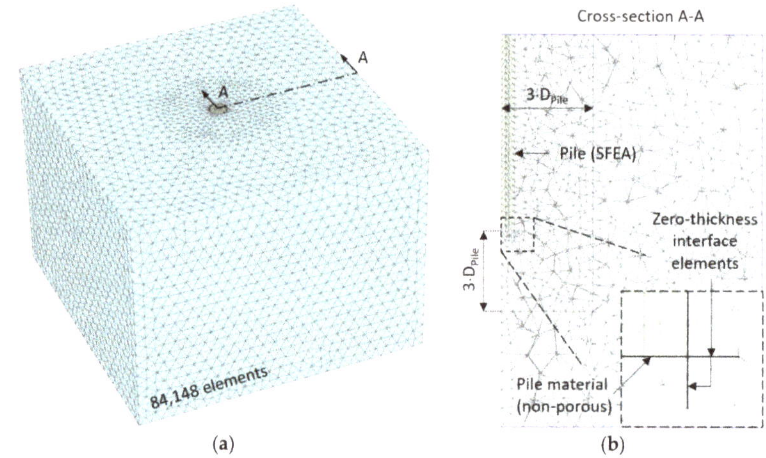

Figure 3. (**a**) 3D view of global SFEA mesh; (**b**) cross-sectional view of SFEA mesh detail.

Contrary to SFEA interfaces, embedded interface elements are not explicitly modelled, but they are internally defined after the global mesh discretization; thus, assigning interface parameters reduces the definition of the ultimate skin and base resistance. For the sake of numerical consistency, the skin resistance is defined as layer-dependent (i.e., dependent of the stress state in the surrounding soil; see Equation (3)). In this way, the embedded interface elements are defined similarly to the interface elements used in the SFEA [12]. In an attempt to capture the SFEA pile capacity with sufficient accuracy, the base resistance is limited by the SFEA normal force developing at the pile base after the final loading step. The mesh sensitivity of both embedded beam formulations is investigated with five different mesh refinements, which are referred to as very fine, fine, medium, coarse, and very coarse (Figure 4). The model boundary conditions are specified equal to the SFEA case.

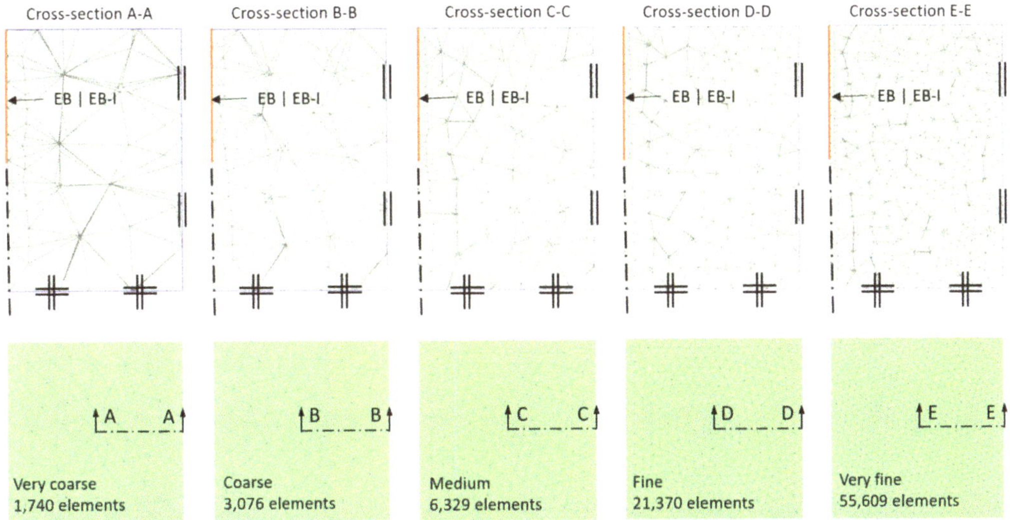

Figure 4. Mesh discretizations considered for EB/EB-I models.

It is noted that both embedded beam formulations are analysed with the same mesh discretizations and considerably less DOFs compared to the SFEA. Since beam elements are superimposed on the solid domain, and therefore overlap the soil, the beam unit weight represents a delta unit weight to the surrounding soil [11]. Noteworthy, this differs from previous interpretations of the embedded beam unit weight documented in [27,30,31,62], where the unit weight is assigned with the actual unit weight of the volume pile; as a consequence, this approach leads to an overestimation of the soil-structure unit weight. To realistically approximate the kinematics at the pile head, the uppermost connection point is considered as free to move and rotate, relative to the surrounding soil. Additional modelling parameters summarized in Table 2 are taken from [16] and resemble on-site conditions as described in [52].

Table 2. Overview of pile parameters applied in the Alzey Bridge model.

Parameter	Symbol	Unit	SFEA		EB/EB-I [1]	
			Pile	Interface	Pile	Interface
Unit weight	γ	kN/m^3	25.0		5.0	
Young's modulus	E	GPa	10.0		10.0	
Poisson's ratio	ν	-	0.2			
Pile diameter	D	m	1.3		1.3	
Pile length	L	m	9.5		9.5	
Base resistance	F_{max}	kN			2300.0	
Interface reduction factor	R_{inter}	-		0.9		0.9
Effective friction angle	φ'	deg		20.0		20.0
Effective cohesion	$c\prime$	kN/m^2		20.0		20.0

[1] Calibrated numerical model documented in [18].

3.3. Constitutive Model and Parameter Determination

Regardless of the pile modelling approach, the use of a proper constitutive model is crucial to cover the complex stress-strain behaviour of soils. Specific to this study, this includes a realistic evolution of potential slip planes along the pile-soil contacts, which is fundamental in the analysis of the pile behaviour. Since piles may undergo a wide range of deformations, leading to considerable strains in the soil, the constitutive model should also be capable of effectively simulating the shear modulus degradation with the evolution of shear strain [63]. Lastly, the constitutive model is supposed to account for the stress dependency of soil stiffness in order to capture the gradual mobilization of skin friction with sufficient accuracy [11].

Concerning these critical modelling aspects, the Hardening Soil Small (HSS) with non-associated flow rule and small-strain stiffness overlay [64], an extension of the Hardening Soil (HS) model [65], is used in this study. While HS parameters and drainage conditions are adopted from [16], HSS-specific model parameters, namely the threshold shear strain for stiffness degradation, $\gamma_{0.7}$, and the initial shear modulus, G_0, are defined using empirical correlations documented in [66] and [67], respectively. The groundwater table is located 3.5 m below ground surface. Soil parameters adopted in the analyses are listed in Table 3.

Table 3. Overview of soil parameters applied in Alzey Bridge model.

Parameter	Symbol	Unit	Value
Drainage conditions	-	-	drained
Depth of groundwater table	-	m	3.5
Unit weight	$\gamma_{sat}, \gamma_{unsat}$	kN/m^3	20.0
Reference deviatoric hardening modulus at p^{ref}	E_{50}^{ref}	kN/m^2	45,000
Reference oedometer stiffness at p^{ref}	E_{oed}^{ref}	kN/m^2	27,150
Reference un-/reloading stiffness at p^{ref}	E_{ur}^{ref}	kN/m^2	90,000
Power index	m	-	1.0
Isotropic Poisson's ratio	ν'_{ur}	-	0.2
Effective friction angle	φ'	deg	20.0
Effective cohesion	c'	kN/m^2	20.0
Pre-overburden pressure	POP	kN/m^2	50.0
Reference pressure	p_{ref}	kN/m^2	100.0
Initial shear modulus at p^{ref}	G_0^{ref}	kN/m^2	116,000
Threshold shear strain	$\gamma_{0.7}$	-	0.00015

4. Numerical Validation

A series of finite element analyses has been performed to assess the performance of the improved embedded beam with interaction surface. Therefore, the latter is compared to the existing embedded beam formulation, as well as the SFEA, which serves as the benchmark. The simulations focus on two specific points: mesh sensitivity studies and displacements induced in the surrounding soil. In addition, practical implications are drawn based on the results.

4.1. *Influence of Mesh Size on Global Pile Behaviour*

The most critical issue in the design of pile supported structures is a reliable prediction of the pile behaviour, in terms of bearing capacity and settlement, under working load conditions [68]. Consequently, the first series of calculations has been performed to investigate the mesh size effect on the mobilization of compressive pile resistance R_c (kN).

As it can be observed in Figure 5, all pile models show a similar behaviour in the first stage of loading, regardless of spatial discretization level and pile modelling approach. However, as the load increases beyond 2000 kN, the results reveal considerable limitations of the EB, including a significant scatter of results and overestimation of bearing capacity, compared to the reference solution. Although it is generally known that coarser meshes yield a stiffer pile response, the wide scatter of load-displacement curves must be regarded as considerable obstacles to produce consistent results. In contrast, a more realistic pile response is obtained when employing the EB-I, thereby transforming the pile-soil line interface to an explicit interaction surface. Up to a relative displacement of 2%, the mobilized pile resistance magnitudes are almost independent of the domain discretization. Even at higher displacement levels, the mesh size effect remains insignificant compared to the EB.

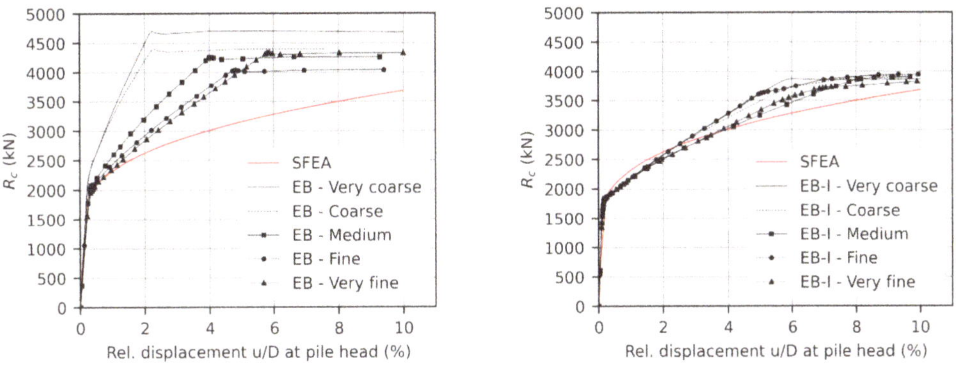

Figure 5. Influence of spatial discretization on mobilization of compressive pile resistance.

In view of the load-displacement behaviour, the EB-I gives a softer global response compared to the EB after reaching the ultimate skin resistance, which is mainly attributed to the different mobilization of base resistance R_b (kN); see Figure 6. The base–resistance curves, deduced from both embedded beam formulations, may be well approximated through bi-linear functions bounded by the max. base resistance F_{max}, which is a direct input to the analysis. However, the EB shows considerably steeper gradients in comparison with the EB-I, which is a direct consequence of the modified embedded base interface stiffness considered in the actual EB-I configuration. As a by-product, EB-I curves are in remarkable agreement with the reference solution. Moreover, the results confirm that the EB-I is less mesh-size dependent; for example, at a relative displacement of 2%, the range of R_c-values obtained with the EB-I (<100 kN) is considerably smaller when compared with the EB (>1000 kN). Nevertheless, both formulations have deficiencies regarding the initial mobilization of base resistance in the first part of loading (see Figure 6); in all cases considered, the base resistance is considerably underestimated and shows a lower mobilization rate compared to the SFEA. This must be regarded as a considerable limitation of the actual embedded beam configurations.

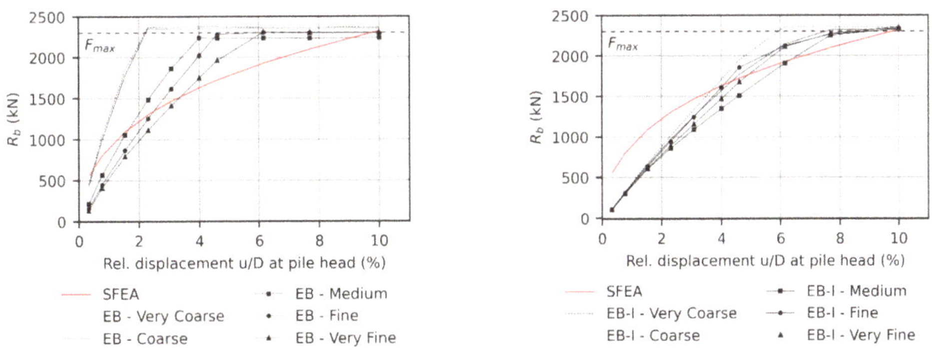

Figure 6. Influence of spatial discretization on the mobilization of base resistance.

Figure 7 provides more insight into the effect of different mesh discretizations on the mobilization of R_c at different pile displacement levels, namely u_{L1}, u_{L2}, and u_{ult}; see Figure 2b. The obtained R_c-values are normalized with respect to the benchmark solution:

$$R_{c,norm} = \frac{R_{c,EB|EB-I}}{R_{c,SFEA}} \cdot 100\ \% \tag{8}$$

where $R_{c,norm}$ (%) is the normalized compressive pile resistance. $R_{c,EB|EB-I}$ (kN) and $R_{c,SFEA}$ (kN) represent the mobilized compressive pile resistance, either obtained with the EB, EB-I, or SFEA at a given pile head displacement. To quantify the refinement level of the considered mesh, independent of the model dimensions, the results are plotted versus the average element size L_{avg} (m), which is internally controlled by the mesh coarseness factor and the model dimensions [57].

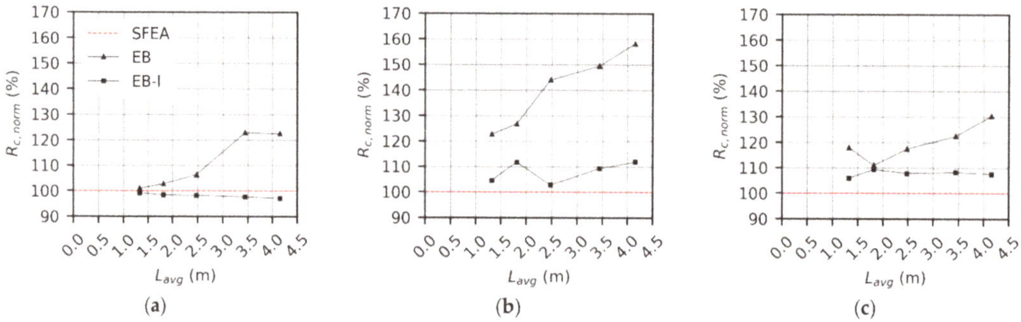

Figure 7. Influence of spatial discretization on the mobilization of pile resistance at different displacement levels, namely (**a**) u_{L1}, (**b**) u_{L2}, (**c**) u_{rel} .

In compliance with the previous results, the EB-I achieves a satisfactory agreement with the reference solution, with the max. deviation being as little as 12%. On the contrary, R_c-magnitudes obtained with the EB are highly mesh-dependent. In all cases considered, the EB yields the highest R_c-values. In particular, for coarse mesh discretizations, the benchmark value is considerably overestimated with a peak deviation of 58%.

In order to describe the mesh size effect by means of a quantitative measure, the mesh dependency ratio MDR is introduced:

$$MDR = \frac{R_{c,max}}{R_{c,min}} \geq 1.0 \tag{9}$$

where $R_{c,max}$ and $R_{c,min}$ (kN) represent the max. and min. compressive pile resistance values obtained at a given pile head displacement, either using the EB or the EB-I. Obviously, a MDR of 1.0 indicates mesh-independency, even though the results may differ from the reference solution. As expected, MDR-values listed in Table 4 are remarkably smaller for the EB-I, thus underlining that the EB-I is clearly superior in the elimination of the mesh-size effect for all cases considered. For example, at the pile head displacement of u_{L1}, the MDR reduces to 1.02 for the EB-I; at the other extreme, the MDR gives 1.22 for the EB.

Table 4. Summary of pile resistance values obtained at different displacement levels.

Mesh Configuration	L_{avg}		**EB**			**EB-I**		**Unit**
	m	$R_{c,\,u_{L1}}$	$R_{c,\,u_{L2}}$	$R_{c,\,u_{ult}}$	$R_{c,\,u_{L1}}$	$R_{c,\,u_{L2}}$	$R_{c,\,u_{ult}}$	
Very coarse	4.15	2355	4642	4667	1867	3284	3855	kN
Coarse	3.44	2363	4386	4385	1877	3208	3876	kN
Medium	2.48	2045	4224	4208	1887	3017	3863	kN
Fine	1.81	1974	3723	3975	1891	3283	3916	kN
Very Fine	1.32	1940	3603	4228	1906	3071	3796	kN
	MDR	1.22	1.29	1.17	1.02	1.09	1.03	-

4.2. *Influence of Mesh Size on Pile Load Transfer Mechanism*

A realistic representation of the load sharing between pile shaft and pile base is critical in the analysis of geotechnical structures, including deep foundation elements. For example, Zhou et al. [69] have reported that the development of differential settlements, developing between the pile and surrounding soil, is more pronounced for end-bearing piles, in comparison with floating piles. In addition, the load-carrying mechanism influences the development of skin tractions along the pile length, hence, this is of paramount importance in the analysis of composite foundations such as piled rafts [70]. As discussed in Section 1, these structures are often modelled using EB, although the load transfer mechanism has been rarely validated in literature.

To close this knowledge gap, a closer inspection of the load separation is given in Figure 8, where relative pile head displacements are plotted versus skin resistance ratios (i.e., percentage of total load resisted by skin friction). This approach is often utilized to characterize the load-bearing behaviour of piles, whereas admissible values range from 0% (end-bearing pile) to 100% (floating pile). In all cases considered, the results infer that the skin resistance ratio decreases with increasing pile displacement, albeit underestimated, at around 20%, in the initial stage of loading. While all curves obtained with the EB-I converge to the reference solution at approximately 40%, the EB indicates mesh sensitive values that vary within higher bounds. This observation complies with previous findings concerning R_c [42]; as the ultimate base resistance is limited by F_{max}, the scatter of magnitudes must be caused by an overestimation of the skin resistance. This tendency, in turn, is presumably triggered by singular skin traction values developing near the pile base, which are likely to occur due to a combination of very high stress gradients in the soil elements around the pile base and constitutive models concerning stress-dependent soil stiffness; a comprehensive discussion on this limitation inherent to EBs is provided in [11].

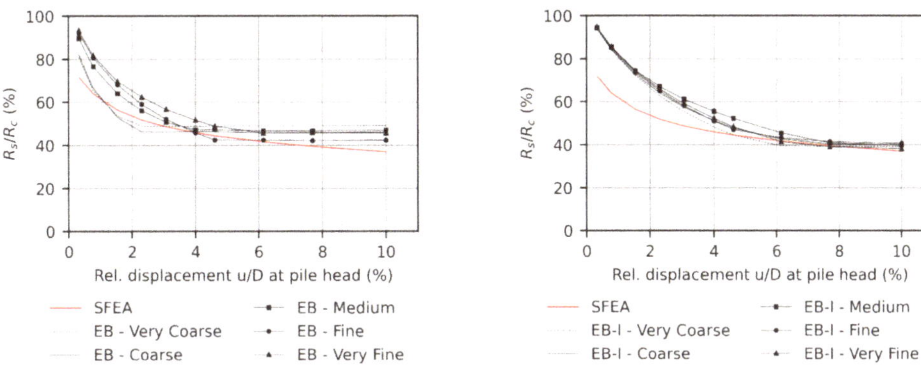

Figure 8. Influence of spatial discretization on load separation plotted by means of skin resistance ratio.

4.3. *Influence of Mesh Size on Pile Stiffness Coefficient*

In traditional serviceability analysis of pile-plate systems, SSI effects are numerically considered as spring constants acting on bending plates [56]. In this context, the pile stiffness coefficient k_s (MN/m), representing the pile stiffness, is expressed by the formula:

$$k_s = \frac{R_{c,i}}{u_i} \tag{10}$$

where $R_{c,i}$ (kN) is the compressive pile resistance, mobilized at the pile head displacement u_i (mm). The current engineering approach assumes k_s from the secant on the resistance-settlement curve of pile load tests working in the initial linear region, commonly adopting empirical data with respect to allowable settlements [69]. Following engineering practice, k_s is calculated as the slope of the secant line connecting the origin to the curve at u_{L2}; see

Figure 3b. Figure 9a presents k_s-values as a function of L_{avg}; in analogy to Equation (8), derived quantities are normalized with respect to the benchmark solution. As one can observe, all EB-I results are in remarkable agreement with the benchmark solution; on the contrary, the EB overestimates k_s up to 26%. With regard to the mesh dependency ratio, the EB-I yields values that are too small to warrant differentiation, while the EB, again, indicates a notable mesh sensitivity; see Figure 9b. From a practical point of view, the EB-I increases the confidence with which k_s may be estimated by reducing the influence of the mesh size.

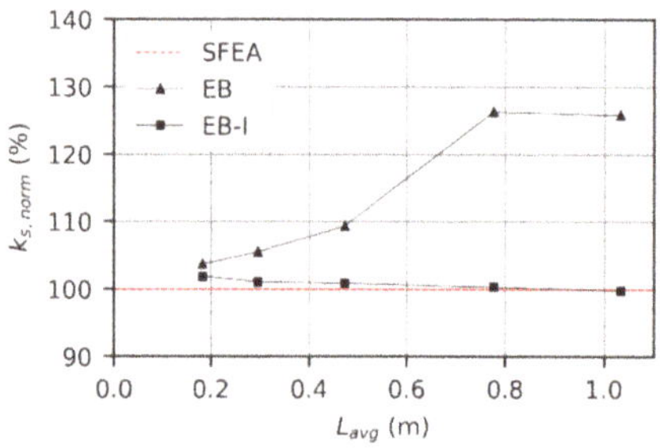

(a)

Mesh configuration	k_s		Unit
	EB	EB-I	
Very coarse	560	444	MN/m
Coarse	562	446	MN/m
Medium	486	449	MN/m
Fine	470	450	MN/m
Very Fine	461	453	MN/m
MDR	1.22	1.02	-

(b)

Figure 9. Mesh size effect on (**a**) normalized pile stiffness factor and (**b**) respective mesh dependency ratio.

4.4. *Influence of Mesh Size on Mobilization of Skin Traction*

Many researchers have focused on the evaluation of normal forces to study the load-transfer mechanism of EB along the pile length (for example, see [13,24,37]). Skin traction profiles, in contrast, have received only subordinate attention. The underlying cause becomes apparent from Figure 10, which compares both approaches at typical working load conditions. Regardless of the pile modelling approach, the normal force distribution indicates a smooth distribution of skin tractions along the pile length. In comparison with the reference solution, increasing deviations towards the pile base are attributed to the spurious mobilization of base resistance, which is underestimated with the EB and EB-I in the initial stage of loading; see Section 4.1.

In contrast, the EB yields numerical oscillations in the predicted skin traction response for all mesh discretizations considered, even though the mean values are almost identical to the benchmark results. To explain the origin for this abnormality, the following points deserve attention:

- EB results of the beam and the line interface are presented at the nodes [71]. Therefore, the normal forces are internally extrapolated from Gaussian beam stress points to the node of interest, leading to smooth profiles. In contrast, Equation (2) is used to work out nodal skin tractions, which are consequently a function of the relative displacement vector field and the embedded stiffness matrix. An additional parametric study (not shown) has revealed that skin traction oscillations also occur with linear elastic soil behaviour, ensuring identical (stress-independent) embedded stiffness values along the pile length. Consequently, the apparent reasoning of the oscillations must be attributed to the relative displacement vector field, which is interpolated using displacement vector fields of different continuity at the element boundaries (C0 for soil displacements; C1 for beam displacements).
- The spurious tendency to produce oscillations is amplified by high stress and strain gradients, predominantly occurring at the EB axis. Since the EB-I evaluates the relative displacements at multiple points over the real pile perimeter, instead of one point located at the pile axis, local effects are significantly reduced. As a consequence, skin traction profiles are considerably smoothened.

- Oscillations of integration point stresses, observed with the SFEA, are caused by steep stress gradients at the pile ends. This was already explained in [58].

In view of skin traction analysis, previous observations demonstrate that the EB-I poses a significant enhancement compared to the EB, as it significantly reduces numerical inconsistencies leading to unrealistic oscillations in skin traction.

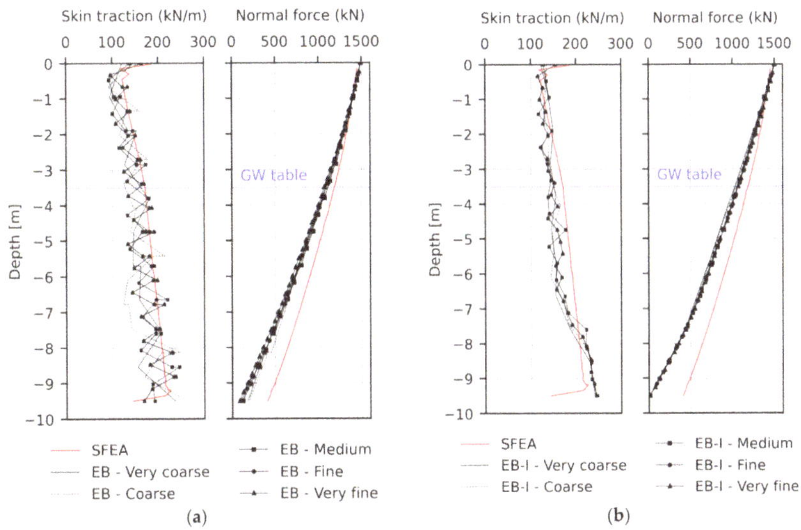

Figure 10. Influence of mesh size on development of skin traction and normal force distribution in an axial direction at a vertical load of 1500 kN: (**a**) EB, (**b**) EB-I.

4.5. Displacements Induced in Surronding Soil

In design situations, where the cost-effectiveness of pile caps are of primary interest, pile spacings should be kept as close as possible [68]. The optimal pile layout, however, should also ensure the effectiveness of resisting piles, which is primarily achieved through sufficiently large pile spacings [72]. The assessment of optimal pile spacings may be based on the evaluation of settlement profiles at different pile depths [52]. Moreover, mixed foundations, involving deep foundation elements, must satisfy serviceability limits, defined in terms of admissible differential settlements [73]. In any of these cases, it is important to realistically capture spatial soil displacements induced by external loads. An insight into the predicted displacement behaviour can be gained by looking at Figure 11, where the final distribution of soil settlements, induced by a vertical load of 1500 kN, are illustrated. At a first glance, it can be observed that:

- Except for the displacement field in close proximity to the pile base, all calculations yield similar results within the soil domain (i.e., zone Ω_1).
- Apparently, the pile domain (i.e., zone Ω_2) experiences settlement concentrations, which are particularly pronounced for the EB. This is a direct consequence of the SSI considered along a line, thereby introducing high displacement gradients in a radial direction. In contrast, the EB-I finds homogenous displacement regions of lower magnitude enclosed by the explicit interaction surface; although vertical displacements are slightly underestimated compared to the SFEA, the EB-I is obviously superior to the EB with regard to the prediction of the general deformation pattern inside Ω_2.

A closer inspection of the settlement profiles, at different pile depths, reveals the apparent origin of previous observations (see Figure 12): With the SFEA, an explicit interface (i.e., series of linked pairs of nodes) is inserted at the physical pile-soil contact. This interface fulfils the role of a kinematic discontinuity, capable of accounting for extremely high strain gradients [74]. Concerning both embedded beam formulations, the fallacy to account for the displacement jump at the pile skin becomes obvious. This is due to the implicit nature of the embedded interface, where displacement jumps are internally considered to describe the nodal connectivity of beam and soil nodes, but do not evolve in the physical mesh.

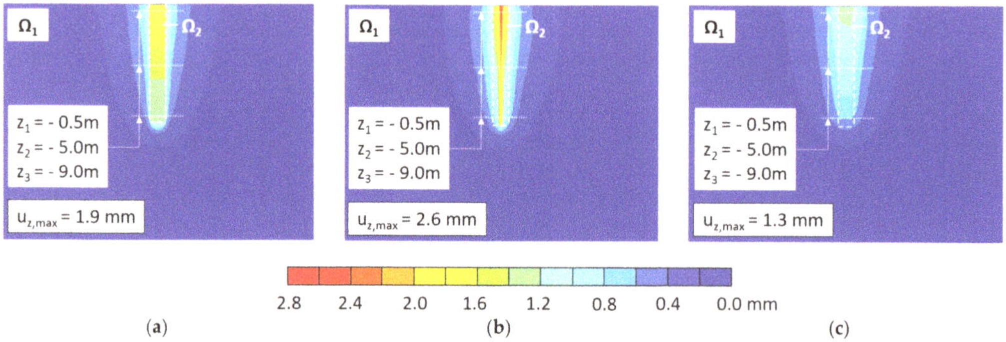

Figure 11. Settlement contour plots at a vertical load of 1500 kN: (**a**) SFEA, (**b**) EB, (**c**) EB-I. The latter models are obtained with the fine mesh configuration. The real pile skin is indicated by dashed lines.

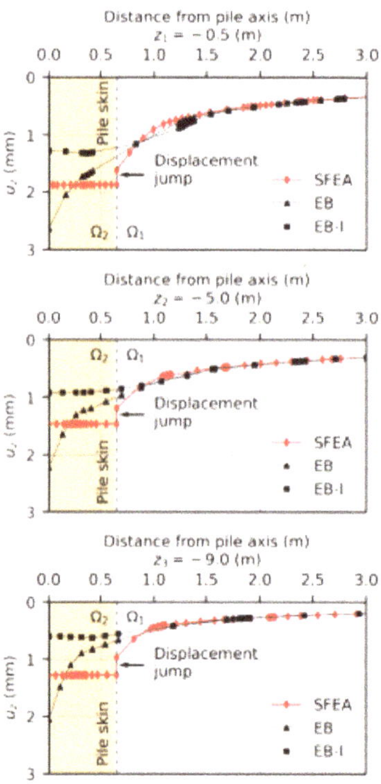

[1] Relative difference with respect to SFEA results.

	SFEA	EB	EB-I	Unit
u_z at pile axis	1.9	2.6	1.3	mm
Rel. difference 1	0	37	-32	%

[1] Relative difference with respect to SFEA results.

	SFEA	EB	EB-I	Unit
u_z at pile axis	1.5	2.2	0.9	mm
Rel. difference 1	0	47	-40	%

[1] Relative difference with respect to SFEA results.

	SFEA	EB	EB-I	Unit
u_z at pile axis	1.3	2	0.6	mm
Rel. difference 1	0	54	-54	%

(**a**) (**b**)

Figure 12. (**a**) Settlement profiles and (**b**) vertical settlements at the pile axis, obtained with a vertical load of 1500 kN at different pile depths.

From an engineering perspective, the settlement curves evolving outside the pile domain vary within acceptable bounds; at a distance greater than two times the pile diameter, the settlement profiles of all pile modelling approaches practically coincide. On the contrary, increasing deviations are observed towards the pile axis. While the EB tends to overestimate the settlements, the EB-I gives lower values. In all cases considered, peak deviations are observed at the lowermost settlement profile (EB: −54%, EB-I: 54%). Further improvements of the spatial deformation behaviour are mainly related to the stiffness definition of the embedded interface, which is at the forefront of ongoing research.

5. Case Study

As part of the restructuring of the Vienna rail node, Tschuchnigg [11] has described the application of 3D finite element simulations to the foundation design of central station "Wien Mitte"; see Figure 13. Due to serviceability requirements, critical zones of the existing slab are supported by jet-grout columns to satisfy serviceability criterions, mainly concerning differential settlements.

Figure 13. Project overview "Wien Mitte".

The performance of previously described embedded beam formulations is assessed by performing the same numerical simulations. In absence of measurement data, the results are, again, numerically validated against a full 3D representation of the problem including SFEA, which is widely recognized as convenient for the numerical analysis of similar problems; for example, see [75,76]. By taking advantage of the symmetry, the model size is defined as 50 × 4 × 48 m; see Figure 14a. The soil layering is similar to the real project. The raft of the simplified model is supported by nine jet-grout columns, which are symmetrically arranged on the foundation axis; see Figure 14b. SSI between soil and raft boundaries are addressed by means of zero-thickness interface elements [58,59] (not shown). The base model boundary condition is defined as fully fixed, whereas the vertical boundaries are set to allow vertical movement of soil layers (i.e., roller supports). Depending on the column modelling approach, the domain is discretized with 102,811 (SFEA) and 41,205 (EB, EB-I) 10-noded tetrahedral elements with quadratic element function.

Table 5 gives the HSS soil parameters adopted in the analyses; in analogy to Section 3.3, HSS-specific soil parameters are obtained using profound empirical correlations [66,67]. Additional material parameters related to the column-raft system, as well as the calculation phase sequence, are described in [11].

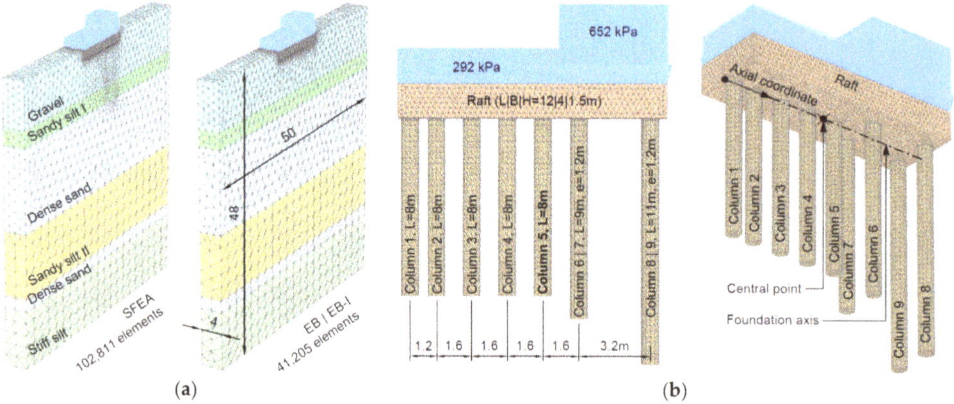

Figure 14. Description of the FE analysis model "Wien Mitte": (**a**) Model configuration and (**b**) side/3D view of pile-raft system.

Table 5. Overview of HSS soil parameters used in the analyses (Wien Mitte).

Parameter	Symbol	Unit	Gravel	Sandy Silt I I II	Sand	Stiff Silt
Depth of groundwater table	-	m	6.0	-	-	-
Layer thickness	t	m	8.0	3.0 I 11.0	14.0 I 2.0	10.0
Unit weight	$\gamma_{sat}, \gamma_{unsat}$	kN/m^3	21.5 I 21.0	20.0	21.0 I 20.0	20.0
Reference deviatoric hardening modulus at p^{ref}	E_{50}^{ref}	kN/m^2	40,000	20,000	25,000	30,000
Reference oedometer stiffness at p^{ref}	E_{oed}^{ref}	kN/m^2	40,000	20,000	25,000	30,000
Reference un-/reloading stiffness at p^{ref}	E_{ur}^{ref}	kN/m^2	120,000	50,000	62,500	90,000
Power index	m	-	0.0	0.8	0.65	0.6
Isotropic Poisson's ratio	ν'_{ur}	-	0.2	0.2	0.2	0.2
Effective friction angle	φ'	deg	35.0	27.5	32.5	27.5
Effective cohesion	$c\prime$	kN/m^2	0.1	20.0 I 30.0	5.0	30.0
Ultimate dilatancy angle	Ψ	deg	5.0	0.0	2.5	0.0
Pre-overburden pressure	POP	kN/m^2	600	600	600	600
Reference pressure	p^{ref}	kN/m^2	100	100	100	100
Initial shear modulus at p^{ref}	G_0^{ref}	kN/m^2	138,000	81,000	93,000	116,000
Threshold shear strain	$\gamma_{0.7}$	-	0.00015	0.00015	0.00015	0.00015

Figure 15a compares the load-settlement curves, obtained with the embedded beam models, against the benchmark solution. It is interesting to note that settlements occurring at the central point (see Figure 14) are practically coincident (<1 mm) up to an applied load of around 40%. Beyond this value, one can observe slightly higher settlements for both embedded beam models, presumably due to lower base resistance mobilization rates in the initial stage of loading; see Section 4.1. The settlement profiles, illustrated in Figure 15b, indicate that this tendency is further amplified towards edge columns 8 and 9, which are exposed to a higher surface load. At final load, corresponding edge settlements are 35 mm (SFEA), 41 mm (EB), and 42 mm (EB-I).

Figure 16a indicates the displacements of the foundation at the final loading stage. Following the definition documented in [73], the displacements along the foundation axis are expressed as the tangent rotation θ (i.e., negative when the tangent of successive nodes points downwards). Again, almost identical results are calculated for both embedded beam formulations. Ignoring the presence of successive peak values, the derived foundation movements appear to be in reasonable agreement with the reference solution. Successive peak values, however, are an important aspect in structural analysis (especially for the design of the raft) and thus, require discussion. In the simulations, all column-raft contacts are imposed with a rigid connection. While the reference solution spreads the column-raft interaction over multiple nodes, both embedded beam formulations consider the column-raft connection at one node only. As the raft is discretized with solid elements occupying no rotational DOFs, the connection type reduces to a hinge for single-node connections, which has a remarkable influence on the magnitude of the tangent rotation developing in direction of the foundation axis.

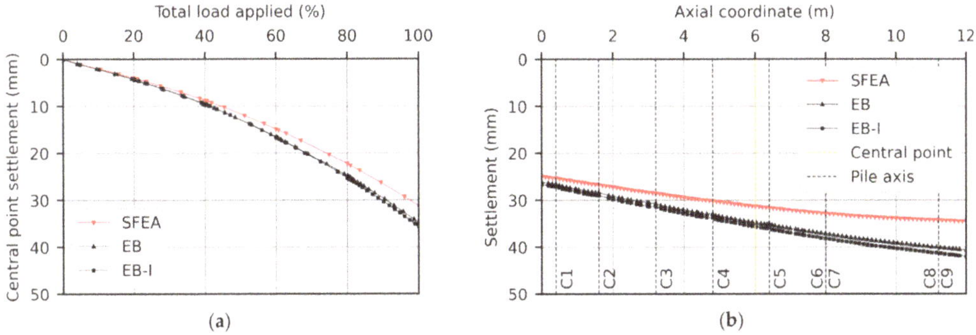

Figure 15. (**a**) Central point settlement plotted as a function of total load applied and (**b**) settlement curve along the foundation axis at final load, obtained with SFEA, EB, and EB-I.

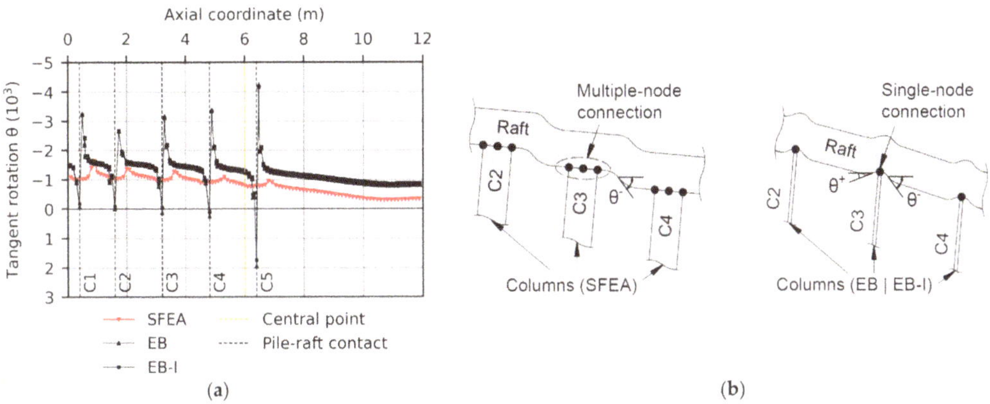

Figure 16. (**a**) Foundation movement at peak load and (**b**) schematic illustration of deformation patterns obtained with different column modelling approaches.

Figure 16b illustrates the observed deformation patterns schematically. In the reference model, the tangent rotation achieves max. values outside the column-raft contacts, indicating offsets between the columns. Since the tangent rotations are purely negative in sign, the tangent between successive nodes (in direction of the axial coordinate) points strictly downwards. In contrast, tangent rotation peak values evolve at the column-raft contacts with the EB and EB-I. In addition, the tangent rotations vary in sign at the column-raft contacts, indicating relatively high curvatures. Taking into account the bending moment–curvature relationship, this is likely to result in the spurious prediction of internal forces and an inconvenient raft design. To circumvent these obstacles, a more realistic representation of foundation movement could be realized by using alternative modelling techniques such as the hybrid modelling approach described by Lődör and Balázs [24], where embedded beams are circumscribed by volume elements at connection areas.

The last section is devoted to examining the load sharing between members of the composite geotechnical structure. In this context, the column raft coefficient α_{cr} at different loading stages is calculated as:

$$\alpha_{cr} = \frac{\sum R_{column}}{R_{total}} \tag{11}$$

where $\sum R_{column}$ (kN) defines the sum of loads carried by the columns and R_{total} (kN) the total load applied. Figure 17a depicts α_{cr} as a function of relative load applied. All models show higher α_{cr}-values with increasing load level, indicating that additional load portions are predominantly carried by the columns. On the one hand, both embedded beam formulations show almost identical results and converge to α_{cr} = 0.46 after the final stage of loading. On the other hand, deviations compared to the benchmark are striking (α_{cr} = 0.71) and require discussion.

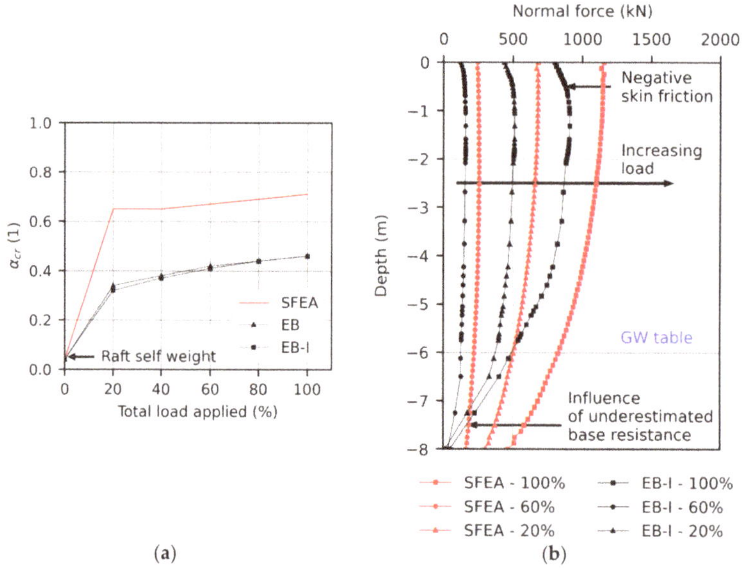

Figure 17. (**a**) Column raft coefficient and (**b**) normal force distribution developing along column 5, plotted as a function of total load applied.

To study the load transfer characteristics in more detail, Figure 17b shows the normal force distribution along (central) column 5. Obviously, arising differences with regard to α_{cr} can be attributed to a combination of remaining issues associated with both embedded beam configurations, namely:

- At load levels, fairly below the ultimate skin resistance, the load carried by the base resistance is significantly underestimated; see Section 4.1. As a consequence, the general column response is too soft.
- The single-node connection causes a combination of unrealistic settlement concentrations and spurious stress path evolutions in the vicinity of column-raft contacts. As a result, column 5 experiences negative skin friction along the upper portion of the shaft (normal force increases up to a depth of around 1.0 m), instead of a direct pile head load. As a consequence, lower α_{cr}-values are obtained.

More research, with the aim of analysing complex boundary value problems, is part of ongoing research.

6. Discussion and Conclusions

Two different embedded beam formulations for finite element modelling of deep foundation elements have been extensively compared in terms of geotechnical and structural performance, namely the embedded beam element (EB) and the recently developed embedded beam element with interaction surface (EB-I). Derived quantities are numerically validated against the widely accepted standard FE approach (SFEA), a full 3D representation of respective structures. To the authors' knowledge, this work poses the first attempt to assess the suitability of the EB-I for practical use. In this respect, a number of critical aspects have been tackled, such as load sharing between base and shaft, evolution of soil displacements in the surrounding soil, and mobilization of skin resistance. In this context, particular emphasis has been given to the influence of the mesh sensitivity of results. The following conclusions are drawn from the FEA of deep foundation elements:

- In the initial phase of loading, both embedded beam formulations yield load-displacement responses which are in remarkable agreement with the SFEA. At load levels beyond the shaft capacity, load-displacement curves obtained with the EB are considerably mesh sensitive, whereas the pile behaves stiffer with increasing mesh-coarseness. The EB-I, in contrast, reduces the mesh size effect tremendously. Moreover, the EB-I achieves a satisfactory agreement with the SFEA.
- Concerning the predicted pile capacity, the EB produces a wide scatter of results which must be regarded as unsatisfactory. This shortcoming is effectively eliminated by the EB-I; in all cases considered, the pile capacity varies, within acceptable bounds, slightly higher than the SFEA target value. Reducing the mesh size effect also allows engineers to deduce pile stiffness coefficients with more confidence.
- At typical working load conditions, skin traction profiles obtained with both embedded beam formulations fit SFEA results qualitatively well. However, the EB produces numerical oscillations about the mean that are significantly reduced with the EB-I.
- Although both embedded beam formulations appear to capture the evolution of spatial soil displacements with sufficient accuracy, major differences occur inside the pile domain. While the EB calculates the highest soil displacement at the pile axis, the EB-I computes almost constant displacement profiles within the pile boundaries, as is the case with the SFEA. However, reproducing displacement jumps at the pile skin lies beyond the capabilities of the actual EB-I configuration.
- When modelling deep foundation elements of composite structures, by means of embedded beams, the connection of the individual structures needs to be considered carefully: specifically, if embedded beams are imposed with a rigid connection. Otherwise, the prediction of structural forces, shaft-base load sharing, and differential settlements may lack physical meaning.

Considering the above findings, the EB-I proves superior to the EB. Nevertheless, several limitation of the current version have been observed which require further research effort. This includes the development of a generally applicable concept capable of predicting SSI effects at the base with sufficient accuracy; in addition, guidelines concerning the definition of the ultimate base resistance are still in demand. Further, future studies should explore whether the elastic zone approach is still required with the EB-I. In the course of this paper, the application of EB-Is was limited to axial loading cases, neglecting passive, as well as lateral, loading; therefore, future studies should focus on more complex loading situations, whereas the credibility of results should be carefully validated using measurement data. Research to resolve remaining issues is ongoing with promising results so far.

Author Contributions: A.-N.G. focused on the conceptualization, methodology, software, investigation, writing of the original draft preparation, formal analysis, and the visualization. F.T. pursued the supervision, editing of the original draft and the project administration. All authors have read and agreed to the published version of the manuscript.

Funding: Open Access Funding by the Graz University of Technology. This research received no external funding.

Institutional Review Board Statement: Not applicable.

Informed Consent Statement: Not applicable.

Acknowledgments: Special thanks should be given to Sandro Brasile and Saman Hosseini for their valuable technical support on this research.

Conflicts of Interest: The authors declare no conflict of interest.

Abbreviations

u_x, u_y, u_z	Translational DOFs
$\varphi_x, \varphi_y, \varphi_z$	Rotational DOFs
$\boldsymbol{t}_{EB\|EB-I}^{skin}$	Skin traction vector
$\boldsymbol{u}_{rel}$	Relative displacement vector
$\boldsymbol{u}^b$, $\boldsymbol{u}^s$	Beam (b) and soil (s) displacement vector
$\mathbf{N}^b$, $\mathbf{N}^s$	Beam (b) and (s) soil interpolation matrix
$\boldsymbol{D}^e$	Elastic stiffness matrix of interface
$t_{axial,max}$	Ultimate shear traction at line interface
c', φ'	Effective shear strength parameters
$\sigma_n^{avg\prime}, \sigma_n'$	Effective normal stress at interface
R, D	Pile radius (R) and diameter (D)
K_{base}	Interface stiffness at base
F_{max}	Ultimate base resistance
$\boldsymbol{H}$	Interpolation matrix for interaction surface
$\boldsymbol{a}^b, \boldsymbol{a}^b$	Nodal beam (b) and soil (s) DOFs
u_{L1}, u_{L2}, u_{ult}	Empirical pile head displacements
$\tau_{max}, \tau_s, \tau_t$	(Max.) shear stress (component) at interface
R_{inter}	Interface reduction factor
$R_{c,norm}, R_c$	(Normalized) compressive pile resistance
$R_{c,min}, R_{c,max}$	Min./max. compressive pile resistance
R_b, R_s	Base (b) and skin (s) resistance
L_{avg}	Average element size
MDR	Mesh dependency ratio
k_s	Pile stiffness coefficient
Ω_1	Soil domain
Ω_2	Column domain
θ	Tangent rotation
R_{column}	Load carried by columns
R_{total}	Total load applied
α_{cr}	Column raft coefficient

Appendix A

For the investigation of boundary effects, it is proposed to use the stiffness multiplier G_m (i.e., state parameter of the HSS model) [77]:

$$G_m = \frac{G}{G_{ur}} \geq 0.8\frac{G_0}{G_{ur}} \tag{A1}$$

in the outermost stress points (i.e., at kinematically constrained model boundaries). In Equation (A1), G denotes the elastic tangent shear modulus, G_0 the initial shear modulus, and G_{ur} the unloading-reloading shear modulus. In this respect, Figure A1 shows the G_m-contour plot obtained with the SFEA, which is considered as a benchmark model in Section 4. The minimum G_m-value (at the kinematically constrained model boundaries) at the end of loading is 2.7, which satisfies the criterion specified in Equation (A1). Hence, it can be concluded that the application domain is large enough in both the horizontal and vertical directions to avoid any boundary cut-off effects (or at least that such effects are reduced to an acceptable limit). The same conclusion holds for all FE models where the piles are idealised with EBs/EB-Is (minimum G_m-value = 2.7, not shown).

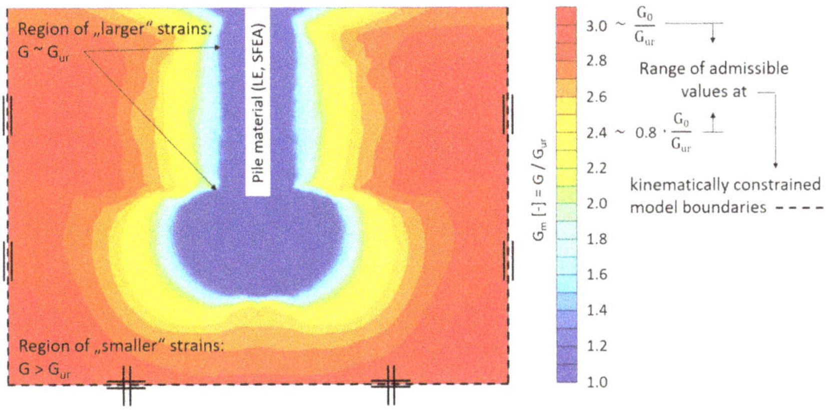

Figure A1. Alzey Bridge pile load test: factorized increase in unloading/reloading stiffness at the end of loading (SFEA).

Appendix B

Mesh sensitivity analyses have been conducted to ensure that acknowledged errors [46], associated with the discretization of the single pile domain, are (practically) eliminated for the benchmark model employed in Section 4. Figure A2 illustrates the mesh discretizations considered. Based on the mesh associated with the benchmark model (i.e., Mesh 1), the solid elements are systematically refined towards the pile axis for Mesh 2 and Mesh 3.

In the present case, compressive pile resistance magnitudes, mobilized at characteristic pile head displacements, are used to demonstrate the suitability of the mesh employed with the benchmark model; see also Figure 2b. As shown in Figure A3, the load settlement curves almost coincide for all investigated mesh discretizations. In comparison with the benchmark model, the max. relative difference in compressive pile resistance is observed at the end of loading, where Mesh 3 gives a slightly lower magnitude (−0.8%).

The results clearly demonstrate that the discretization of the benchmark model is sufficiently fine to approximate the boundary value problem with high accuracy, which allows the FE model to be used as a reference model to numerically validate both embedded beam formulations in Section 4.

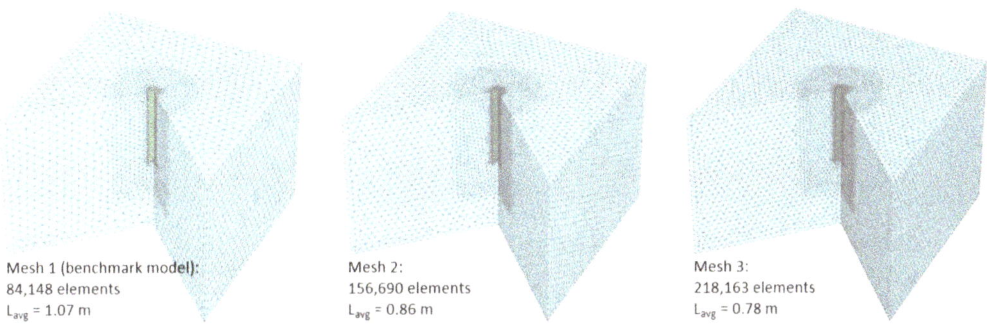

Figure A2. Alzey Bridge pile load test: 3D view of mesh discretizations considered in mesh sensitivity analyses (SFEA).

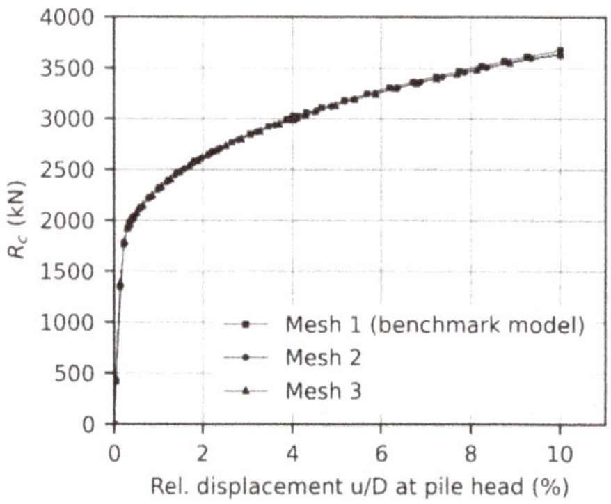

(**a**)

[1] Relative difference with respect to Mesh 1.

	Mesh 1	Mesh 2	Mesh 3	Unit
R_c at u/D = 0.34%	1969	1971	1970	kN
Rel. difference [1]	-	0.1	0.1	%
R_c at u/D = 4%	3009	2996	2991	kN
Rel. difference [1]	-	−0.4	−0.6	%
R_c at u/D = 10%	3675	3652	3646	kN
Rel. difference [1]	-	−0.6	−0.8	%

(**b**)

Figure A3. Alzey Bridge pile load test: influence of spatial discretization on (**a**) vertical pile head settlements at the pile axis and (**b**) mobilization of compressive pile resistance at different pile head settlements.

References

1. Maheshwari, B.K.; Watanabe, H. Nonlinear Dynamic Behavior of Pile Foundations: Effects of Separation at the Soil-Pile Interface. *Soils Found* **2006**, *46*, 437–448. [CrossRef]
2. Poulos, H.G.; Davis, E.H. *Pile Foundation Analysis and Design*, 1st ed.; Poulos, H.G., Davis, E.H., Eds.; Wiley: New York, NY, USA, 1980; ISBN 0471020842.
3. von Wolffersdorff, P.-A.; Henke, S. Möglichkeiten und Grenzen numerischer Methoden in der Geotechnik. *Bautechnik* **2021**, *98*, 1–17. [CrossRef]
4. Potts, D.M.; Zdravkovic, L. *Finite Element Analysis in Geotechnical Engineering: Application*, 1st ed.; Thomas Telford Ltd.: London, UK, 2001; ISBN 0727728121.
5. Engin, H.K. Modelling of Installation Effects: A Numerical Approach. Ph.D. Thesis, Delft University of Technology, Delft, The Netherlands, 2013.
6. Schmüdderich, C.; Shahrabi, M.M.; Taiebat, M.; Lavasan, A.A. Strategies for numerical simulation of cast-in-place piles under axial loading. *Comput. Geotech.* **2020**, *125*, 1–19. [CrossRef]
7. Wehnert, M.; Vermeer, P.A. Numerische Simulation von Probebelastungen an Großbohrpfählen. In Proceedings of the Bauen in Boden und Fels: 4. Kolloquium, Ostfildern, Germany, 20–21 January 2004; TAE: Esslingen, Germany, 2004; pp. 555–565, ISBN 3924813558.
8. Han, F.; Salgado, R.; Prezzi, M.; Lim, J. Shaft and base resistance of non-displacement piles in sand. *Comput. Geotech.* **2017**, *83*, 184–197. [CrossRef]
9. Chen, Y.-J.; Lin, S.-S.; Chang, H.-W.; Marcos, M.C. Evaluation of side resistance capacity for drilled shafts. *J. Mar. Sci. Technol.* **2011**, *19*, 210–221. [CrossRef]
10. Turello, D.F.; Sánchez, P.J.; Blanco, P.J.; Pinto, F. A variational approach to embed 1D beam models into 3D solid continua. *Comput. Struct.* **2018**, *206*, 145–168. [CrossRef]
11. Tschuchnigg, F. 3D Finite Element Modelling of Deep Foundations Employing an Embedded Pile Formulation. Ph.D. Thesis, Graz University of Technology, Graz, Austria, 2013.
12. Tschuchnigg, F.; Schweiger, H.F. The embedded pile concept—Verification of an efficient tool for modelling complex deep foundations. *Comput. Geotech.* **2015**, *63*, 244–254. [CrossRef]
13. Tradigo, F.; Pisanò, F.; Di Prisco, C. On the use of embedded pile elements for the numerical analysis of disconnected piled rafts. *Comput. Geotech.* **2016**, *72*, 89–99. [CrossRef]
14. Ninić, J.; Stascheit, J.; Meschke, G. Beam-solid contact formulation for finite element analysis of pile-soil interaction with arbitrary discretization. *Int. J. Numer. Anal. Meth. Geomech.* **2014**, *38*, 1453–1476. [CrossRef]
15. Sadek, M.; Shahrour, I. A three dimensional embedded beam element for reinforced geomaterials. *Int. J. Numer. Anal. Meth. Geomech.* **2004**, *28*, 931–946. [CrossRef]
16. Engin, H.; Brinkgreve, R.; Septanika, E. Improved embedded beam elements for the modelling of piles. In *Numerical Models in Geomechanics, Proceedings of the 10th International Symposium on Numerical Models in Geomechanics, Rhodes, Greece, 25–27 April 2007*; Pande, G.N., Ed.; Taylor & Francis: London, UK, 2007; ISBN 978-0-415-44027-1.
17. Oliveria, D.; Wong, P.K. Use of embedded pile elements in 3D modelling of piled-raft foundations. *Aust. Geomech. J.* **2011**, *46*, 9–19.

18. Smulders, C.M. An Improved 3D Embedded Beam Element with Explicit Interaction Surface. Master's Thesis, Delft University of Technology, Delft, The Netherlands, 2018.
19. Ukritchon, B.; Faustino, J.C.; Keawsawasvong, S. Numerical investigations of pile load distribution in pile group foundation subjected to vertical load and large moment. *Geomech. Eng.* **2016**, *10*, 577–598. [CrossRef]
20. Engin, H.K.; Brinkgreve, R.B.J. Investigation of Pile Behaviour Using Embedded Piles. In Proceedings of the 17th International Conference on Soil Mechanics and Geotechnical Engineering, Alexandria, Egypt, 5–9 October 2009; Hamza, M., Shahien, M., El-Mossallamy, Y., Eds.; IOS Press: Amsterdam, The Netherlands, 2009; pp. 1189–1192.
21. Jongpradist, P.; Haema, N.; Lueprasert, P. Influence of pile row under loading on existing tunnel. In *Tunnels and Underground Cities: Engineering and Innovation Meet Archaeology, Architecture and Art*; Peila, D., Viggiani, G., Celestino, T., Eds.; CRC Press: London, UK, 2019; pp. 5711–5719, ISBN 9781003031857.
22. Engin, H.; Septanika, E.; Brinkgreve, R. Estimation of Pile Group Behavior using Embedded Piles. In Proceedings of the 12th International Conference on Computer Methods and Advances in Geomechanics, Goa, India, 1–6 October 2008; Jadhav, M.N., Ed.; Curran: New York, NY, USA, 2008, ISBN 9781622761760.
23. Engin, H.; Brinkgreve, R.; Bonnier, P.; Septanika, E. Modelling piled foundation by means of embedded piles. In *Geotechnics of Soft Soils: Focus on Ground Improvement, Proceedings of the 2nd International Workshop, Glasgow, Scotland, 3–5 September 2008*; Karstunen, M., Leoni, M., Eds.; Taylor & Francis: London, UK, 2008; pp. 131–136, ISBN 978-0-415-47591-4.
24. Lődör, K.; Balázs, M. Finite element modelling of rigid inclusion ground improvement. In *Numerical Methods in Geotechnical Engineering, Proceedings of the 9th European Conference on Numerical Methods in Geotechnical Engineering, Porto, Portugal, 25–27 June 2018*; Cardoso, A.S., Borges, J.L., Costa, P.A., Gomes, A.T., Marques, J.C., Vieira, C.S., Eds.; CRC Press: London, UK, 2018; pp. 1–8, ISBN 9781138544468.
25. Tschuchnigg, F.; Schweiger, H.F. Numerical study of simplified piled raft foundations employing an embedded pile formulation. In Proceedings of the 1st International Symposium on Computational Geomechanics. ComGeo I, Juan-les-Pins, France, 29 April–1 May 2009; IC2E: Rhodes, Greece, 2009; pp. 743–752.
26. Lee, S.W.; Cheang, W.; Swolfs, W.M.; Brinkgreve, R. Modelling of piled rafts with different pile models. In *Numerical Methods in Geotechnical Engineering, Proceedings of 7th European Conference on Numerical Methods in Geotechnical Engineering, Trondheim, Norway, 2–4 June 2010*; Benz, T., Nordal, S., Eds.; CRC Press: London, UK, 2010; pp. 1–6, ISBN 9780429206191.
27. Elshehahwy, E.R.; Eltahrany, A.; Dif, A. Numerical Analysis Validation Using Embedded Pile. In *Advances in Numerical Methods in Geotechnical Engineering, Proceedings of 2nd GeoMEast International Congress and Exhibition on Sustainable Civil Infrastructures, Cairo, Egypt, 10–15 November 2019*; Shehata, H., Desai, C.S., Eds.; Springer: Berlin/Heidelberg, Germany, 2019; pp. 199–208, ISBN 978-3-030-01925-9.
28. Watcharasawe, K.; Jongpradist, P.; Kitiyodom, P.; Matsumoto, T. Measurements and analysis of load sharing between piles and raft in a pile foundation in clay. *Geomech. Eng.* **2021**, *24*, 559–572. [CrossRef]
29. Banerjee, R.; Bandyopadhyay, S.; Sengupta, A.; Reddy, G.R. Settlement behaviour of a pile raft subjected to vertical loadings in multilayered soil. *Geomech. Geoengin.* **2020**, 1–15. [CrossRef]
30. Abbas, Q.; Kim, G.; Kim, I.; Kyung, D.; Lee, J. Lateral Load Behavior of Inclined Micropiles Installed in Soil and Rock Layers. *Int. J. Geomech.* **2021**, *21*, 1–13. [CrossRef]
31. Abbas, Q.; Choi, W.; Kim, G.; Kim, I.; Lee, J. Characterizing uplift load capacity of micropiles embedded in soil and rock considering inclined installation conditions. *Comput. Geotech.* **2021**, *132*, 1–12. [CrossRef]
32. El-Sherbiny, M.M.; El-Sherbiny, R.M.; El-Mamlouk, H. Three Dimensional Effect of Grouted Discontinuous Berms for Passive Support of Diaphragm Walls. In Proceedings of the Grouting 2017: Jet Grouting, Diaphragm Walls, and Deep Mixing, Honolulu, HI, USA, 9–12 July 2017; Gazzarrini, P., Richards, J.T.D., Bruce, D.A., Byle, M.J., El Mohtar, C.S., Johnsen, L.F., Eds.; American Society of Civil Engineers: Reston, VA, USA, 2017; pp. 571–583, ISBN 9780784480809.
33. Marjanović, M.; Vukićević, M.; König, D.; Schanz, T.; Schäfer, R. Modeling of laterally loaded piles using embedded beam elements. In *Contemporary Achievements in Civil Engineering, Proceedings of 4th International Conference, Subotica, Serbia, 22–23 April 2016*; Faculty of Civil Engineering: Novi Sad, Serbia, 2016; pp. 349–358.
34. Abo-Youssef, A.; Morsy, M.S.; El Ashaal, A.; El Mossallamy, Y.M. Numerical modelling of passive loaded pile group in multilayered soil. *Innov. Infrastruct. Solut.* **2021**, *6*, 1–13. [CrossRef]
35. Al-abboodi, I.; Sabbagh, T.T. Numerical Modelling of Passively Loaded Pile Groups. *Geotech. Geol. Eng.* **2019**, *37*, 2747–2761. [CrossRef]
36. Scarfone, R.; Morigi, M.; Conti, R. Assessment of dynamic soil-structure interaction effects for tall buildings: A 3D numerical approach. *Soil Dyn. Earthq. Eng.* **2020**, *128*, 1–14. [CrossRef]
37. Di Prisco, C.; Flessati, L.; Porta, D. Deep tunnel fronts in cohesive soils under undrained conditions: A displacement-based approach for the design of fibreglass reinforcements. *Acta Geotech.* **2020**, *15*, 1013–1030. [CrossRef]
38. Ninic, J. Computational Strategies for Predictions of the Soil-Structure Interaction during Mechanized Tunneling. Ph.D. Thesis, Ruhr University Bochum, Bochum, Germany, 2015.
39. Granitzer, A.; Tschuchnigg, F.; Summerer, W.; Galler, R.; Stoxreiter, T. Construction of a railway tunnel above the main drainage tunnel of Stuttgart using the cut-and-cover method. *Bauingenieur* **2021**, *96*, 156–164. [CrossRef]

40. Lo, S. A study in an attempt to use embedded pile structure elements to simulate soil nail structures in PLAXIS 2D 2012 and 3D 2012. In *Numerical Methods in Geotechnical Engineering, Proceedings of the 8th European Conference on Numerical Methods in Geotechnical Engineering, Delft, The Netherlands, 18–20 June 2014*; Hicks, M., Brinkgreve, R., Rohe, A., Eds.; CRC Press: London, UK, 2014; pp. 765–769, ISBN 978-1-138-00146-6.
41. Turello, D.F.; Pinto, F.; Sánchez, P.J. Embedded beam element with interaction surface for lateral loading of piles. *Int. J. Numer. Anal. Meth. Geomech.* **2016**, *40*, 568–582. [CrossRef]
42. Smulders, C.M.; Hosseini, S.; Brinkgreve, R. Improved embedded beam with interaction surface. In Proceedings of the 17th European Conference on Soil Mechanics and Geotechnical Engineering, Reykjavík, Iceland, 1–6 September 2019; Sigursteinsson, H., Erlingsson, S., Bessason, B., Eds.; COC: Reykjavík, Iceland, 2019; pp. 1048–1055.
43. Turello, D.F.; Pinto, F.; Sánchez, P.J. Three dimensional elasto-plastic interface for embedded beam elements with interaction surface for the analysis of lateral loading of piles. *Int. J. Numer. Anal. Meth. Geomech.* **2017**, *41*, 859–879. [CrossRef]
44. Turello, D.F.; Pinto, F.; Sánchez, P.J. Analysis of lateral loading of pile groups using embedded beam elements with interaction surface. *Int. J. Numer. Anal. Meth. Geomech.* **2019**, *43*, 272–292. [CrossRef]
45. Jauregui, R.; Silva, F. Numerical Validation Methods. In *Numerical Analysis—Theory and Application*; Awrejcewicz, J., Ed.; InTech: Toyama, Japan, 2011; ISBN 978-953-307-389-7.
46. Oberkampf, W.L.; Trucano, T.G.; Hirsch, C. Verification, validation, and predictive capability in computational engineering and physics. *Appl. Mech. Rev.* **2004**, *57*, 345–384. [CrossRef]
47. Ferreira, D.; Manie, J. DIANA—Finite Element Analysis: DIANA Documentation—Release 10.4. Available online: https://dianafea.com/manuals/d104/Diana.html (accessed on 25 February 2021).
48. Truty, A.; Zimmermann, T.; Podles, K.; Obrzud, R. ZSOIL.PC 2020: User Manual—Theory. Available online: https://www.zsoil.com/zsoil_manual/TM-Man.pdf (accessed on 26 July 2021).
49. Rocscience Online Help: Online Documentation. Available online: https://www.rocscience.com/about (accessed on 25 February 2021).
50. FLAC3D 7.0 Documentation. Available online: http://docs.itascacg.com/flac3d700/common/sel/doc/manual/sel_manual/piles/piles.html?node903 (accessed on 26 February 2021).
51. PLAXIS Scientific Manual: 3D—Connect Edition V21. Available online: https://communities.bentley.com/cfs-file/__key/communityserver-wikis-components-files/00-00-00-05-58/PLAXIS3DCE_2D00_V21.01_2D00_04_2D00_Scientific.pdf (accessed on 26 July 2021).
52. Sommer, H.; Hambach, P. Großpfahlversuche im Ton für die Talbrücke Alzey. *Bauingenieur* **1974**, *49*, 310–317.
53. Hirany, A.; Kulhawy, F.H. Interpretation of Load Tests on Drilled Shafts—Part 1: Axial Compression. In Proceedings of the Foundation Engineering—Current Principles and Practices, Evanston, IL, USA, 25–29 June 1989; Kulhawy, F.H., Ed.; American Society of Civil Engineers: New York, NY, USA, 1989; pp. 1132–1149, ISBN 0872627047.
54. Chen, Y.-J.; Fang, Y.-C. Critical Evaluation of Compression Interpretation Criteria for Drilled Shafts. *J. Geotech. Geoenvironmental Eng.* **2009**, *135*, 1056–1069. [CrossRef]
55. Hirany, A.; Kulhawy, F.H. On the Interpretation of Drilled Foundation Load Test Results. In *International Deep Foundations Congress, Proceedings of the Deep Foundations: An International Perspective on Theory, Design, Construction, and Performance, Orlando, FL, USA, 14–16 February 2002*; O'Neill, M.W., Townsend, F.C., Eds.; American Society of Civil Engineers: New York, NY, USA, 2002; pp. 1018–1028, ISBN 9780784406014.
56. German Geotechnical Society. *Recommendations on Piling (EA-Pfähle)*, 2nd ed.; Wilhelm Ernst & Sohn: Berlin, Germany, 2014; ISBN 9783433030189.
57. PLAXIS Reference Manual: 3D—Connect Edition V21. Available online: https://communities.bentley.com/cfs-file/__key/communityserver-wikis-components-files/00-00-00-05-58/PLAXIS3DCE_2D00_V21.01_2D00_02_2D00_Reference.pdf (accessed on 16 August 2021).
58. Day, R.A.; Potts, D.M. Zero thickness interface elements—Numerical stability and application. *Int. J. Numer. Anal. Meth. Geomech.* **1994**, *18*, 689–708. [CrossRef]
59. Goodman, R.; Taylor, R.; Brekke, T. A Model for the Mechanics of Jointed Rock. *J. Soil Mech. Found. Div.* **1968**, *94*, 637–659. [CrossRef]
60. van Langen, H.; Vermeer, P.A. Interface elements for singular plasticity points. *Int. J. Numer. Anal. Meth. Geomech.* **1991**, *15*, 301–315. [CrossRef]
61. Kulhawy, F.H. Limiting tip and side resistance: Fact or Fallacy? In Proceedings of the Analysis and Design of Pile Foundations, San Francisco, CA, USA, 1–5 October 1984; Meyer, J.R., Ed.; American Society of Civil Engineers: New York, NY, USA, 1984; pp. 80–98.
62. Sheil, B.; McCabe, B.A. Predictions of friction pile group response using embedded piles in PLAXIS. In Proceedings of the 3rd International Conference on New Developments in Soil Mechanics and Geotechnical Engineering, Nicosia, Turkey, 28–30 June 2012; Near East University: Nicosia, Turkey, 2012; pp. 679–686.
63. Fleming, W.G.K.; Weltman, A.; Randolph, M.F.; Elson, K. *Piling Engineering*, 3rd ed.; Taylor & Francis: London, UK, 2020; ISBN 9780367659387.
64. Benz, T.; Schwab, R.; Vermeer, P. Small-strain stiffness in geotechnical analyses. *Bautechnik* **2009**, *86*, 16–27. [CrossRef]

65. Schanz, T.; Vermeer, P.A.; Bonnier, P.G. The hardening soil model: Formulation and verification. In *Beyond 2000 in Computational Geotechnics*, 1st ed.; Brinkgreve, R.B.J., Ed.; Routledge: London, UK, 2019; pp. 281–296, ISBN 9781315138206.
66. Vucetic, M.; Dobry, R. Effect of Soil Plasticity on Cyclic Response. *J. Geotech. Eng.* **1991**, *117*, 89–107. [CrossRef]
67. Alpan, I. The geotechnical properties of soils. *Earth-Sci. Rev.* **1970**, *6*, 5–49. [CrossRef]
68. Viggiani, C.; Mandolini, A.; Russo, G. *Piles and Pile Foundations*, 1st ed.; Taylor & Francis: London, UK, 2012; ISBN 978-0-415-49066-5.
69. Urbonas, K.; Slizyte, D.; Mackevicius, R. Influence of the pile stiffness on the ground slab behaviour. *J. Civ. Eng. Manag.* **2016**, *22*, 690–698. [CrossRef]
70. Hanisch, J.; Katzenbach, R.; König, G. (Eds.) *Kombinierte Pfahl-Plattengründungen*; Ernst & Sohn: Berlin, Germany, 2001; ISBN 3433016062.
71. Schreppers, G.-J. Bond-slip Reinforcements and Pile Foundations. Available online: https://dianafea.com/sites/default/files/2018-04/Bondslip_reinforcements_and_Pile_foundations.pdf (accessed on 26 July 2021).
72. Schmidt, H.-H.; Buchmaier, R.F.; Vogt-Breyer, C. *Grundlagen der Geotechnik*, 5th ed.; Springer Fachmedien Wiesbaden: Wiesbaden, Germany, 2017; ISBN 978-3-658-14930-7.
73. British Standards Institution. *BS EN 1997-1:2004: Eurocode 7: Geotechnical Design—Part 1: General Rules*; BSI: London, UK, 2004.
74. de Gennaro, V.; Frank, R.; Said, I. Finite element analysis of model piles axially loaded in sands. *Riv. Ital. Geotec.* **2008**, *2*, 44–62.
75. Cho, J.; Lee, J.-H.; Jeong, S.; Lee, J. The settlement behavior of piled raft in clay soils. *Ocean Eng.* **2012**, *53*, 153–163. [CrossRef]
76. Poulos, H.G. Piled raft foundations: Design and applications. *Géotechnique* **2001**, *51*, 95–113. [CrossRef]
77. Benz, T. Small-Strain Stiffness of Soils and Its Numerical Consequences. Ph.D. Thesis, University of Stuttgart, Stuttgart, Germany, 2007.

processes

MDPI

Article

Implementation of the Non-Associated Elastoplastic $MSDP_u$ Model in FLAC3D and Application for Stress Analysis of Backfilled Stopes

Feitao Zeng, Li Li *, Michel Aubertin and Richard Simon

Research Institute on Mines and Environment, Department of Civil, Geological and Mining Engineering, École Polytechnique de Montréal, C.P. 6079 Succursale Centre-Ville, Montréal, QC H3C 3A7, Canada; feitao.zeng@polymtl.ca (F.Z.); michel.aubertin@polymtl.ca (M.A.); richard.simon@polymtl.ca (R.S.)
* Correspondence: li.li@polymtl.ca

Abstract: The multiaxial Mises-Schleicher and Drucker-Prager unified ($MSDP_u$) criterion has been shown to exhibit several specific features compared to other yield and failure criteria, including a nonlinear mean stress dependency, influence of the Lode angle, use of independent uniaxial compressive and tensile strength values and absence of an apex (singularity) on the envelope surface in the negative stress quadrant. However, $MSDP_u$ has been seldom used in practice to solve geotechnical and geomechanical engineering problems mainly because it had not yet been fully implemented into three-dimensional (3D) numerical codes. To fill this gap, a 3D elastoplastic $MSDP_u$ formulation is developed and implemented into FLAC3D. The proposed $MSDP_u$ elastic-perfectly plastic (EPP) constitutive model is then validated against existing analytical solutions developed for calculating the stress and displacement distributions around cylindrical openings. The FLAC3D $MSDP_u$-EPP model is then applied to evaluate the vertical and horizontal stress distributions in a three-dimensional vertical backfilled stope. The numerical results obtained with the $MSDP_u$-EPP model are compared with those obtained with the Mohr-Coulomb EPP model, to highlight key features of the new formulation.

Keywords: 3D nonlinear yield criterion; elastoplastic model; numerical modeling; circular opening; backfill; FLAC3D

Citation: Zeng, F.; Li, L.; Aubertin, M.; Simon, R. Implementation of the Non-Associated Elastoplastic $MSDP_u$ Model in FLAC3D and Application for Stress Analysis of Backfilled Stopes. *Processes* **2022**, *10*, 1130. https://doi.org/10.3390/pr10061130

Academic Editor: Blaž Likozar

Received: 15 May 2022
Accepted: 2 June 2022
Published: 5 June 2022

1. Introduction

Elastoplastic constitutive models are widely used in geotechnical engineering to assess the mechanical response of geomaterials. The elastoplastic framework typically involves a yield criterion, a flow rule (with a plastic potential) and, in some cases, a hardening or softening function. Over the years, many elastoplastic constitutive models have been proposed and applied to analyze the complex mechanical responses of rocks and soils; the main ones are included in state-of-the-art review publications [1–9].

Elastic-perfectly plastic (EPP) models, with a fixed yield surface, are probably the most often used in practice to solve engineering problems involving geomaterials, due in a large part to their relative simplicity and ease of application. Several EPP models with different plastic (yield) criteria, such as Mohr-Coulomb (MC), Drucker-Prager (DP) and Hoek-Brown (HB), have already been built in commercial codes and applied in geotechnical engineering. In these models, the plastic criterion F is used to determine the limit of the stress state associated with plastic behavior. Criterion F also defines the plastic potential Q when an associated flow rule is used, while it can serve as a basis for Q ($\neq F$) in the case of a non-associated flow rule [9,10].

In addition to those mentioned above, a large number (>100) of failure and yield criteria have been proposed over the years [11–14], each having its advantages and limitations. The most commonly used criteria for soil and rock, MC and HB, are generally expressed with only two principal stresses and thus neglect the effect of the intermediate principal stress. A few well-known 3D criteria (e.g., DP) simplify the effect of the stress geometry by neglecting the influence of the Lode angle (defined below). The open surface defined by function F in the principal stress space along the hydrostatic axis is another limitation of many existing criteria, which cannot describe the volumetric yield behavior of geomaterials under high mean stresses [14–16].

Initially proposed for intact rocks [17,18] and later modified and extended for various geomaterials including rock and rock mass, concrete and mine backfill [19–22], the multiaxial Mises-Schleicher and Drucker-Prager unified ($MSDP_u$) criterion takes into account the effect of the three principal stresses, with a nonlinear (rounded) surface on the negative side of the mean stress axis and a cap on the positive (compressive) side. The $MSDP_u$ criterion exhibits four essential characteristics to define yielding or failure of cohesive/cemented geomaterials:

(i) a nonlinear mean stress dependency;
(ii) influence of the Lode angle to distinguish triaxial compression and extension behavior;
(iii) independence of the uniaxial compressive and tensile strengths;
(iv) absence of an apex (singularity) on the surface in the negative stress quadrant.

Additional features of $MSDP_u$ will be presented in the next section, after recalling the criterion basic formulation. Its applicability to describe the failure and yielding of a large variety of materials has been demonstrated by Aubertin et al. [20,21], Li et al. [14,22,23] and Aubertin and Li [12].

Despite its advantageous features, the practical use of the $MSDP_u$ criterion has been limited because it was only partially implemented in a numerical code. The implementation has been done within the 2D finite difference code FLAC through its user-defined model option with an external language, called "FISH" [24]. The ensuing FLAC2D $MSDP_u$-EPP model can be used to analyze geotechnical problems under plane strain conditions only. Some of the key features and advantages of $MSDP_u$ thus cannot be exploited, particularly when facing three-dimensional problems. In addition, the implementation of the $MSDP_u$-EPP model in FLAC2D (presented by Li et al. [24]) was based on an associated flow rule, which tends to overestimate the volumetric strains of geomaterial (and the related mean stresses). Moreover, the implementation of user-defined models in FLAC with the "FISH" language is no longer recommended by Itasca [25]. There is thus a need to implement a more complete 3D version of the $MSDP_u$-EPP model, with a non-associated flow rule, in FLAC3D, a commercial code widely applied for the solution of three-dimensional problems in geotechnical engineering [26–30].

In this paper, the three-dimensional $MSDP_u$-EPP model is formulated in terms of stress and strain increments, following the guidelines provided by Itasca [26] as well as Desai and Siwardane [1] and Chen and Baladi [2]. Its implementation in FLAC3D is done through its user-defined model option with the programing language C++. The model is then compiled into a DLL (dynamic link library) module that can be loaded and run as a built-in plug-in of FLAC3D. The FLAC3D $MSDP_u$-EPP model is partially validated against some existing analytical solutions developed for evaluating the stresses and displacements distributions around cylindrical openings in plane strain. The new 3D model is applied to evaluate the vertical and horizontal stress distributions in three-dimensional backfilled stopes, which are compared to those obtained with the commonly used Mohr-Coulomb EPP (MC-EPP) model. The results comparison highlights some of the new model's distinctive features such as a better fit to the yield (failure) surface for a wide range of geomaterials and a more representative volumetric yield behavior at relatively high mean stress.

2. The $MSDP_u$ Nonlinear Multiaxial Criterion

The general form of a plastic (yield) criterion for isotropic materials can be expressed in terms of commonly used stress invariants [1,31,32]:

$$F(I_1, J_2, J_3) = const \tag{1}$$

where I_1 is the is the first invariant of the stress tensor σ_{ij}; J_2 and J_3 are respectively the second and third invariants of the deviator stress tensor $S_{ij} = \sigma_{ij} - p\delta_{ij}$; δ_{ij} is the Kronecker delta ($\delta_{ij} = 1$ if $i = j$ and $\delta_{ij} = 0$ if $i \neq j$); the mean stress p is defined as follows:

$$p = \frac{I_1}{3} \tag{2}$$

The three invariants I_1, J_2 and J_3 can be expressed using the general stress tensor components [1,8], or in terms of the major (σ_1), intermediate (σ_2) and minor (σ_3) principal stresses:

$$I_1 = \sigma_1 + \sigma_2 + \sigma_3 \tag{3}$$

$$J_2 = \frac{1}{6}\left[(\sigma_1 - \sigma_2)^2 + (\sigma_2 - \sigma_3)^2 + (\sigma_3 - \sigma_1)^2\right] \tag{4}$$

$$J_3 = (\sigma_1 - p)(\sigma_2 - p)(\sigma_3 - p) \tag{5}$$

The $MSDP_u$ criterion can then be written as

$$F = J_2 - F_0^2 F_\pi^2 = 0 \tag{6}$$

where F_0 is associated with the nonlinear surface for conventional triaxial compression condition ($\sigma_1 > \sigma_2 = \sigma_3$), while F_π defines the surface in the π-plane. These two functions are usually expressed as follows [21]:

$$F_0 = \left[\alpha^2\left(I_1^2 - 2a_1 I_1\right) + a_2^2 - a_3\langle I_1 - I_c\rangle^2\right]^{1/2} \tag{7}$$

$$F_\pi = \frac{b}{\left[b^2 + (1 - b^2)\sin^2(45^\circ - 1.5\theta)\right]^{1/2}} \tag{8}$$

The Macaulay brackets $\langle x \rangle$ (= $(x + |x|)/2$, where x is a variable) are used in Equation (7) to avoid a negative term; α (taken from the DP criterion), a_1 and a_2 are material parameters related to shear failure or yielding:

$$\alpha = \frac{2\sin\phi}{\sqrt{3}(3 - \sin\phi)} \tag{9}$$

$$a_1 = \frac{C_0 - T_0}{2} - \frac{C_0^2 - \left(\frac{T_0}{b}\right)^2}{6\alpha^2(C_0 + T_0)} \tag{10}$$

$$a_2 = \left\{\left[\frac{C_0 + \left(\frac{T_0}{b^2}\right)}{3(C_0 + T_0)} - \alpha^2\right] C_0 T_0\right\}^{1/2} \tag{11}$$

where ϕ is internal friction angle, C_0 is uniaxial compressive strength (UCS), T_0 is uniaxial tensile strength (UTS, positive value) and b is a shape parameter defining the surface in the π-plane. Material parameter a_3 is related to the volumetric yield (cap) surface:

$$a_3 = \frac{\alpha^2\left(I_{1n}^2 - 2a_1 I_{1n}\right) + a_2^2}{(I_{1n} - I_c)^2} \tag{12}$$

where I_c is the I_1 value where the cap starts and I_{1n} is I_1 value where it intersects the hydrostatic axis ($\sigma_1 = \sigma_2 = \sigma_3$) in the principal stress space.

The Lode angle θ is defined as

$$\theta = \frac{1}{3}\sin^{-1}\frac{3\sqrt{3}J_3}{2\sqrt{J_2^3}}, -30^\circ \leq \theta \leq 30^\circ \tag{13}$$

From this definition, one can deduce $\theta = 30°$ for conventional triaxial (axisymmetric) compression (CTC, with $\sigma_1 > \sigma_2 = \sigma_3$) and $\theta = -30°$ for reduced triaxial (axisymmetric) extension (RTE, with $\sigma_1 = \sigma_2 > \sigma_3$). Under CTC testing conditions, $\theta = 30°$ and the commonly defined deviatoric stress $q = (3J_2)^{1/2} = \sigma_1 - \sigma_3$.

Figure 1 shows a typical representation of the MSDP$_u$ criterion in the I_1–$J_2^{1/2}$ plane (Figure 1a) and π-plane (Figure 1b). In the I_1–$J_2^{1/2}$ plane, the MSDP$_u$ envelope can be decomposed into two main parts. In the first part when $I_1 \leq I_c$, $J_2^{1/2}$ increases nonlinearly with I_1. In the second part when $I_1 > I_c$, the cap controls the yield surface and $J_2^{1/2}$ tends to decrease with an increasing I_1. For dense geomaterials such as hard rocks, the values of I_c and I_{1n} can be very high so the cap can be neglected for most applications; this is not the case however for porous materials such as soils, backfills and some weak rocks. In the π-plane, the usual shape of the MSDP$_u$ envelope defined above by Equation (8) depends on the value of parameter b. The typical value of b goes from 1 to about 0.7 and the corresponding envelope evolves from a circle (for $b = 1$) to a rounded triangle ($b < 1$) in the π-plane, as shown in Figure 1b.

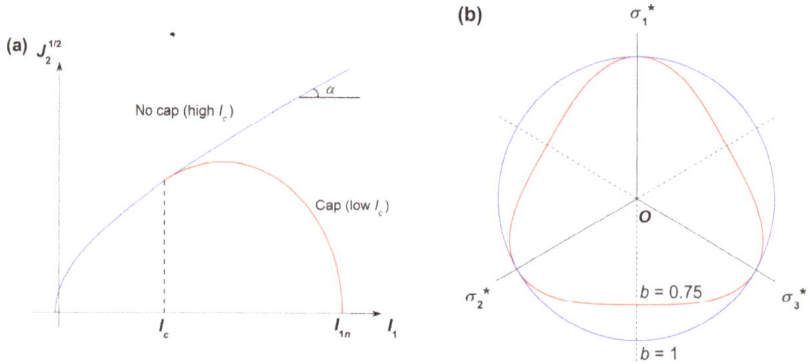

Figure 1. MSDP$_u$ failure surface in the (**a**) I_1–$J_2^{1/2}$ plane and (**b**) π-plane (σ_1 *, σ_2 *, σ_3 * are the projections of the principal stress axes in the π-plane).

The main characteristics of the MSDP$_u$ criterion can be summarized as follows:

- At low mean stress, the criterion reduces to the Mises-Schleicher criterion (for $b = 1$). As the mean stress increases, the MSDP$_u$ surface in the I_1–$J_2^{1/2}$ plane becomes linear with a slope angle α (Figure 1a), similarly to the DP criterion (for $b = 1$).
- The MSDP$_u$ surface has no sharp apex (corner) for negative I_1 values. This curved shape produces a natural tension cut-off that is physically and experimentally more representative than many other criteria. It is also advantageous from a numerical point of view because no extra algorithm operation is needed to handle the corner or apex [2,33–35].
- The rounded surface depends on the Lode angle θ so it can take into account the effect of the loading geometry in the π-plane (again without corners).
- The limiting pressures for the cap onset I_c and closure I_{1n} can be determined explicitly from experimental testing. The cap curvature can also be related to experimental data and then used to define the volumetric strain (through the plastic potential Q).
- The criterion becomes insensitive to the mean stress when $\alpha = 0$ or $\phi = 0°$; unconsolidated-undrained (UU) conditions can thus be simulated.
- The criterion can be applied to a wide variety of geomaterials, from low porosity rocks, to stiff soils and to high porosity materials (e.g., porous rocks, loose soils and backfill); it has also been applied to other types of engineering materials such as metals and ceramics.

Additional details can be found in Aubertin et al. [20,21], Aubertin and Li [12] and Li et al. [23].

3. Implementation of the Multiaxial MSDP$_u$-EPP Model in FLAC3D

The constitutive model formulation in FLAC3D can be expressed in terms of the principal stresses $\sigma_1, \sigma_2, \sigma_3$ and the principal strains $\varepsilon_1, \varepsilon_2, \varepsilon_3$, using procedures described by Itasca [25]. The total strain increment $\Delta\varepsilon_i$ can be divided into elastic $\Delta\varepsilon_i{}^e$ and plastic $\Delta\varepsilon_i{}^p$ strain increments (with i = 1, 2, 3)

$$\Delta\varepsilon_i = \Delta\varepsilon_i^e + \Delta\varepsilon_i^p \tag{14}$$

The principal stress increments associated with the elastic strain increments can then be expressed as follows:

$$\begin{aligned}
\Delta\sigma_1 &= S_1\left(\Delta\varepsilon_1^e, \Delta\varepsilon_2^e, \Delta\varepsilon_3^e\right) = \beta_1\Delta\varepsilon_1^e + \beta_2\left(\Delta\varepsilon_2^e + \Delta\varepsilon_3^e\right) \\
\Delta\sigma_2 &= S_2\left(\Delta\varepsilon_1^e, \Delta\varepsilon_2^e, \Delta\varepsilon_3^e\right) = \beta_1\Delta\varepsilon_2^e + \beta_2\left(\Delta\varepsilon_1^e + \Delta\varepsilon_3^e\right) \\
\Delta\sigma_3 &= S_3\left(\Delta\varepsilon_1^e, \Delta\varepsilon_2^e, \Delta\varepsilon_3^e\right) = \beta_1\Delta\varepsilon_3^e + \beta_2\left(\Delta\varepsilon_1^e + \Delta\varepsilon_2^e\right)
\end{aligned} \tag{15}$$

where S_1, S_2, S_3 are linear functions for the Hooke's law; β_1 and β_2 are material parameters (constants) defined in terms of the isotropic shear modulus G and bulk modulus K,

$$\begin{aligned}
\beta_1 &= K + \tfrac{4}{3}G \\
\beta_2 &= K - \tfrac{2}{3}G
\end{aligned} \tag{16}$$

The plastic strain increment is given by the flow rule

$$\Delta\varepsilon_i^p = \lambda\frac{\partial Q}{\partial\sigma_i} \tag{17}$$

where σ_i is the current stress state component (or initial stress state); λ is plastic coefficient; Q is the plastic potential (function), defined as follows

$$Q = J_2 - \xi F_0^2 F_\pi^2 \tag{18}$$

where coefficient ξ serves to control the plastic deviatoric and volumetric strain components. When ξ = 1, the plastic potential function is the same as the yield criterion (Q = F) so the flow rule is associated. When $\xi \ll 1$ (very small value), the plastic potential leads to quasi-isovolumetric plastic strains, typical of critical state [15,36].

Combining Equations (14) and (17) leads to

$$\Delta\varepsilon_i^e = \Delta\varepsilon_i - \lambda\frac{\partial Q}{\partial\sigma_i} \tag{19}$$

Substituting Equation (19) into Equation (15) gives

$$\Delta\sigma_i = S_i(\Delta\varepsilon_1, \Delta\varepsilon_2, \Delta\varepsilon_3) - \lambda S_i\left(\frac{\partial Q}{\partial\sigma_1}, \frac{\partial Q}{\partial\sigma_2}, \frac{\partial Q}{\partial\sigma_3}\right) \tag{20}$$

Under varying stress conditions, the new stress state $\sigma_i{}^N$ corresponding to the total strain increment $\Delta\varepsilon_i$ is expressed by

$$\sigma_i^N = \sigma_i + \Delta\sigma_i \tag{21}$$

In the user-defined model, the induced (postulated) elastic stresses $\sigma_i{}^I$ are obtained by adding a "trial" stress increment to the current (initial) stress state σ_i. The "trial" stress increment is computed by using the incremental form of Hooke's law and the total strain increment $\Delta\varepsilon_i$. The induced elastic stresses $\sigma_i{}^I$ can be expressed as follows:

$$\sigma_i^I = \sigma_i + S_i(\Delta\varepsilon_1, \Delta\varepsilon_2, \Delta\varepsilon_3) \tag{22}$$

Substituting Equations (20) and (22) into Equation (21) leads to

$$\sigma_i^N = \sigma_i^I - \lambda S_i\left(\frac{\partial Q}{\partial \sigma_1}, \frac{\partial Q}{\partial \sigma_2}, \frac{\partial Q}{\partial \sigma_3}\right) \tag{23}$$

Equation (23) is called the plastic correction in FLAC3D.

The induced elastic stresses are initially taken as the new stress state and then adjusted if required. When the new stress state is within the elastic domain (i.e., $F < 0$), the new stress state is updated by directly using the incremental expression of Hooke's law. When the new stress state would lead to $F > 0$, the plastic coefficient λ is calculated from the MSDP$_u$ yield criterion to bring the new stress state $\sigma_i{}^N$ determined by Equation (23) on the yield surface ($F = 0$), which results in

$$F\left(\sigma_1^N, \sigma_2^N, \sigma_3^N\right) = J_2^N - \left(F_0^N\right)^2\left(F_\pi^N\right)^2 = 0 \tag{24}$$

Substituting Equation (23) into Equation (24) gives the quadratic equation:

$$A\lambda^2 + B\lambda + C = 0 \tag{25}$$

The correct value of λ corresponds to the root with smaller absolute value of the two obtained after solving Equation (25).

Since F_0 includes two pieces and depends on the value of I_1, the expressions of A, B and C in Equation (25) depend on whether (or not) $I_1 \leq I_c$. Hereafter, their expressions are introduced with "s" denoting a shear response (for $I_1 \leq I_c$) and "v" denoting the contribution of a volumetric response related to the cap ($I_1 > I_c$).

When $I_1 \leq I_c$, F_0 becomes

$$F_0^s = \left[\alpha^2\left(I_1^2 - 2a_1 I_1\right) + a_2^2\right]^{1/2} \tag{26}$$

The corresponding plastic potential is then expressed as

$$Q^s = J_2 - \xi(F_0^s)^2 F_\pi^2 \tag{27}$$

The expressions for A^s, B^s and C^s are given in Appendix A.

When $I_1 > I_c$, F_0 is written as

$$F_0^v = \left[\alpha^2\left(I_1^2 - 2a_1 I_1\right) + a_2^2 - a_3(I_1 - I_c)^2\right]^{1/2} \tag{28}$$

and

$$Q^v = J_2 - \xi(F_0^v)^2 F_\pi^2 \tag{29}$$

The expressions for A^v, B^v and C^v are given in Appendix A.

To simplify the calculations of partial derivatives, variation of Lode angle θ is postulated to have a negligible effect on F_π to obtain the new (updated) stress state (from the induced elastic stresses). This simplification leads to the following function in the π-plane:

$$F_\pi^N \approx F_\pi^I = \frac{b}{\left[b^2 + (1 - b^2)\sin^2(45^\circ - 1.5\theta^I)\right]^{1/2}} \tag{30}$$

with

$$\theta^N \approx \theta^I = \frac{1}{3}\sin^{-1}\frac{3\sqrt{3}J_3^I}{2\sqrt{\left(J_2^I\right)^3}} \tag{31}$$

The computational procedure of the MSDP$_u$-EPP model in FLAC3D v6.0 (Itasca, 2017) starts with adding the stress components, which are computed from the incremental

Hooke's law by using the total strain increments to the current (initial) stress state σ_i and the induced elastic stresses σ_i^I are obtained (Equation (22)). Then, the σ_i^I values are substituted into the yield function F to determine if these are in the elastic domain ($F < 0$) or the plastic domain ($F \geq 0$). If in the elastic domain, the new (updated) stress state σ_i^N equals to σ_i^I. If in the plastic domain, the new stress state σ_i^N is updated by applying the plastic correction (Equation (23)) to σ_i^I. The "corrected" new principal stresses σ_1^N, σ_2^N and σ_3^N can then be used to update the stress tensor σ_{ij} in the system of reference axes, assuming that the principal directions have not been affected by the occurrence of a plastic correction.

Figure 2 shows the computational scheme for the implementation of the non-associated MSDP$_u$-EPP model in FLAC3D 6.0 [25]. As indicated above, FLAC3D v6.0 provides the option to load and run user-written model in DLL (dynamic link library). The implementation of the MSDP$_u$-EPP model was hence performed by creating a user-written DLL, which was created by compiling a program written with C++ language in Visual Studio 2017.

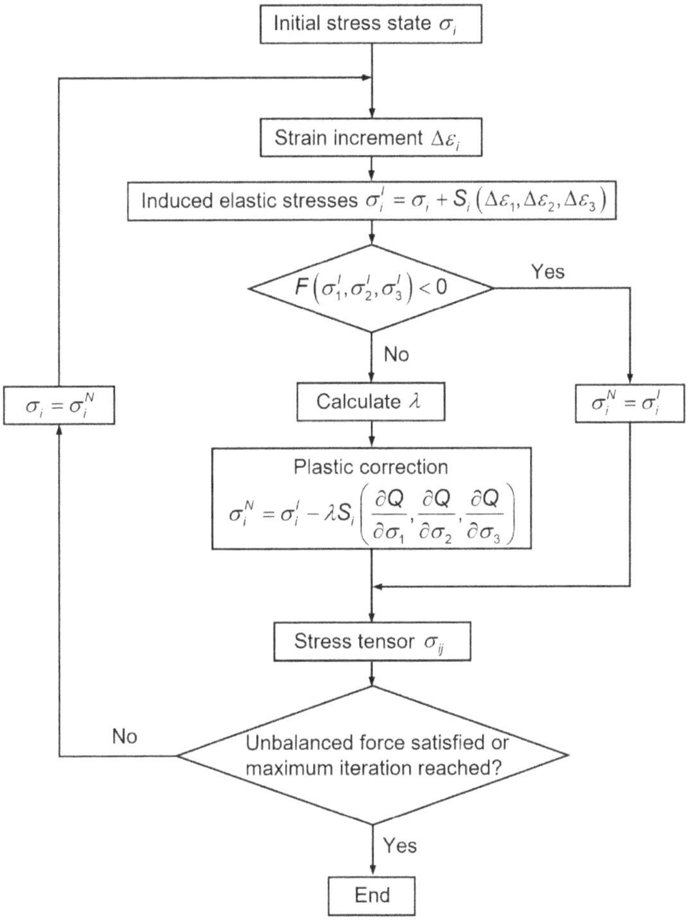

Figure 2. Computational scheme for the implementation of the non-associated MSDP$_u$-EPP model in FLAC3D.

4. Validation of the FLAC3D MSDP$_u$-EPP Model

After its implementation in FLAC3D, preliminary simulations with the user-defined MSDP$_u$-EPP model were conducted to verify that the formulation and code programming were correctly done and to validate (in part) the numerical results. The main results from this assessment are summarized here.

The first validation is made against the analytical solution for the stresses around a cylindrical opening in a $MSDP_u$-EPP material developed by Li et al. [23] for the open $MSDP_u$ surface (without cap) and the solution of Li and Aubertin [37] with the closed $MSDP_u$ yield surface (with cap). It should be noted that the analytical solutions were developed for a constant Lode angle $\theta = 0°$, even though the actual Lode angle tends to vary slightly in the plastic region (between about $-19°$ and $-25°$) along radial coordinate r around the cylindrical opening, while $\theta = 0°$ in the elastic region [23,37].

Figure 3a shows a cylindrical opening in an elastic-perfectly plastic material subjected to a hydrostatic far-field stress P_0; an internal pressure p_0 is applied to the wall of the cylindrical opening (in plane strain condition). In the figure, r_0 is the radius of the opening, R is the radius of the interface between the plastic and elastic regions, σ_r and σ_ψ are the radial and tangential stresses respectively and r and ψ are the corresponding cylindrical coordinates. The calculations are made for $r_0 = 1$ m, $P_0 = 30$ MPa and $p_0 = 2$ MPa. The $MSDP_u$-EPP model parameters are given in Table 1.

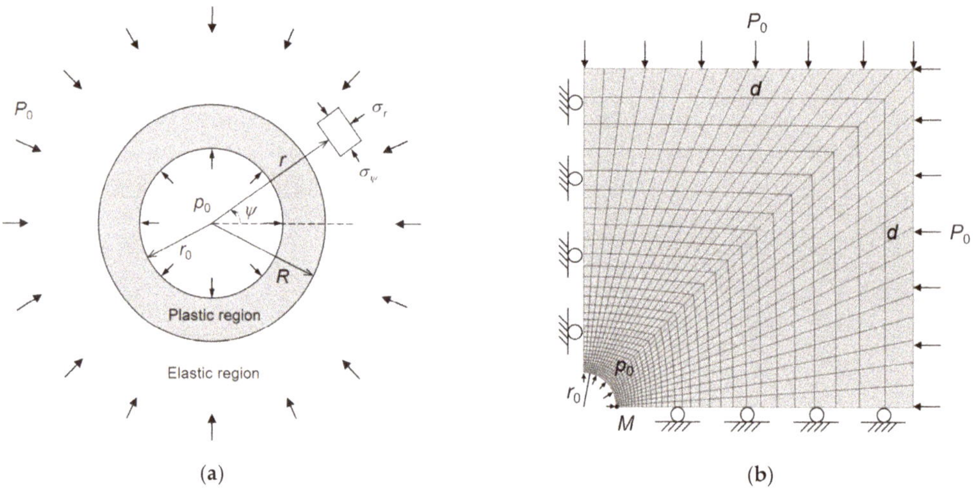

Figure 3. (**a**) A cylindrical opening under far-field isotropic stresses outside and an internal pressure applied to the inside wall, (**b**) a numerical model built with FLAC3D after taking advantage of the symmetry along two axes.

Table 1. Parameters used for comparing the results obtained with the FLAC3D $MSDP_u$-EPP model and analytical solutions (taken from Li and Aubertin [37]).

Parameter	G (GPa)	K (GPa)	C_0 (MPa)	T_0 (MPa)	ϕ (°)	I_c (MPa)	a_3
With cap	20	60	7	0.2	27	10	0.06
No cap	20	60	7	0.2	27	-	0

Figure 3b shows a numerical model with domain size $d \times d$ and the boundary conditions. The model built with FLAC3D takes advantage of the quarter-symmetry geometry. The analytical solutions were developed for a plane-strain condition, so the numerical simulations are conducted in 2D even if the $MSDP_u$-EPP model implemented in FLAC3D is three-dimensional. A commonly used method to do that is to isolate a thin domain in the direction perpendicular to the axis, with the displacements fixed in the direction parallel to the opening axis but allowed in the directions perpendicular to the axis. The thickness t of the modeling domain is taken as one-fifth of the cylinder radius, i.e., $t = 0.2$ m.

A sensitivity analysis was conducted to obtain an optimal configuration of the numerical model, which is based on an optimal domain size and optimal mesh. The optimal domain corresponds to the smallest size of the model to minimize the time of calculation and large enough to ensure stable and reliable numerical results. Similarly, an optimal mesh size m is associated with the coarsest elements (blocs) to minimize the time of calculation, with elements that are fine enough to ensure stable and reliable numerical results.

Figure 4 shows the variation of the radial displacements (Figure 4a) and stresses (Figure 4b) at a reference point M (shown in Figure 2) as a function of the mesh size (for domain size d = 10 m). In this figure, the mesh size is defined from the minimum size of the first layer of the grid around the opening, which increases at a constant ratio from the opening wall to the domain boundary (based on radial meshing; Itasca [25]). The results shown in Figure 4 indicate that, for both versions of the MSDP$_u$-EPP model (with and without the cap), the numerical results tend to become stable once the mesh size is equal to or smaller than 0.02 m. The optimal mesh size corresponds to m = 0.02 m.

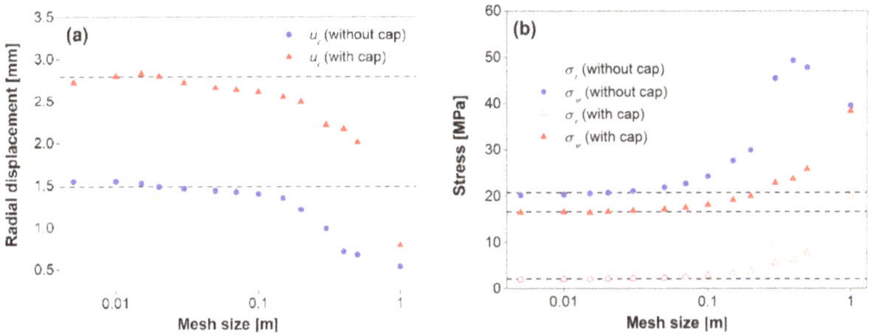

Figure 4. Variation of the (**a**) radial displacements and (**b**) stresses at point M (see Figure 3b) as a function of the mesh size (with d = 10 m), obtained from the MSDP$_u$-EPP model with and without the cap.

Figure 5 shows the variation of the radial displacements (Figure 5a) and stresses (Figure 5b) as a function of the domain size (with m = 0.02 m). It can be seen that d = 10 m is the smallest value, which remains large enough to ensure stable and reliable numerical results. The optimal numerical model is then constructed with m = 0.02 m and d = 10 m, which are used to conduct the numerical simulations.

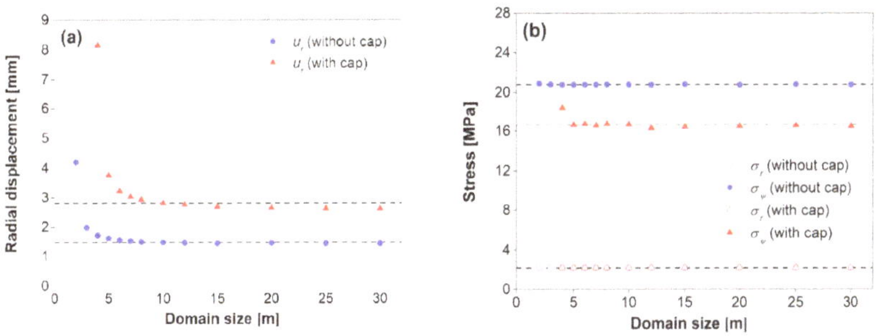

Figure 5. Variation of the (**a**) radial displacements and (**b**) stresses at point M (see Figure 3b) as a function of domain size (with m = 0.02 m) obtained from the MSDP$_u$-EPP model with and without the cap.

Figure 6 shows the radial (σ_r) and tangential (σ_ψ) stress distributions obtained from the numerical modeling and analytical solution for the cases of a closed MSDP$_u$ surface (Figure 6a) and an open MSDP$_u$ surface (Figure 6b). The agreements between the numerical and analytical results are excellent in all cases. Additional calculations were also made for other material parameters and loading conditions (results not shown here) to help assess the validity of the MSDP$_u$-EPP model and its numerical implementation in FLAC3D. The good agreements obtained between the numerical and analytical results support the validation and indicate that the analytical solutions developed by considering a constant Lode angle provide a good approximation of the stress distribution around a cylindrical opening.

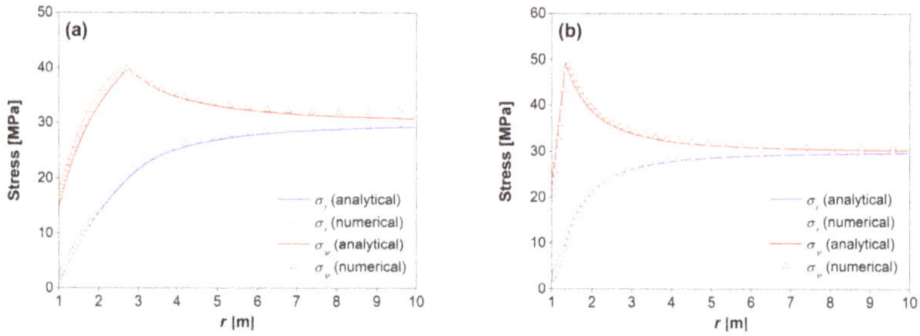

Figure 6. Distributions of the radial and tangential stresses, obtained from the numerical simulations and analytical solutions developed by Li and Aubertin [37] (**a**) for a closed MSDP$_u$ surface (with cap) and (**b**) for an open MSDP$_u$ surface (no cap).

The second component of this validation of the FLAC3D MSDP$_u$-EPP model is performed against the analytical solution of Salençon [38,39], developed for evaluating the distribution of stresses and radial displacements around a cylindrical opening in a MC-EPP material (considering an associated flow rule). Parameter ξ = 1 (associated flow rule) and a_3 = 0 (without cap) are thus taken for the FLAC3D MSDP$_u$-EPP model. As the MSDP$_u$ is nonlinear and MC is linear in the I_1–$J_2^{1/2}$ plane, the MC strength parameters ϕ and c were chosen so the two yield surfaces would correlate to each other (as well as possible) in the stress domain of interest around the cylindrical opening (for 50 MPa $\leq I_1 \leq$ 100 MPa in this case). The resulting MC strength parameters are ϕ = 32° and c = 3.9 MPa for matching the open MSDP$_u$ surface with the parameters given in Table 1. The two yield surfaces in the I_1–$J_2^{1/2}$ plane (for θ = 0°) are very close to each other for the stress states of interest, as shown in Figure 7. A constant Lode angle θ = 0° is again considered an acceptable approximation for the stresses around the cylindrical opening.

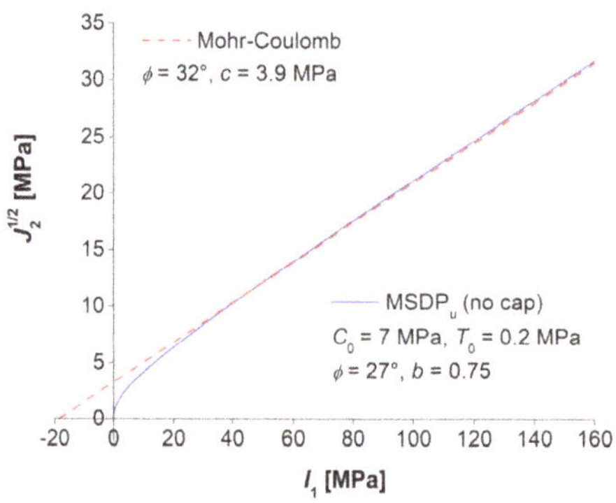

Figure 7. The open MSDP$_u$ yield surface (no cap) and corresponding Mohr-Coulomb surface in I_1–$J_2^{1/2}$ plane (for θ = 0°).

Figure 8 illustrates the distributions of the stresses (Figure 8a) and radial displacements (Figure 8b), obtained from FLAC3D MSDP$_u$-EPP model and calculated with the analytical solutions of Salençon [38,39]. The good agreement between the two types of solutions indicates that the FLAC3D MSDP$_u$-EPP model correctly represents the problem at hand.

This (partial) validation supports the use of the FLAC3D MSDP$_u$-EPP model to analyze the elastoplastic response of geomaterials around underground openings.

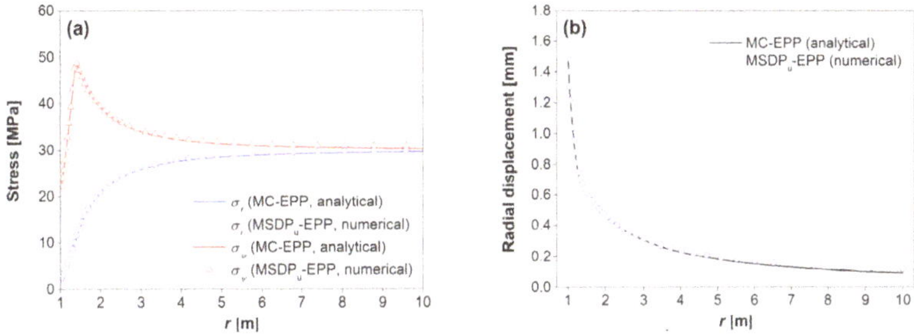

Figure 8. Distributions of the (**a**) stresses and (**b**) radial displacements, obtained from the numerical simulation with FLAC3D MSDP$_u$-EPP model and calculated with the analytical solutions of Salençon [38,39].

5. Applications of the FLAC3D MSDP$_u$-EPP Model

In order to further assess the features of the new FLAC3D MSDP$_u$-EPP model, simulations are conducted to analyze the mechanical response of cemented backfill in a 3D vertical mine stope.

Backfill is commonly used in underground mine excavations to ensure safe working conditions and improve ore recovery. Underground backfilling is also gaining momentum as a mine waste management approach [40]. Evaluating the stresses in backfilled stopes is of great importance and interest for underground mine stability. Until recently however, most of the numerical analyses were done under plane strain (2D) conditions. Only a few investigations have included 3D numerical analyses with the Mohr-Coulomb EPP model [27–29,41–43].

Figure 9 shows the conceptual model of a three-dimensional vertical mine stope in a semi-infinite rock mass before (Figure 9a) and after (Figure 9b) the addition of backfill. This stope is located at a depth of 500 m below the ground surface. It is excavated to a height of 45 m; the width is 6 m in the two horizontal directions (Figure 9a). After excavation, the stope is filled by a cemented (cohesive) backfill to a final height of 44.5 m, leaving a void space of 0.5 m at the top (Figure 9b). The stress distribution in the backfill and surround rock mass then depends on the fill settlement under its own weight and interaction with the rock walls. The stability of the rock walls of the empty stope is first evaluated using both MSDP$_u$-EPP and MC-EPP models, respectively.

Figure 10 shows the corresponding numerical model of the stope, built by FLAC3D after excavation, before backfilling. Half of the stope is modelled to take advantage of the symmetric geometry. As indicated in the figure, the boundary conditions applied to the model domain are defined with a free surface on the top boundary and a fixed surface at the base. The conditions imposed along the four external vertical surfaces allow displacements within their respective plane, but not in the perpendicular direction. The rock mass is considered homogeneous, isotropic and perfectly elastoplastic; it is characterized by the following properties: E_r = 30 GPa (Young's modulus), ν_r = 0.3 (Poisson's ratio), γ_r = 27 kN/m^3 (unit weight). These parameters are used for both of the MC-EPP (with a tension cut-off at T_0) and MSDP$_u$-EPP models. The strength parameters are ϕ = 32°, c = 3.9 MPa and T_0 = 0.2 MPa for the MC criterion and ϕ = 27°, C_0 = 7 MPa, T_0 = 0.2 MPa and a_3 = 0 for the open MSDP$_u$ criterion. The rock mass is subjected to three-dimensional in-situ stresses, including the vertical in-situ stress σ_v (z-direction) generated by the overburden (i.e., $\sigma_v = \gamma_r h$; h is the depth below ground surface), maximum horizontal in-situ stress in the x-direction σ_H taken as two times the vertical in-situ stress (i.e., $\sigma_H = 2\sigma_z$) and minimum horizontal in-situ stress in the y-direction σ_h (varying values in the different simulations).

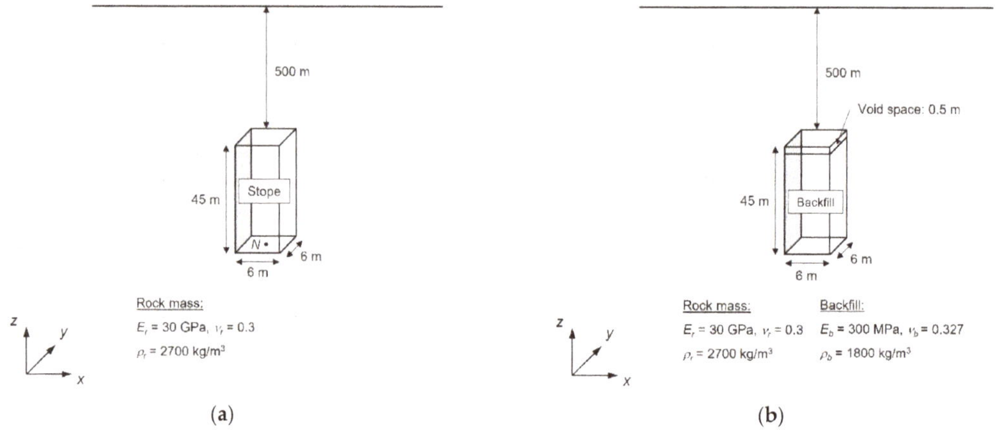

Figure 9. Conceptual model of a three-dimensional vertical stope in rock mass (**a**) before and (**b**) after being backfilled.

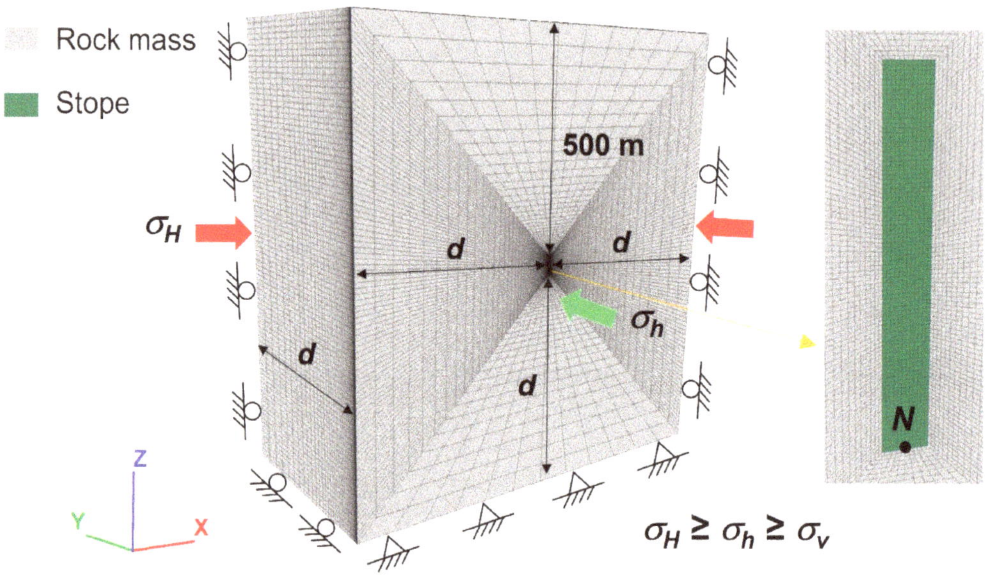

Figure 10. A three-dimensional numerical model of the vertical stope in rock mass built by FLAC3D.

Sensitivity analyses are performed to ensure stable and reliable numerical results and obtain optimal numerical model for each case. For a numerical model such as the one shown in Figure 10, the sensitivity analysis gives the optimal mesh m and optimal domain size d.

The procedure is illustrated in Figure 11, which shows the variation of vertical displacement (z-direction) and horizontal stress σ_{xx} at a reference point N (at the center of the stope base, see Figs. 9a and 10) as a function of the mesh size (with d = 500 m) for simulations conducted with the MSDP$_u$-EPP model, for an in-situ stress state $\sigma_h = \sigma_H = 2\sigma_z$. The results indicate that stable numerical results are achieved with m = 0.5 m. Figure 12 shows the variation of the vertical displacement (z-direction) and σ_{xx} as a function of the domain size (with m = 0.5 m). It is seen that the numerical results become stable when the domain size d is equal to or larger than 500 m. The results thus indicate that the optimal numerical model has a mesh size of 0.5 m and a domain size of 500 m.

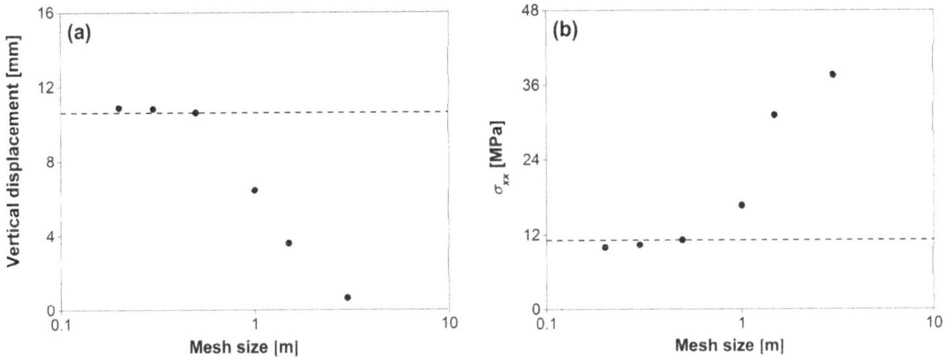

Figure 11. Variation of the (**a**) vertical displacement (z-direction) and (**b**) horizontal stress σ_{xx} at point N (see Figures 9a and 10) as a function of the mesh size m (for d = 500 m).

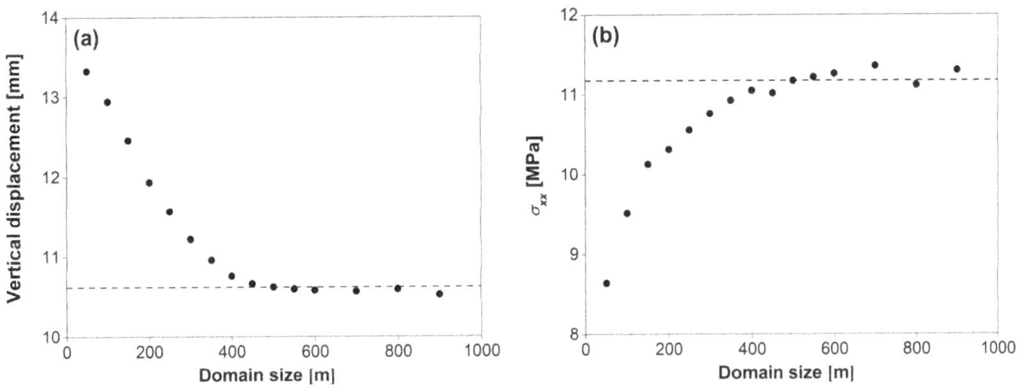

Figure 12. Variation of the (**a**) vertical displacement (z-direction) and (**b**) horizontal stress σ_{xx} at point N (see Figures 9a and 10) as a function of the domain size d (for m = 0.5 m).

Figure 13 shows the yielded areas around the stope along a vertical cross cut section in the xz symmetric plane, obtained from the numerical simulations conducted with the MSDP$_u$-EPP and MC-EPP models, respectively. The results have been obtained for σ_h equals to 1, 1.4 and 2 times σ_v, while σ_v and σ_H are kept unchanged. It can be seen that the size of yielded area obtained with the MSDP$_u$-EPP model significantly increases as the intermediate in-situ stress σ_h increases from σ_v to $2\sigma_v$, indicating that the intermediate in-situ stress plays an important role in the response and stability of openings. With the MC-EPP model however, the yielded area stays almost unchanged when σ_h increases from σ_v to $2\sigma_v$ because the MC criterion is based on the 2D formulation and thus fails to consider explicitly the effect of the intermediate principal stress in the 3D extension used in FLAC3D.

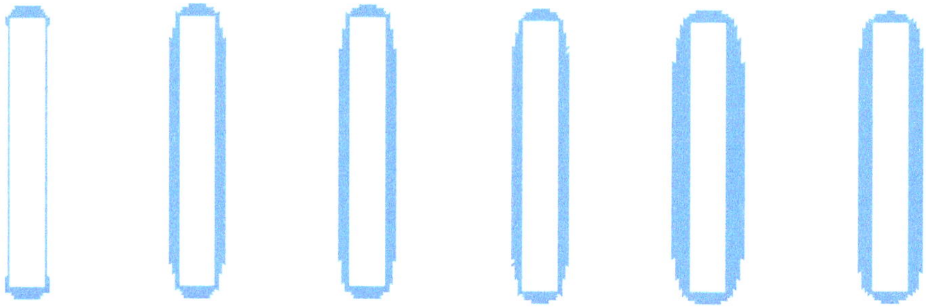

Figure 13. Yielded areas around the stope along a vertical cross cut section in the xz symmetric plane, obtained from the numerical simulations with the MSDP$_u$-EPP and MC-EPP models, respectively, considering (**a**) $\sigma_h = \sigma_v$, (**b**) $\sigma_h = 1.4\sigma_v$ and (**c**) $\sigma_h = 2\sigma_v$.

Figure 14 shows the corresponding numerical model after the placement of backfill in the stope, built with FLAC3D. As was the case for the numerical model shown in Figure 10, half of the stope is modelled to take advantage of the symmetric geometry. The boundary conditions are defined by a free top boundary surface, a fixed bottom boundary surface and four vertical external surfaces whose displacements are allowed within their respective plane, but not allowed in the direction perpendicular to that plane. The stope backfilling is performed in multiple layers after the convergence of the rock walls has been completed. The thickness of each layer is 5 m (except for the top layer having a thickness of 4.5 m). A mesh size m of 0.5 m is used for the backfill; the domain size d = 500 m, determined for the rock wall stability analysis shown in Figure 10, is also used here. As the backfill is placed in the open stope after all the elastic and plastic strains have been released, the mechanical response of the backfill is almost independent on the in-situ stresses and rock model. The rock mass is thus considered homogeneous, isotropic and linearly elastic (without yield), characterized by E_r = 30 GPa, ν_r = 0.3 and γ_r = 27 kN/m^3. The vertical in-situ stress σ_v is generated by the overburden and the horizontal in-situ stress is isotropic, with $\sigma_h = \sigma_H = 2\sigma_z$.

The stresses in the backfilled stope are evaluated using the MSDP$_u$-EPP and MC-EPP models. The weakly cemented backfill is characterized by E_b = 300 MPa (Young's modulus), ν_b = 0.327 (Poisson's ratio), ρ_b = 1800 kg/m^3 (density). The MSDP$_u$ criterion is defined by ϕ = 30°, C_0 = 10 kPa, T_0 = 0.2 kPa, b = 0.75, I_c = 100 kPa and a_3 = 0.06. A value of ξ = 0.01 is used to express the plastic potential Q for the non-associated flow rule to represent the quasi-isovolumetric plastic strain of the backfill following settlement and mobilization of the frictional stress along the vertical walls. The MC yield parameters were selected so the surface would be close to the MSDP$_u$ yield surface for comparative purposes of the stresses obtained from numerical simulations; the corresponding MC strength parameters are ϕ; = 31°, c = 3.8 kPa and T_0 = 0.2 kPa. A non-associated flow rule was imposed with ϕ_d = 0° (dilation angle). It is noted that the values of the Poisson's ratio ν_b and internal friction angle ϕ are interrelated through $\nu_b = (1 - \sin\phi)/(2 - \sin\phi)$ to ensure a consistent at-rest earth pressure coefficient K_0 [28,29,44–46]. This results in a value of Poisson's ratio $\nu_b = (1 - \sin\phi) / (2 - \sin\phi) = (1 - \sin31°) / (2 - \sin31°) = 0.327$ and a Jaky's [47] earth pressure coefficient at rest $K_0 = 1 - \sin\phi = 1 - \sin31° = 0.485$; the same value of K_0 is obtained from the equation based on Poisson's ratio. The theoretical horizontal stress due to the overburden can then be calculated as $\sigma_{xx} = K_0\sigma_{zz}$.

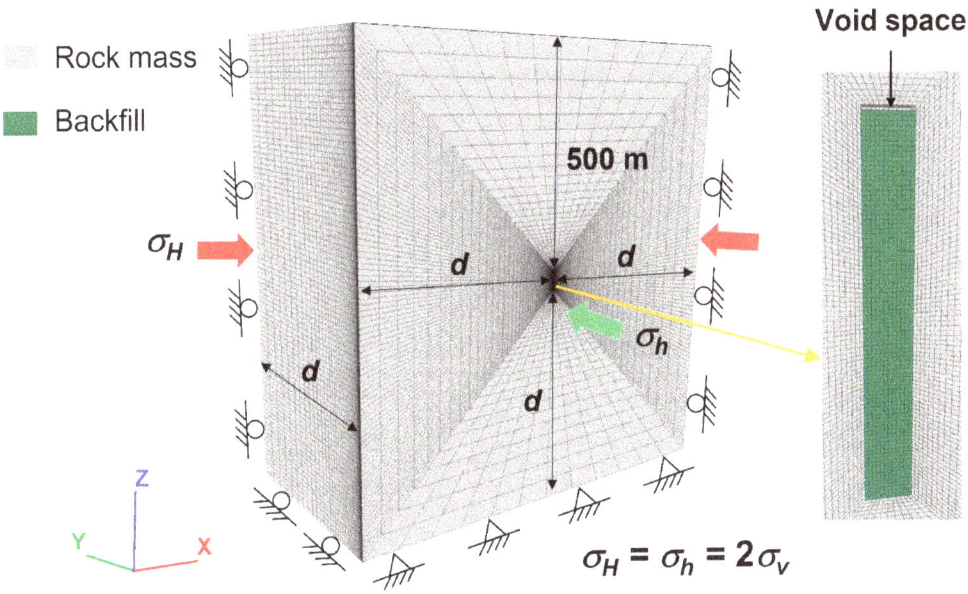

Figure 14. The numerical FLAC3D model of the three-dimensional vertical backfilled stope in an elastic rock mass.

Figure 15 shows the vertical (σ_{zz}) and horizontal (σ_{xx}) stress distributions along the centerline (CL) of the backfilled stope obtained with the two models. It is seen that the general tendencies of the stress distributions obtained by numerical modeling with the MSDP$_u$-EPP model and MC-EPP model are similar. The vertical and horizontal stresses follow the overburden stresses (marked as dashed lines in Figure 15) at shallow depth, but the overburden stresses significantly exceed the stresses in the backfill deeper in the opening due to the well-known arching effect associated with the frictional stress transfer along the rock walls.

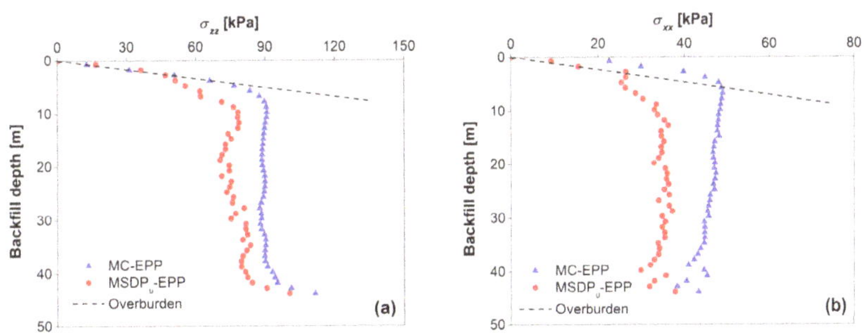

Figure 15. Stress distributions along the centerline (CL) of the backfilled stope with the MSDP$_u$-EPP model and the MC-EPP model obtained by FLAC3D for (**a**) vertical stress σ_{zz} and (**b**) horizontal stress σ_{xx}.

The trends obtained here are consistent with those given by previous numerical simulations and also by analytical solutions developed with the MC-EPP model [27,29,45,48,49]. The results shown here nonetheless indicate that the vertical and horizontal stresses obtained by the MSDP$_u$-EPP model are smaller than those obtained by the MC-EPP model, due to the differences in the MSDP$_u$ and MC yield surfaces. Representative field experimental data are necessary to evaluate which one is more representative of the actual stress state.

One of the specific features of the MSDP$_u$-EPP model is the introduction of a cap on the yield surface, which starts at the key parameter I_c (not included in other criteria and models). As indicated above, the value of I_c controls the cap position on the yield surface, which then departs from the (quasi) linear strength increase given by the most other models. Additional simulations were conducted to investigate the effect of I_c on the stress distributions in the backfilled stope. Three values (I_c = 20 kPa, 30 kPa, 100 kPa) were used with the MSDP$_u$-EPP model, while the other parameters were kept constant for all cases (E_b = 300 MPa, ν_b = 0.327, ρ_b = 1800 kg/m^3, ϕ = 30°, C_0 = 10 kPa, T_0 = 0.2 kPa, b = 0.75, a_3 = 0.06 and ξ = 0.01).

Figure 16 shows the vertical (σ_{zz}) and horizontal (σ_{xx}) stress distributions along the CL obtained for different I_c. The results indicate that the vertical and horizontal stresses along the CL decrease with an increasing I_c, due to the higher shear strength (larger yield surface) of the backfill. The results also indicate that the arching effect tends to become more significant for stronger cemented (cohesive) backfill. These calculations highlight the importance of including the effect of the cap through I_c, which is a unique characteristic of the MSDP$_u$-EPP model.

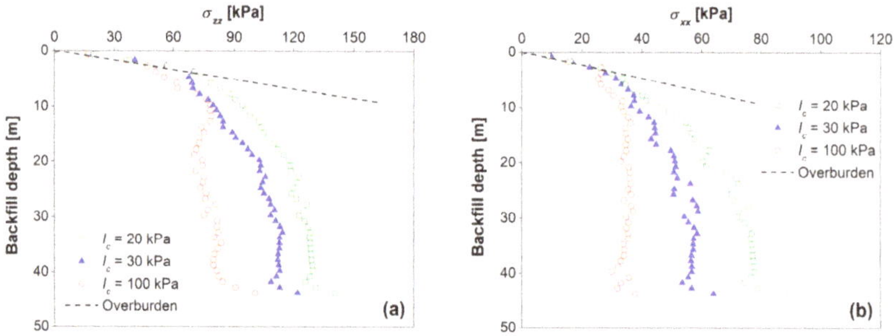

Figure 16. Stress distributions along the CL of the backfilled stope obtained from the numerical simulation conducted with the MSDP$_u$-EPP model by considering different I_c for (**a**) vertical stress σ_{zz} and (**b**) horizontal stress σ_{xx}.

To further illustrate the features of the FLAC3D MSDP$_u$-EPP model with associated and non-associated flow rules, the stresses in the backfilled stope shown in Figures 9b and 14 are analyzed by considering ξ = 1 for associated flow rule.

Figure 17 shows the vertical (σ_{zz}) and horizontal (σ_{xx}) stress distributions along the CL of the backfilled stope obtained by numerical modeling with the FLAC3D MSDP$_u$-EPP model by considering ξ = 1 (associated flow) and ξ = 0.01 (non-associated flow), respectively. It can be seen that the vertical stress along the CL is much lower for the associated flow rule than that for the non-associated flow rule (Figure 17a), due to the significant volumetric strains (dilatancy) which tend to increase the arching effect. For the same reason, the horizontal stresses along the CL for the associated flow rule significantly exceed the overburden horizontal stress at shallow depth between around 0 to 5 m (Figure 17b). Similar results have been shown by Li and Aubertin [27] with the MC-EPP model for 2D backfilled stopes. Figure 18 shows the vertical stress distribution along the CL of the 3D backfilled stope (Figures 9b and 14) obtained by numerical modeling with the MC-EPP model by considering a non-associated (dilation angle ϕ_d = 0°) and associated (ϕ_d = ϕ = 31°) flow rules. The results suggest once again that the numerical modeling with an associated flow rule tends to lead an underestimation of the stresses in backfilled stopes.

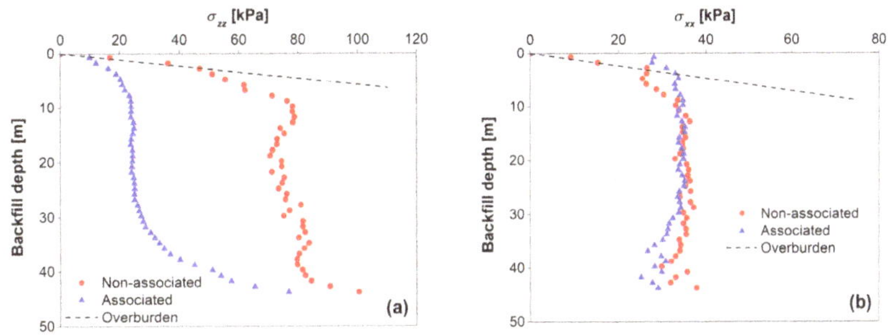

Figure 17. Stress distributions along the CL of the backfilled stope obtained from the numerical simulation conducted with the MSDP$_u$-EPP model by considering the non-associated ($\xi = 0.01$) and associated ($\xi = 1$) flow rules for (**a**) vertical stress σ_{zz} and (**b**) horizontal stress σ_{xx}.

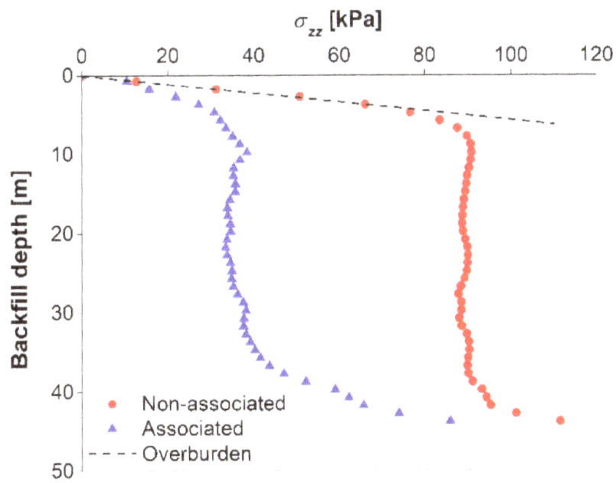

Figure 18. Vertical stress σ_{zz} distributions along the CL of the backfilled stope obtained from numerical simulation conducted with the MC-EPP model by considering the non-associated (dilation angle $\phi_d = 0°$) and associated ($\phi_d = \phi = 31°$) flow rules; other material parameters are $\phi = 31°$, $c = 3.8$ kPa and $T_0 = 0.2$ kPa.

6. Discussion

The non-associated MSDP$_u$-EPP model recently implemented in FLAC3D is applied here for the analysis of underground backfilled stopes, to illustrate some of the features of this three-dimensional nonlinear model. The MSDP$_u$ criterion used here was previously shown to capture the essential characteristics of yielding and failure of a large variety of geomaterials, including soils, rocks and rockfills. The MSDP$_u$-EPP model uses a closed yield surface with a controllable cap at high mean stress. It employs a non-associated flow rule to better describe the deviatoric and volumetric plastic strains. The availability of the FLAC3D MSDP$_u$-EPP model will favor its application to help solve various geotechnical engineering problems such as those related to the behavior of slopes, tunnels, dams and barricades in underground mines.

Despite the various advantages of the model, there are a few limitations with the FLAC3D MSDP$_u$-EPP model. For instance, the model has not yet been applied to cohesionless (granular) materials such as sand, rockfill, waste rock and uncemented backfill. This can be done by expressing the MSDP$_u$ criterion with $C_0 = T_0 = 0$ in Equations (10) and (11), which leads to $a_1 = a_2 = 0$. Equation (7) then becomes:

$$F_0 = \left[\alpha^2 I_1^2 - a_3 \langle I_1 - I_c \rangle^2\right]^{1/2} \tag{32}$$

This specific version of the criterion has not yet been validated in the context of the FLAC3D MSDP$_u$-EPP model.

Another limitation of the model, common to all perfectly plastic models, is the absence of strain hardening and softening in the formulation. This is acceptable for many types of application, but not for some specific ones [3,4,8]. Additional work is considered for an evolving yield surface in the principal stress space with the MSDP$_u$-EPP model.

Also, the model has not yet been validated for coupled problems involving pore water pressures (particularly for transient conditions).

Despite such limitations, the FLAC3D MSDP$_u$-EPP model offers a powerful alternative to simulate the complex mechanical response of geomaterials as a built-in model in FLAC3D (and other commercial codes).

7. Conclusions

The non-associated MSDP$_u$ elastic-perfectly plastic constitutive model (MSDP$_u$-EPP model) has been implemented in FLAC3D with a user-written DLL under the assistance of the C++ plug-in option. The resulting FLAC3D MSDP$_u$-EPP model is validated, in part, by comparing simulation results with existing analytical solutions developed for evaluating the stress and displacement distributions around cylindrical openings. The FLAC3D MSDP$_u$-EPP model is then used to analyze a specific three-dimensional geotechnical problem, taking into account the nonlinear stress-strain behavior, yielding and volumetric strains associated with the cap surface and considering an associated or non-associated flow rule. Application of the FLAC3D MSDP$_u$-EPP model is illustrated with simulations conducted to analyze the stability of a 3D open stope and stress distribution in the backfilled stope. The results were compared with those obtained with the commonly used Mohr-Coulomb (MC-EPP) model. The numerical results obtained with the MSDP$_u$-EPP model indicate that the size of yielded area around the empty stope significantly increases as the intermediate in-situ stress increases, indicating that the intermediate in-situ stress plays an important role in the response and stability of openings. This effect is not as clearly perceived with the MC-EPP model, however, as the simulated yielded area stays almost unchanged when the intermediate in-situ stresses are increased, hence showing that this model largely neglects this aspect. Other particular features of the FLAC3D MSDP$_u$-EPP model related to the closed yield envelope (with a cap) are shown with the stress analysis of backfilled stopes. These results show that when a higher mean stress is necessary to reach backfill volumetric yield surface, defined by the closed (cap) envelope, the backfill becomes stronger, leading to more pronounced arching effect and smaller stresses in the backfilled stope. This is a unique characteristic obtained with numerical simulations conducted with the FLAC3D MSDP$_u$-EPP model. It can thus be considered as a very useful numerical tool to analyze and solve geotechnical engineering problems. Nevertheless, improvements are still being considered for the model to minimize its limitations. For example, the model has not yet been applied to cohesionless (granular) materials; strain hardening and softening are not included in the formulation; the model has not been validated for couple problems involving pore water pressures. Work is underway to address these specific issues.

Author Contributions: F.Z.: formulation, programming, numerical modeling, literature and writing of the original draft. L.L.: project administration, supervision and editing of the original draft. M.A.: co-supervision and editing of the original draft. R.S.: co-supervision and editing of the original draft. All authors have read and agreed to the published version of the manuscript.

Funding: This work was financially supported by the Natural Sciences and Engineering Research Council of Canada (RGPIN89749-04, RGPIN-2018-06902, ALLRP-566888-21), Fonds de recherche du Québec-Nature et Technologies (2017-MI-202860), Mitacs (IT12569) and industrial partners of the Research Institute on Mines and the Environment (RIME UQAT-Polytechnique; http://rime-irme.ca/ accessed on 1 May 2022). The authors are grateful for their support.

Institutional Review Board Statement: Not applicable.

Informed Consent Statement: Not applicable.

Data Availability Statement: Data are contained within the article.

Acknowledgments: The authors acknowledge the financial support from the Natural Sciences and Engineering Research Council of Canada, Fonds de recherche du Québec-Nature et Technologies, Mitacs and industrial partners of the Research Institute on Mines and the Environment.

Conflicts of Interest: The authors declare no conflict of interest.

Appendix A. Expressions for A^S, B^S and C^S

When $I_1 \leq I_c$, the expressions for A, B and C in Equation (25) are suffixed with a "s" to denote a shear response and expressed as follows:

$$\begin{aligned} A^s = &(\beta_1-\beta_2)^2\left[\left(\frac{\partial Q^s}{\partial\sigma_1}-\frac{\partial Q^s}{\partial\sigma_2}\right)^2+\left(\frac{\partial Q^s}{\partial\sigma_2}-\frac{\partial Q^s}{\partial\sigma_3}\right)^2+\left(\frac{\partial Q^s}{\partial\sigma_3}-\frac{\partial Q^s}{\partial\sigma_1}\right)^2\right] \\ &-6\alpha^2(\beta_1+2\beta_2)^2\left(F_\pi^I\right)^2\left(\frac{\partial Q^s}{\partial\sigma_1}+\frac{\partial Q^s}{\partial\sigma_2}+\frac{\partial Q^s}{\partial\sigma_3}\right)^2 \end{aligned} \tag{A1}$$

$$\begin{aligned} B^s = &12\alpha^2(\beta_1+2\beta_2)\left(I_1^I-2a_1\right)\left(F_\pi^I\right)^2\left(\frac{\partial Q^s}{\partial\sigma_1}+\frac{\partial Q^s}{\partial\sigma_2}+\frac{\partial Q^s}{\partial\sigma_3}\right) \\ &-2(\beta_1-\beta_2)\left[\left(2\sigma_1^I-\sigma_2^I-\sigma_3^I\right)\frac{\partial Q^s}{\partial\sigma_1}+\left(2\sigma_2^I-\sigma_3^I-\sigma_1^I\right)\frac{\partial Q^s}{\partial\sigma_2}+\left(2\sigma_3^I-\sigma_1^I-\sigma_2^I\right)\frac{\partial Q^s}{\partial\sigma_3}\right] \end{aligned} \tag{A2}$$

$$C^s = J_2^I-\left(F_0^{sI}\right)^2\left(F_\pi^I\right)^2 \tag{A3}$$

where

$$\begin{aligned} \frac{\partial Q^s}{\partial\sigma_1} &= \frac{1}{3}(2\sigma_1-\sigma_2-\sigma_3)-2\xi F_\pi^2\left[\alpha^2(I_1-a_1)+\frac{3}{4}\left(\frac{1}{b^2}-1\right)\left(F_0^s\right)^2F_\pi^2\cos 3\theta\frac{\partial\theta}{\partial\sigma_1}\right] \\ \frac{\partial Q^s}{\partial\sigma_2} &= \frac{1}{3}(2\sigma_2-\sigma_1-\sigma_3)-2\xi F_\pi^2\left[\alpha^2(I_1-a_1)+\frac{3}{4}\left(\frac{1}{b^2}-1\right)\left(F_0^s\right)^2F_\pi^2\cos 3\theta\frac{\partial\theta}{\partial\sigma_2}\right] \\ \frac{\partial Q^s}{\partial\sigma_3} &= \frac{1}{3}(2\sigma_3-\sigma_1-\sigma_2)-2\xi F_\pi^2\left[\alpha^2(I_1-a_1)+\frac{3}{4}\left(\frac{1}{b^2}-1\right)\left(F_0^s\right)^2F_\pi^2\cos 3\theta\frac{\partial\theta}{\partial\sigma_3}\right] \end{aligned} \tag{A4}$$

and

$$\begin{aligned} \frac{\partial\theta}{\partial\sigma_1} &= \frac{\sqrt{3}(\sigma_1-\sigma_2)(\sigma_1-\sigma_3)(\sigma_2-\sigma_3)^2}{6J_2\sqrt{4J_2^3-27J_3^2}} \\ \frac{\partial\theta}{\partial\sigma_2} &= \frac{\sqrt{3}(\sigma_2-\sigma_3)(\sigma_2-\sigma_1)(\sigma_1-\sigma_3)^2}{6J_2\sqrt{4J_2^3-27J_3^2}} \\ \frac{\partial\theta}{\partial\sigma_3} &= \frac{\sqrt{3}(\sigma_3-\sigma_1)(\sigma_3-\sigma_2)(\sigma_1-\sigma_2)^2}{6J_2\sqrt{4J_2^3-27J_3^2}} \end{aligned} \tag{A5}$$

Appendix B. Expressions for A^v, B^v and C^v

When $I_1 > I_c$, the expressions for A, B and C in Equation (25) are suffixed with a "v" to denote volumetric response and expressed as follows:

$$\begin{aligned} A^v = &(\beta_1-\beta_2)^2\left[\left(\frac{\partial Q^v}{\partial\sigma_1}-\frac{\partial Q^v}{\partial\sigma_2}\right)^2+\left(\frac{\partial Q^v}{\partial\sigma_2}-\frac{\partial Q^v}{\partial\sigma_3}\right)^2+\left(\frac{\partial Q^v}{\partial\sigma_3}-\frac{\partial Q^v}{\partial\sigma_1}\right)^2\right] \\ &-6\left(\alpha^2-a_3\right)(\beta_1+2\beta_2)^2\left(F_\pi^I\right)^2\left(\frac{\partial Q^v}{\partial\sigma_1}+\frac{\partial Q^v}{\partial\sigma_2}+\frac{\partial Q^v}{\partial\sigma_3}\right)^2 \end{aligned} \tag{A6}$$

$$\begin{aligned} B^v = &12(\beta_1+2\beta_2)\left[\alpha^2\left(I_1^I-a_1\right)-a_3\left(I_1^I-I_c\right)\right]\left(F_\pi^I\right)^2\left(\frac{\partial Q^v}{\partial\sigma_1}+\frac{\partial Q^v}{\partial\sigma_2}+\frac{\partial Q^v}{\partial\sigma_3}\right) \\ &-2(\beta_1-\beta_2)\left[\left(2\sigma_1^I-\sigma_2^I-\sigma_3^I\right)\frac{\partial Q^v}{\partial\sigma_1}+\left(2\sigma_2^I-\sigma_3^I-\sigma_1^I\right)\frac{\partial Q^v}{\partial\sigma_2}+\left(2\sigma_3^I-\sigma_1^I-\sigma_2^I\right)\frac{\partial Q^v}{\partial\sigma_3}\right] \end{aligned} \tag{A7}$$

$$C^v = J_2^I-\left(F_0^{vI}\right)^2\left(F_\pi^I\right)^2 \tag{A8}$$

where

$$\begin{aligned} \frac{\partial Q^v}{\partial\sigma_1} &= \frac{1}{3}(2\sigma_1-\sigma_2-\sigma_3)-2\xi F_\pi^2\left[\alpha^2(I_1-a_1)-a_3(I_1-I_c)+\frac{3}{4}\left(\frac{1}{b^2}-1\right)\left(F_0^v\right)^2F_\pi^2\cos 3\theta\frac{\partial\theta}{\partial\sigma_1}\right] \\ \frac{\partial Q^v}{\partial\sigma_2} &= \frac{1}{3}(2\sigma_2-\sigma_1-\sigma_3)-2\xi F_\pi^2\left[\alpha^2(I_1-a_1)-a_3(I_1-I_c)+\frac{3}{4}\left(\frac{1}{b^2}-1\right)\left(F_0^v\right)^2F_\pi^2\cos 3\theta\frac{\partial\theta}{\partial\sigma_2}\right] \\ \frac{\partial Q^v}{\partial\sigma_3} &= \frac{1}{3}(2\sigma_3-\sigma_1-\sigma_2)-2\xi F_\pi^2\left[\alpha^2(I_1-a_1)-a_3(I_1-I_c)+\frac{3}{4}\left(\frac{1}{b^2}-1\right)\left(F_0^v\right)^2F_\pi^2\cos 3\theta\frac{\partial\theta}{\partial\sigma_3}\right] \end{aligned} \tag{A9}$$

Notations:

The following symbols are used in the paper:

A, B, C	coefficients of the quadratic equation for solving λ
A^s, B^s, C^s	coefficients of the quadratic equation for solving λ ($I_1 \leq I_c$)
A^v, B^v, C^v	coefficients of the quadratic equation for solving λ ($I_1 > I_c$)
a_1, a_2, a_3, α	material parameters for $MSDP_u$ criterion
b	material parameter associated with shape function
c	cohesion
C_0	uniaxial compressive strength
d	domain size
E_r, E_b	Young's modulus for rock mass and backfill
F	failure function or yield function
F_0	function of mean stress dependence
F_0^s, F_0^v	functions of mean stress dependence ($I_1 \leq I_c$ and $I_1 > I_c$)
F_0^I, F_0^N	functions of mean stress dependence in terms of elastic guess and new stress state
F_0^{sI}, F_0^{vI}	functions of mean stress dependence in terms of elastic guess ($I_1 \leq I_c$ and $I_1 > I_c$)
F_π	shape function for π-plane
F_π^I, F_π^N	shape functions in terms of elastic guess and new stress state
G	shear modulus
h	depth below ground surface
I_1	first invariant of stress state
$I_1{}^I$	first invariant of stress state in terms of elastic guess
I_c	I_1 value where the cap starts
I_{1n}	I_1 value where the cap intersects the hydrostatic axis ($\sigma_1 = \sigma_2 = \sigma_3$) in the principal stress space
J_2	second invariant of deviator stress
$J_2{}^I, J_2{}^N$	second invariants of deviator stress in terms of elastic guess and new stress state
J_3	third invariant of deviator stress
$J_3{}^I$	third invariant of deviator stress in terms of elastic guess
K	bulk modulus
K_0	at-rest earth pressure coefficient
M, N	reference points
m	mesh size
P_0	hydrostatical far-field stress
p	mean stress
p_0	internal pressure in the cylindrical opening
Q	plastic potential function
Q^s, Q^v	plastic potential functions ($I_1 \leq I_c$ and $I_1 > I_c$)
R	radius at the interface between plastic and elastic region
r, ψ	cylindrical coordinates
r_0	radius of the cylindrical opening
S_1, S_2, S_3	linear functions for the Hooke's law
S_{ij}	deviator stress tensor component
T_0	uniaxial tensile strength
t	domain thickness
u_r	radial displacement
β_1, β_2	elastic constants associated with K and G
$\varepsilon_1, \varepsilon_2, \varepsilon_3$	principal strains
$\Delta\varepsilon_i$	total principal strain increment
$\Delta\varepsilon_i{}^e$	principal elastic strain increment
$\Delta\varepsilon_i{}^p$	principal plastic strain increment
ϕ	internal friction angle
ϕ_d	dilation angle
λ	plastic coefficient
ν_r, ν_b	Poisson's ratio for rock mass and backfill
θ	Lode angle

θ^I, θ^N	Lode angles in terms of elastic guess and new stress state
ρ_r, ρ_b	density for rock mass and backfill
$\sigma_1, \sigma_2, \sigma_3$	principal stresses
$\sigma_1^*, \sigma_2^*, \sigma_3^*$	projections of the principal stress axes in π-plane
$\sigma_1^I, \sigma_2^I, \sigma_3^I$	principal stresses in terms of elastic guess
$\sigma_1^N, \sigma_2^N, \sigma_3^N$	principal stresses in terms of new stress state
σ_H, σ_h	Maximum and minimum horizontal in-situ stresses
σ_v	vertical in-situ stress
σ_{ij}	stress tensor component
$\Delta\sigma_i$	principal stress increment
σ_{xx}	horizontal stress
σ_{zz}	vertical stress
σ_r	radial stress
σ_ψ	tangential stress
ξ	coefficient in plastic potential for non-associated flow rule

References

1. Desai, C.S.; Siwardane, H.J. *Constitutive Laws for Engineering Materials with Emphasis on Geologic Materials*; Prentice-Hall: Englewood Cliffs, NJ, USA, 1984.
2. Chen, W.F.; Baladi, G.Y. *Soil Plasticity: Theory and Implementation*; Elsevier Science: New York, NY, USA, 1985.
3. Potts, D.M.; Zdravkovic, L. *Finite Element Analysis in Geotechnical Engineering. Theory*; Thomas Telford: London, UK, 1999.
4. Potts, D.M.; Zdravkovic, L. *Finite Element Analysis in Geotechnical Engineering. Application*; Thomas Telford: London, UK, 2001.
5. Muir Wood, D. *Geotechnical Modelling*; CRC Press: Boca Raton, FL, USA, 2004.
6. Lade, P.V. Overview of constitutive model for soils. In *Soil Constitutive Models: Evaluation, Selection, and Calibration*; Yamamuro, J.A., Kaliakin, V.N., Eds.; ASCE: Reston, VA, USA, 2005; pp. 1–34.
7. Jaeger, J.C.; Cook, N.G.; Zimmerman, R. *Fundamentals of Rock Mechanics*; John Wiley & Sons: Hoboken, NJ, USA, 2007.
8. Yu, H.S. *Plasticity and Geotechnique*; Springer: Berlin/Heidelberg, Germany, 2007.
9. Pietruszczak, S. *Fundamentals of Plasticity in Geomechanics*; CRC Press: Boca Raton, FL, USA, 2010.
10. Vermeer, P.A.; De Borst, R. Non-associated Plasticity for Soils, Concrete and Rock. *HERON* **1984**, *29*, 51–64.
11. Yu, M. Advances in strength theories for materials under complex stress state in the 20th century. *Appl. Mech. Rev.* **2002**, *55*, 169–218. [CrossRef]
12. Aubertin, M.; Li, L. A porosity-dependent inelastic criterion for engineering materials. *Int. J. Plast.* **2004**, *20*, 2179–2208. [CrossRef]
13. Labuz, J.F.; Zeng, F.; Makhnenko, R.; Li, Y. Brittle failure of rock: A review and general linear criterion. *J. Struct. Geol.* **2018**, *112*, 7–28. [CrossRef]
14. Li, L.; Aubertin, M.; Simon, R. The MSDP$_u$ multiaxial criterion for the strength of rocks and rock masses. In *Rock Mechanics and Engineering*; Feng, X.T., Ed.; CRC Press: Boca Raton, FL, USA, 2017; Volume 1, pp. 396–423.
15. Roscoe, K.H.; Schofield, A.; Wroth, A.P. On the yielding of soils. *Géotechnique* **1958**, *8*, 22–53. [CrossRef]
16. DiMaggio, F.L.; Sandler, I.S. Material model for granular soils. *J. Eng. Mech.* **1971**, *97*, 935–950. [CrossRef]
17. Aubertin, M.; Simon, R. A multiaxial failure criterion that combines two quadric surfaces. In *Rock Mechanics-Tools and Techniques*; Aubertin, M., Hassani, F., Mitri, H., Eds.; A.A. Balkema: Rotterdam, The Netherlands, 1996; Volume 2, pp. 1729–1736.
18. Aubertin, M.; Simon, R. A damage initiation criterion for low porosity rocks. *Int. J. Rock Mech. Min. Sci.* **1997**, *34*, 17.e1–17.e15. [CrossRef]
19. Aubertin, M.; Simon, R. Un critère de rupture multiaxial pour matériaux fragiles. *Can. J. Civ. Eng.* **1998**, *25*, 277–290. [CrossRef]
20. Aubertin, M.; Li, L.; Simon, R.; Khalfi, S. Formulation and application of a short-term strength criterion for isotropic rocks. *Can. Geotech. J.* **1999**, *36*, 947–960. [CrossRef]
21. Aubertin, M.; Li, L.; Simon, R. A multiaxial stress criterion for short-and long-term strength of isotropic rock media. *Int. J. Rock Mech. Min. Sci.* **2000**, *37*, 1169–1193. [CrossRef]
22. Li, L.; Gamache, M.; Aubertin, M. Parameter determination for nonlinear stress criteria using a simple regression tool. *Can. Geotech. J.* **2000**, *37*, 1332–1347. [CrossRef]
23. Li, L.; Aubertin, M.; Simon, R.; Bussière, B. Formulation and application of a general inelastic locus for geomaterials with variable porosity. *Can. Geotech. J.* **2005**, *42*, 601–623. [CrossRef]
24. Li, L.; Aubertin, M.; Shirazi, A. Implementation and application of a new elastoplastic model based on a multiaxial criterion to assess the stress state near underground openings. *Int. J. Geomech.* **2010**, *10*, 13–21. [CrossRef]
25. Itasca. *FLAC3D 6.0, FLAC3D Modeling*; Itasca Consulting Group: Minneapolis, MN, USA, 2017.
26. Chen, J.F.; Han, J.; Oztoprak, S.; Yang, X.M. Behavior of single rammed aggregate piers considering installation effects. *Comput. Geotech.* **2009**, *36*, 1191–1199. [CrossRef]
27. Li, L.; Aubertin, M. Numerical investigation of the stress state in inclined backfilled stopes. *Int. J. Geomech.* **2009**, *9*, 52–62. [CrossRef]

28. Jahanbakhshzadeh, A.; Aubertin, M.; Li, L. Numerical simulations to assess the stress state in backfilled stopes with inclined walls. In Proceedings of the Canadian Geotechnical Conference, GeoQuebec, Quebec City, QC, Canada, 21–23 September 2015.
29. Jahanbakhshzadeh, A.; Aubertin, M.; Li, L. Three-dimensional stress state in inclined backfilled stopes obtained from numerical simulations and new closed-form solution. *Can. Geotech. J.* **2018**, *55*, 810–828. [CrossRef]
30. Li, Z.W.; Yang, X.L. Active earth pressure for retaining structures in cohesive backfills with tensile strength cut-off. *Comput. Geotech.* **2019**, *110*, 242–250. [CrossRef]
31. Nadai, A. *Theory of Flow and Fracture of Solids*; McGraw-Hill: New York, NY, USA, 1950; Volume 1.
32. Davis, R.O.; Selvadurai, A.P. *Plasticity and Geomechanics*; Cambridge University Press: Cambridge, UK, 2005.
33. Ortiz, M.; Popov, E.P. Accuracy and stability of integration algorithms for elastoplastic constitutive relations. *Int. J. Numer. Methods Eng.* **1985**, *21*, 1561–1576. [CrossRef]
34. Sloan, S.W.; Booker, J.R. Removal of singularities in Tresca and Mohr-Coulomb yield functions. *Commun. Appl. Numer. Methods* **1986**, *2*, 173–179. [CrossRef]
35. Jeremić, B.; Sture, S. Implicit integrations in elastoplastic geotechnics. *Mech. Cohes. Frict. Mater.* **1997**, *2*, 165–183. [CrossRef]
36. Wroth, C.P.; Houlsby, G.T. Soil mechanics-property characterization and analysis procedures. In Proceedings of the 11th International Conference on Soil Mechanics and Foundation Engineering, San Francisco, CA, USA, 12–16 August 1985; pp. 1–55.
37. Li, L.; Aubertin, M. An elastoplastic evaluation of the stress state around cylindrical openings based on a closed multiaxial yield surface. *Int. J. Numer. Anal. Methods Geomech.* **2009**, *33*, 193–213. [CrossRef]
38. Salençon, J. Expansion quasi-statique d'une cavité à symétrie sphérique ou cylindrique dans un milieu élastoplastique. *Ann. Des. Ponts Chaussées* **1966**, *3*, 175–187.
39. Salençon, J. Contraction quasi-statique d'une cavité à symétrie sphérique ou cylindrique dans un milieu élastoplastique. *Ann. Des. Ponts Chaussées* **1969**, *4*, 231–236.
40. Benzaazoua, M.; Bussière, B.; Demers, I.; Aubertin, M.; Fried, É.; Blier, A. Integrated mine tailings management by combining environmental desulphurization and cemented paste backfill: Application to mine Doyon, Quebec, Canada. *Miner. Eng.* **2008**, *21*, 330–340. [CrossRef]
41. Yan, B.; Zhu, W.; Hou, C.; Guan, K. A three-dimensional analytical solution to the arching effect in inclined backfilled stopes. *Geomech. Geoeng.* **2019**, *14*, 136–147. [CrossRef]
42. Wang, R.; Zeng, F.; Li, L. Stability analyses of side-exposed backfill considering mine depth and extraction of adjacent stope. *Int. J. Rock Mech. Min. Sci.* **2021**, *142*, 104735. [CrossRef]
43. Wang, R.; Zeng, F.; Li, L. Applicability of constitutive models to describing the compressibility of mining backfill: A comparative study. *Processes* **2021**, *9*, 2139. [CrossRef]
44. Falaknaz, N.; Aubertin, M.; Li, L. Numerical investigation of the geomechanical response of adjacent backfilled stopes. *Can. Geotech. J.* **2015**, *52*, 1507–1525. [CrossRef]
45. Jahanbakhshzadeh, A.; Aubertin, M.; Li, L. Analysis of the stress distribution in inclined backfilled stopes using closed-form solutions and numerical simulations. *Geotech. Geol. Eng.* **2018**, *36*, 1011–1036. [CrossRef]
46. Yang, P.; Li, L.; Aubertin, M. Theoretical and numerical analyses of earth pressure coefficient along the centerline of vertical openings with granular fills. *Appl. Sci.* **2018**, *8*, 1721. [CrossRef]
47. Jaky, J. Pressure in silos. In Proceedings of the 2nd International Conference on Soil Mechanics and Foundation Engineering, Balkema, Rotterdam, 21–30 June 1948; Volume 1, pp. 103–107.
48. Li, L.; Aubertin, M.; Belem, T. Formulation of a three dimensional analytical solution to evaluate stresses in backfilled vertical narrow openings. *Can. Geotech. J.* **2005**, *42*, 1705–1717. [CrossRef]
49. Aubertin, M.; Li, L.; Arnoldi, S.; Belem, T.; Bussière, B.; Benzaazoua, M.; Simon, R. Interaction between backfill and rock mass in narrow stopes. In *Proceedings of Soil and Rock America*; Culligan, P.J., Einstein, H.H., Whittle, A.J., Eds.; Verlag Glückauf Essen (VGE): Essen, Germany, 2003; Volume 1, pp. 1157–1164.

processes

MDPI

Article

Analysis of the Surface Subsidence Induced by Mining Near-Surface Thick Lead-Zinc Deposit Based on Numerical Simulation

Yifan Zhao *, Xingdong Zhao *, Jiajia Dai and Wenlong Yu

Geomechanics Research Center, Northeastern University, Shenyang 110819, China; daijiajia@stumail.neu.edu.cn (J.D.); 2010396@stu.neu.edu.cn (W.Y.)
* Correspondence: zyf_neu@126.com (Y.Z.); zhaoxingdong@mail.neu.edu.cn (X.Z.)

Abstract: This paper describes a case study of surface subsidence in the Hongling Lead-Zinc Mine. Hongling Lead-Zinc Mine is located in Inner Mongolia, China, about 240 km away from the border between China and Mongolia. There is a batch of outcrops of the near-surface thick steep-dip metamorphic orebody. The large-scale surface subsidence induced by underground excavation has brought some impact on the safety of herdsmen and their daily husbandry activities nearby. The requirements of reclamation for subsidence areas in the relevant laws and regulations, raise enormous pressure and risk on safe and economic operation. In this paper, a 3D numerical model of this mine was built by 3DMine and FLAC3D to analyse the excavation procedure and mechanism. The results of the simulation were in good agreement with the field subsidence data collected by satellites and unmanned aerial vehicles from 2009 to 2019. The analysis showed that the current mining method—an integrated underground method of stoping and caving—accelerated the surface subsidence, and some measures of monitoring, controlling and management were expected to take in order to improve economic and ecological benefits.

Keywords: near-surface thick deposit; surface subsidence; numerical simulation; unmanned aerial survey; accurate model

Citation: Zhao, Y.; Zhao, X.; Dai, J.; Yu, W. Analysis of the Surface Subsidence Induced by Mining Near-Surface Thick Lead-Zinc Deposit Based on Numerical Simulation. *Processes* **2021**, 9, 717. https://doi.org/10.3390/pr9040717

Academic Editor: Orlando Vaselli

Received: 19 March 2021
Accepted: 14 April 2021
Published: 19 April 2021

1. Introduction

Mining subsidence is the continuous deformation, and discontinuous failure of strata and surface by the destroy of in-situ stress field induced by the underground excavation [1]. The first acts and laws related to subsidence were published in Belgium between the fifteenth and sixteenth century [2]. Plenty of comprehensive researches on theory and control technology of mining subsidence have been conducted in Soviet Union, Poland, Germany, Australia, Britain, Canada, Japan and the United States, and there are a great number of results and findings on the excavation under buildings, railway, water body and above confined water [3–5]. In the long term, many scholars try to accurately calculate the subsidence displacement induced by underground mining. In 1838, Doris first put forward the "vertical line theory" of surface subsidence. Twenty years later, this theory was developed into "normal line theory" by Gonot, which was revised by Dumon to the calculation model of subsidence $w = m \times \cos\alpha$ [6]. In Germany, "bisection theory" and "natural slope theory" were proposed by Jlcinsky (1876) and Oesterr (1882), respectively [7,8]. Besides these theories, some other early studies including Fayol's (1885) [9] "circular arch theory" and Hausse's (1885–1897) "bisection theory", analyzed the relationship between deformation and movement of overburden and surface subsidence, and established several geometric theoretical models [10]. In 1903, Halban deduced that the surface strain is inversely proportional to the radius of curvature [1]. Schimizx (1923), Keinhost (1925) and Bals (1932) studied the area of mining subsidence and formed the concept of influence function [11]. In 1947, С. Г. Авершин's monograph "Strata Movement in Underground Coal Mining", built the differential relationship between the vertical and horizontal directions of the surface movement vector, pointed out that the subsidence section equation is in exponential form, and introduced the famous theory that the horizontal movement is inversely proportional to the ground tilt [12]. However, the theory complicated surface movement calculation formulas were of difficulty to be applied to field analysis. J. Litwaniszyn applied discontinuous medium mechanics to study the relation of strata and surface movement and introduced the stochastic medium theory, which was later developed into the widely used probability integral method [2,13,14]. Smolarski deduced the surface movement function of a stochastic medium theory under the condition of inclined coal seam mining from the viewpoint of medium symmetry [15]. Salamon combined the influence function with continuum mechanics and applied the elastic theory to put forward the surface element principle, which presented the foundation of boundary element method [16,17]. Berry derived the surface movement and deformation formula in isotropic plane, transversely isotropic plane and three-dimensional conditions based on elastic theory [18–21].

With the continuous and rapid development of computer and information technology, a variety of numerical methods and artificial intelligence algorithms have been applied to the study of surface subsidence [22–24]. The finite difference software—FLAC3D—developed by Itasca Inc. is widely used for the analysis of mining induced surface subsidence [25,26]. Due to the complexity of model built in FLAC3D, the combination of SURPAC and FLAC3D is popular in building the numerical models of mines [27]. Additionally, 3DMine software can be used to establish the 3D geological model of orebody which can be converted into the file containing blocky and nodes by Midas, and then the Midas file can be imported into FLAC3D model through interface program [28,29]. In this paper, the 3Dmine, Midas second method was used to build the 3D numerical model, and the field mining procedure and sequence were simulated by FLAC3D, which is employed to analyze the surface subsidence in Hongling Lead-Zinc Mine.

2. Motivation and the Study Objective

If the large-scale surface subsidence suddenly collapses in mine void, high ground pressure can be induced and cause the permanent loss and dilution of ore, even the mine is closed; strong airwave can form air blast to threaten the safety of deep underground mining, more enlarge the scope of surface subsidence. The surface is grazed by nearby herders. In order to ensure the safety of herdsmen and the normal operation of the mine, it is necessary to study the mine collapse. Thus, it was of great importance for decreasing the risk of large-scale collapse by analyzing and predicting the characteristics of surface subsidence induced by underground mining.

3. Engineering Background

3.1. Geological Setting

Hongling Lead-Zinc Mine is located in Chifeng City (Figure 1), Inner Mongolia, China, and 240 km south away from the border between China and Mongolia, with a mining area of 3.394 km^2. Hongling Lead-Zinc Mine is a skarn type polymetallic deposits dominated by lead and zinc, at the north slope of the watershed in the southwest of Greater Khingan, with the surface elevation between +976 m and +1220 m. Hongling Lead-Zinc Mine has a production capacity of 5000 t/day. The surface boundary of deposit is in 17# exploratory line, and its deep boundary of deposit is in 13# exploratory line divided by the feldspar porphyry belt. The eastern side of the feldspar porphyry belt is named 1# mineralization zone, and the deposit is divided two veins by invasive marble in the west side of the feldspar porphyry belt, other northwestern deposit is called 2# mineralization zone (Figure 2). The 1# mineralization zone is 1350 m long along the strike, with an average width of 10 m, and its strikes 55–59° and has a NW dip of 65–85°, and there is a batch of outcrops vertical to the surface. The 2# mineralization zone is 975 m long along the strike, with an average width of 7 m.

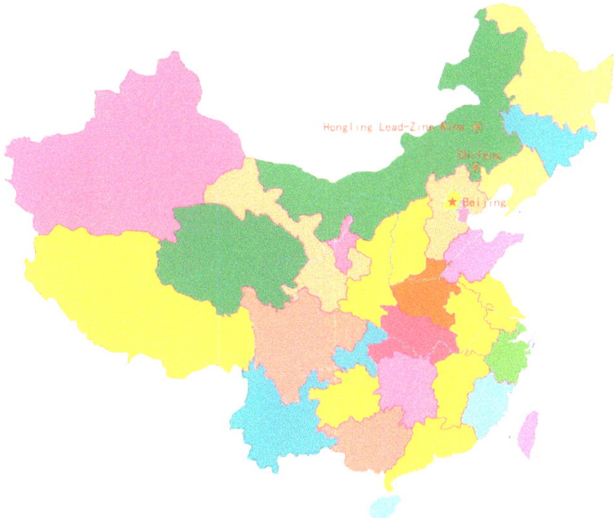

Figure 1. The location of Hongling lead-zinc Mine.

Figure 2. The aerial view of Hongling Lead-Zinc Mine (taken in 2016).

As illustrated in the cross-section of 17# exploratory line (Figure 3), the mineralization zone occurs in the skarn belt, and the bed rock is skarn with small horizontal thickness. The lithological units of the mineralization zone include slate, marble and skarn. The surrounding rock is the argillite, siltstone, metasandstone and marble. The red area shown in Figure 3 is the ore body that will be mined. It is clear that mining activity starts at the surface. The gravity anomaly map of Hongling Lead-Zinc Mine is shown in Figure 4.

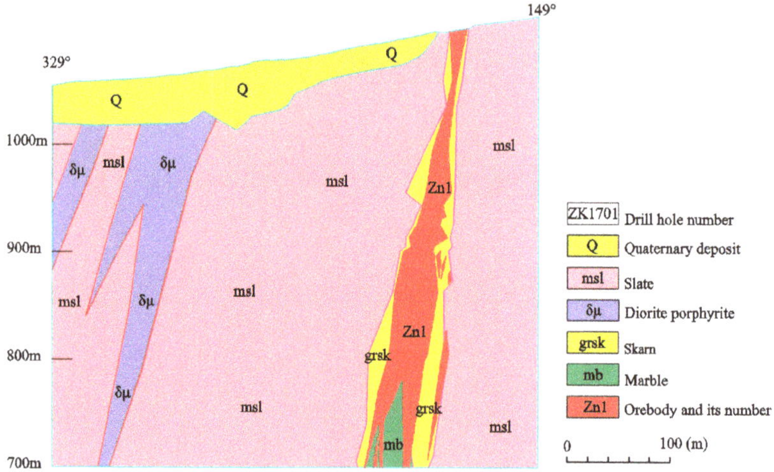

Figure 3. The cross-section of 17# exploring line.

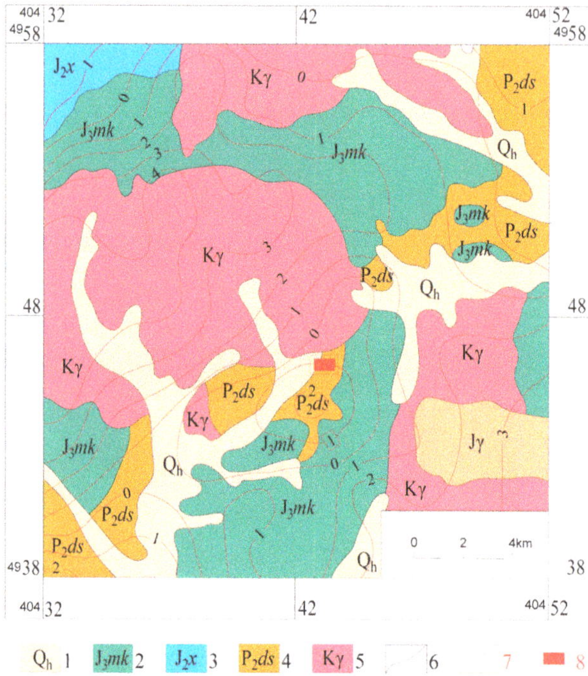

Figure 4. The gravity anomaly map. 1—Loess sandy soil, 2—Acid lava, acid lithic crystalline tuff, tuffaceous glutenite, etc. 3—Metamorphic acid tuff, tuffaceous conglomerate, glutenite, etc. 4—argillaceous slate intercalated with silty slate, marble, metamorphic sandstone, siltstone, etc. 5—Biotite granite, 6—Geological boundary, 7—Gravity contour, 8—Mine site.

3.2. Mining Method

The ore is mined depending on the thickness of the ore body. In the Hongling Lead-zinc Mine, the room and pillar method has been adopted, and the typical design of mining method is shown in Figure 5a. The level interval of this mine is 50 m, the height of one stope is 40 m with 10 m thickness crown pillar. The length of one stope can be determined according to the horizontal thickness of the orebody. When the horizontal thickness of the orebody is less than 30 m, and the stope layout is along the strike of orebody, the length of one stope is 32 m, and the span of one stope is the thickness of the deposit. When the horizontal thickness of the orebody is more than 30 m, and the stope layout is perpendicular to the strike of the orebody; the length of one stope is the horizontal thickness of the orebody, and the span of one stope is 30 m. The width of the barrier pillar is 18 m. The square panel pillar is of a 5 m × 2 m cross-section, and its pattern is 5 m × 10 m in long and wide. Stoping is carried out using 175 mm diameter long hole in room; the caving method is adopted to extract the pillar, and the blasting sequence is illustrated in Figure 5b. There is no backfill for mined-out stopes.

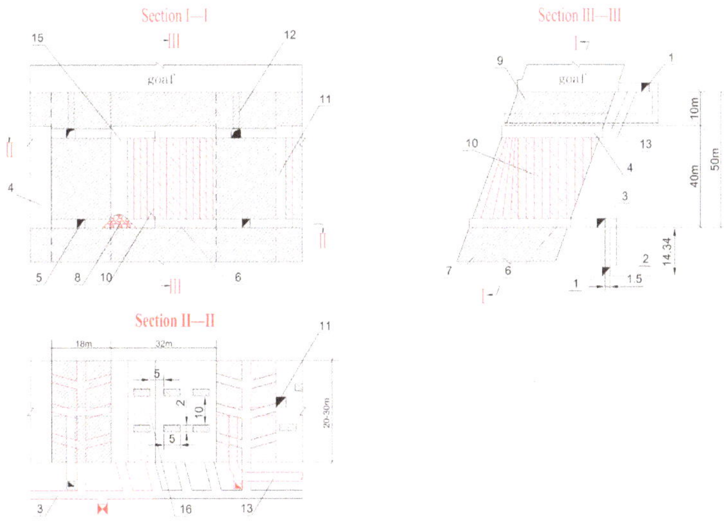

(**a**) The layout of room and pillar method

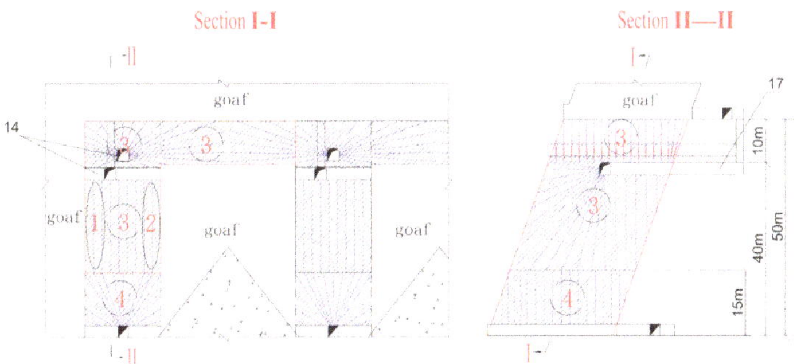

(**b**) The sequence of extracting pillars

Figure 5. The blasting sequence of extracting the pillar. 1—haulage drift; 2—ore pass; 3—loading drift for rooms; 4—goaf; 5—loading drift for pillars; 6—undercutting level; 7—orebody; 8—broken ore; 9—crown pillar; 10—longhole; 11—slot raise; 12—service and ventilation raise; 13—ramp; 14—drilling drift; 15—cutting slot; 16—drilling chamber; 17—access.

3.3. Current Surface Subsidence

Due to continuous underground mining, extracting the pillar, and no backfill materials fill the mine void. The giant goaf has gradually formed with the development of underground mining for more than ten years, which has induced the massive surface collapse and shown in Figure 5. The overlying rock mass gradually subsided under the action of gravity and mining-induced stress, resulting in new cracks occurrence on the surface. Field surveying and data analysis have been done at +1035 m level, and above in 2016, the length of the collapse zone is 684 m, and the maximum width is 150 m, the surface collapse area is more than 60,000 m^2 (Figure 6). Continuous surface cracks, ground sliding and collapse have been observed in or near the subsided pit, and the cracks were usually parallel to the edges of the surface collapse. Some cracks' width is more than 1 m, and they were identified as the primary cracks. The surface collapse caused by underground mining in this mine also caused a slope landslide with a length of 240 m and a width of 60 m.

Figure 6. The aerial view of surface collapse.

4. Preliminary Study

33 rock samples were collected from four different levels (+805 m, +855 m, +905 m and +955 m) of Hongling Lead-zinc Mine with five kinds of rock (ore, marble, argillaceous slate, skarn and feldspar porphyry) to ensure that the range of lithology and alteration types were included. 111 standard rock specimens (Figure 7) including cylindrical specimens with the height-to-diameter ratio of 2:1 and the size of Φ50 × 100 mm, disc specimens with the size of Φ50 × 25 mm and specimens with diameter of 50 mm and height ranged from 64 mm to 100 mm, were processed from the in-situ samples (Figure 8). The rocks' physical and mechanical properties were determined from laboratory testing on intact rock specimens following ISRM (1981) recommended methods (Figures 9–11), and the test results are given in Table 1. Uniaxial compression tests, Brazil splitting tests and direct shear tests were performed to obtain rock mechanical parameters. The uniaxial compressive strengths (UCS, σc), Young's modulus (E) and Poisson's ratios (μ) of specimens were estimated by YE4-100 hydraulic pressure testing machine with the maximum loading of 100 t, and stress sensors and UCOM-60A static acquisition instrument were applied to monitor stress distributions

and axial and lateral strains (Figure 9). The tensile strengths (σt) and shear strengths were determined by YAW-300 microcomputer controlled electro hydraulic servo direct shear testing machine. The densities (ρ) of specimens were calculated by measured weights and volumes. All the laboratory tests were carried out under the Standards Press of China GB/T50266-99. The cohesion (c) and internal friction angle (φ) of rocks were obtained by the means of linear regression with the shear stresses and normal stresses measured in the direct shear tests under the Mohr-Coulomb criterion.

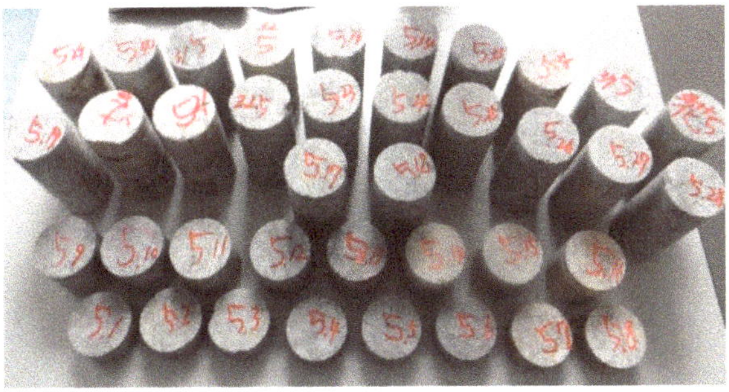

Figure 7. Partial rock specimens.

Figure 8. Processing rock samples.

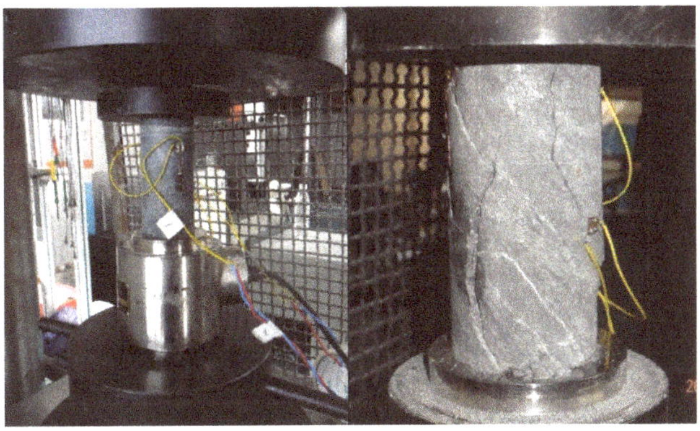

Figure 9. Equipment and failure modes of uniaxial compression test.

Figure 10. Equipment and failure modes of Brazilian splitting test.

Figure 11. Equipment and failure modes of direct shear test.

Table 1. Physical and mechanical parameters of intact rock.

Rock Type	ρ (kg/m^3)	σ_c (MPa)	μ	E (GPa)	σ_t (MPa)	c (MPa)	φ (°)
Marble	2663	58.51	0.21	35.64	3.98	6.52	36.50
Slate	2770	116.00	0.27	75.60	12.15	9.75	48.68
Ore	3565	142.68	0.18	55.16	7.86	6.31	58.24
Skarn	3086	87.05	0.30	36.10	5.64	5.31	50.15
Feldspar Porphyry	2730	65.32	0.24	18.30	7.26	5.85	42.76

The quantitative value of GSI was deeply studied by Bertuzzi et al. In this paper, the GSI value of rock mass was determined according to the results of geological survey based on the research results of Bertuzzi et al. [30]. The geological survey work includes the collection of borehole data, structural plane measurements (weathering, roughness, water, etc.). The value of D was determined depended on experience in the mine according to the reference of Hoek et al.'s paper in 2002 [31].

According to the China GB50218-94 Standard for engineering classification of rock masses [32], the samples of skarn, slate, feldspar porphyry and ore were hard rock ($\sigma_c \geq 60$ MPa), and the marble were medium hard rock (30 MPa $\leq \sigma_c \leq$ 60 MPa). The tensile strength and shear strength of contact surrounding rock of orebody—skarn were the weakest, which is not conducive to the mined-out area's stability. Because the bedrock in the mining zone is slate, and the development and extension of collapses and subsidence were close to slate distribution and mechanical properties.

5. Numerical Model Set-Up

The current underground mining of Hongling Lead-Zinc Mine has caused surface subsidence. As the orebody is mined, which will induce subsidence, and it is necessary to retain pillars and fill mine void during mining. To solve this series of problems, a 3D numerical model was built to analyze the stability of the surface subsidence area in

Hongling Lead-Zinc Mine, judge the impact of surface collapse extraction on the, and predict the development of the subsidence scope.

Build a 3D numerical model for FLAC3D to calculate and analyze the surface subsidence, and there were eight steps in this numerical model set-up progress:

① Establish geological database including drilling sheet, survey sheet, geological sheet and mineral grade sheet in 3DMine software;
② Generate a 3D model (Figure 12) in 3DMine and export the surface elevation data;
③ Import the surface elevation data into MidasGTSNX and generate the 3D surface and surrounding rock model;
④ Import the orebody data and optimize the 3D orebody model in MidasGTSNX;
⑤ Connect the surrounding rock model with the orebody model and optimize topological relations;
⑥ Mesh the orebody and surrounding rock model;
⑦ Export the grid node and element data;
⑧ Convert data into flac3d format and import to FLAC3D (Figure 13).

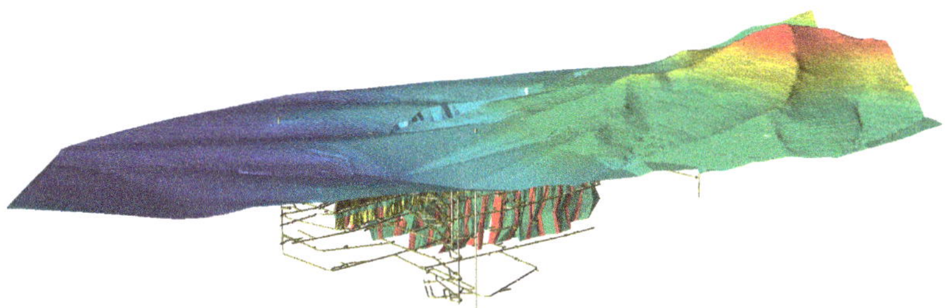

Figure 12. A 3D model of mine structure, orebody and surface.

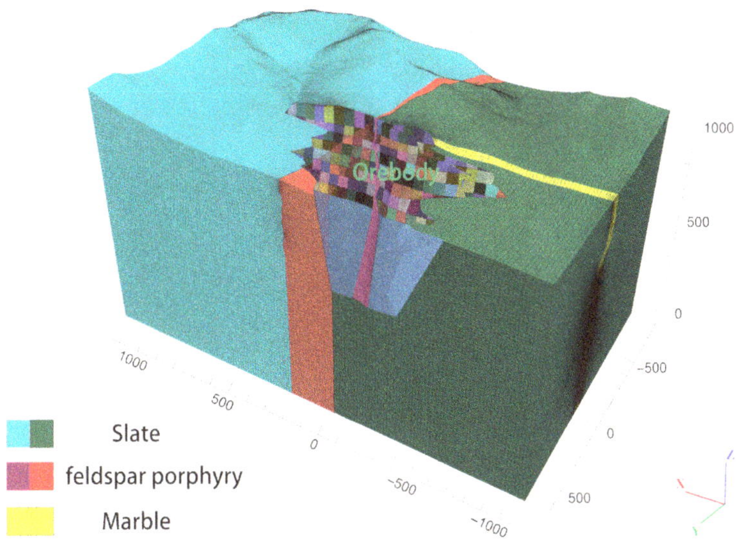

Figure 13. The FLAC3D model.

When generating a 3D model in FLAC3D, the original coordinate point is taken as 49,444,800, 40,443,100. Taking the original coordinate point as the rotation center, the orebody was rotated 31 degrees clockwise, so that the dip direction of the orebody was the positive direction of the y axis of the coordinate system, and the trend is the x-axis. The model extended from point (−1075 m, −830 m) to point (1160 m, 660 m) with the size of 2235 m × 1490 m and height ranging from 980 m to 1300 m.

6. Numerical Modelling and Analysis

6.1. Numerical Model Introduction

In order to better understand the size and position relationship of different rock masses, 6 sections were made for the model (Figure 14). The orebody was cut off by feldspar porphyry. The orebodies are roughly distributed between sections b and e (Figure 15). Section b shows that the marble near the surface is supported by orebodies. The main rock mass is slate.

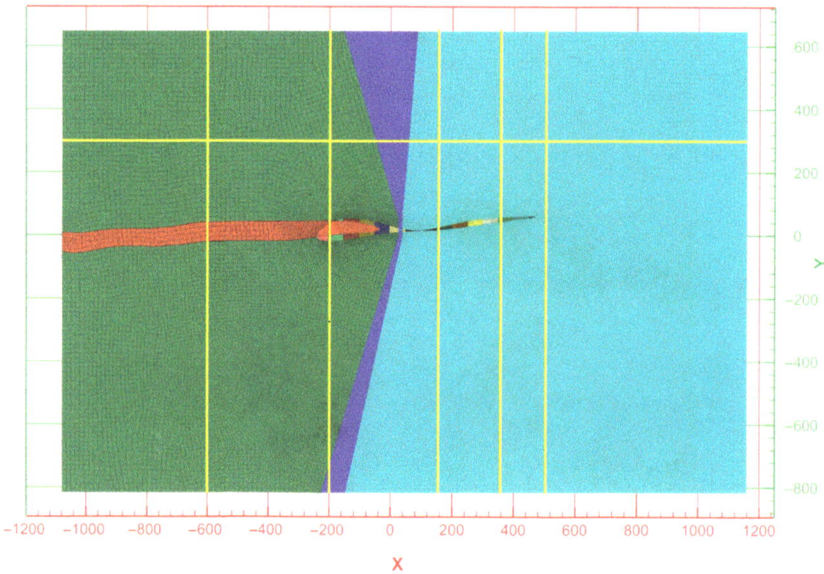

Figure 14. Top view of the numerical model.

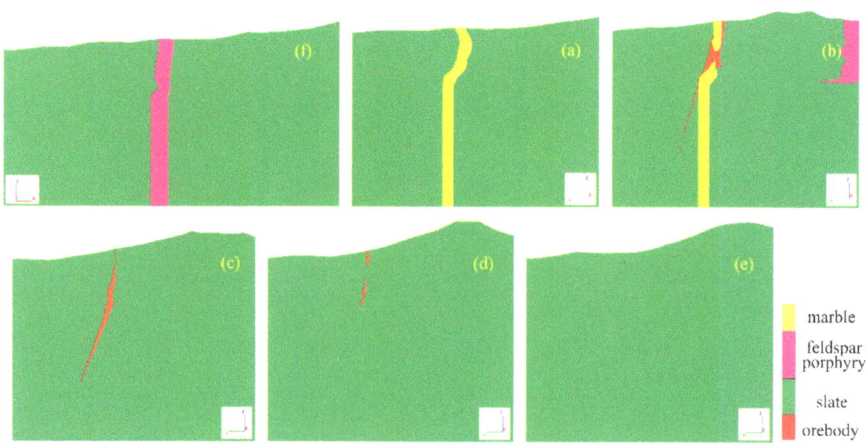

Figure 15. Sectional view of the numerical model.

6.2. In-Situ Stresses and Boundary Restrains

According to in-situ stress testing results by the acoustic emission method, the maximum horizontal principal stress, the minimum horizontal principal stress, and the vertical stress varied almost linearly with the depth. The in-situ stress was described by the following equations [33]:

$$\sigma_z = 0.0264h + 1.8118 \quad (1)$$

$$\sigma_{h\max} = 0.026h + 10.008 \quad (2)$$

$$\sigma_{h\min} = 0.023h + 1.8024 \quad (3)$$

where $\sigma_{h\max}$ is the maximum horizontal principal stress, $\sigma_{h\min}$ is the minimum horizontal principal stress, σ_z is the vertical stress, and h is the depth from the surface.

The in-situ stress field of the model was calculated by Equations (1)–(3). The results of fieldwork showed that the direction of maximum principal stress ($\sigma_1 = \sigma_{h\max}$) ranged from 5° to 22°, and the direction of minimum principal stress ($\sigma_3 = \sigma_{h\min}$) ranged from 185° and 202°. The intermediate principal stress ($\sigma_2 = \sigma_z$) was in an almost vertical direction. The boundary restraints are shown in Table 2.

Table 2. The boundary restrains of model.

Restrain X	Left and Right Boundaries
Restrain Y	Front and Back Boundaries
Restrain XYZ	Bottom Boundary
Free	Top Boundary

6.3. Parameters and Failure Criterion for Calculation

In 1980, Hoek and Brown proposed the original Hoek-Brown criterion considering the strength and integrity of rock mass [34]. The generalized Hoek-Brown criterion for the estimation of rock mass strength was introduced with some improved parameters in 1992 [35]. In 2002, the introduction of disturbance factor D improved the comprehensive and specific application of the Hoek-Brown criterion to the nonlinear failure of various rock masses [31]. In 2006, Hoek et al. revised the calculation of elastic modulus, as shown in Equation (4) [36].

$$E_m = E_i\left[0.02 + \frac{(1 - D/2)}{1 + e^{(60+15D-GSI)/11}}\right] \quad (4)$$

The basic parameters input in FLAC3D were composed of the physical and mechanical properties of rock mass. The value of *mi* was determined according to the reference of Hoek et al.'s paper in 2000 [37]. The parameters of rock mass were calculated by the Hoek-Brown criterion from its corresponding laboratory test and field results, as shown in Table 3.

Table 3. The rock masses mechanical parameters of model.

Rock Type	D	GSI	m_i	σ_{cm} (MPa)	σ_{tm} (MPa)	E_m (GPa)	c_m(MPa)	φ_m (°)	μ	ρ (kg/m³)
Marble	0.8	65	9	4.08	−0.26	8.11	3.04	25.41	0.21	2663
Slate	0.8	62	9	6.43	−0.39	14.54	3.81	28.95	0.27	2770
Ore	0.8	65	24	9.96	−0.24	12.56	6.01	40.07	0.18	3565
Skarn	0.8	70	24	8.91	−0.23	10.57	5.56	38.66	0.30	3086
Feldspar Porphyry	0.8	65	30	4.56	−0.09	4.17	4.74	35.89	0.24	2730

In the numerical simulation of rock engineering, the Mohr-Coulomb model's safety factor is a little larger than the practical result. The introduction of the strain-softening Mohr-Coulomb model results in larger deformation of surrounding rock and deeper plastic zone and unstable zone. The strain-softening Mohr-Coulomb model can more reasonably reflect the stable state and rock mass's deformation tendency [38]. The strain-softening

Mohr-Coulomb model in FLAC3D is based on the Mohr-Coulomb model. The difference is that the cohesion, friction, dilation, and tensile strength will soften after the onset of plastic yield. In the Mohr-Coulomb model, those properties are assumed to remain constant.

6.4. Mining Sequences

The mining sequence of the mine is downward from the center of the ore body to the two wings. The upper level is ahead of one stope in the neighbour mining level, as shown in Figure 16. The simulation steps are determined according to the progress of the mine. The same color area is shown in Figure 16 for each excavation. The displacement, damage and surface collapse as the mining process are analyzed by FLAC3D.

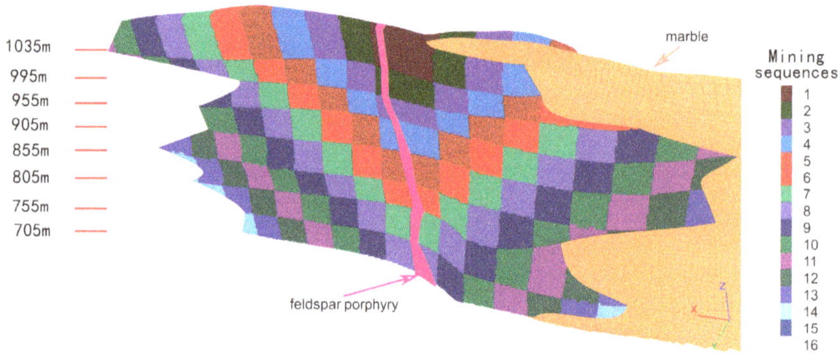

Figure 16. The diagram of mining sequences.

6.5. Results of Numerical Simulation

There were two kinds of fractures in the subsidence area of Hongling Lead-Zinc Mine, namely tensile fracture and collapse-induced fracture. Tensile fractures were caused by tensile stress exceeding the tensile strength of surface soil or rock mass, with characteristics of small width and depth, same strike with orebody and development ahead of mining faces. Collapse-induced fractures were caused by collapses of mine-out areas, surrounding rock and surface, with characteristics of longitudinal subsidence, large depth and development later than mining faces [39]. The results of numerical simulation illustrated the evolution of stress field, displacements and plastic zones with the mining process. The displacement and failure of surrounding rocks (feldspar porphyry, marble and slate) were analyzed from the perspectives of displacement, plastic zone, maximum principal stress and minimum principal stress.

6.5.1. Analysis of Feldspar Porphyry

As shown in Figure 17, there was a significant increase of maximum principal stress and obvious stress concentration at the intersection with the orebody, with the stress value of 32 MPa. There were two zones where the hanging wall feldspar porphyry was in tension state and even in tensile failure: ① Adjacent to the hanging wall of the orebody, the length in the dip direction is about 30 m, almost in tensile plastic (as shown in Figure 18); ② It is located in 225–447 m away from hanging wall of the orebody and about 202 m in dip direction, The tensile plastic zone is located 341–447 m away from the hanging wall of the ore body. The feldspar porphyry at the intersection of orebodies was in plastic state with a small part in tensile plastic and the majority in shear yielding. The comprehensive analysis revealed that: the intersection of the feldspar porphyry was fully plastic, and the hanging wall of 30 m range of non-feldspar porphyry was also fully plastic; there was an overall collapse in this area, with strike toward the orebody and a largely wide fracture on the surface; in zone ②, there would be tensile fractures distributed in the hanging wall of ore body about 341–447 m.

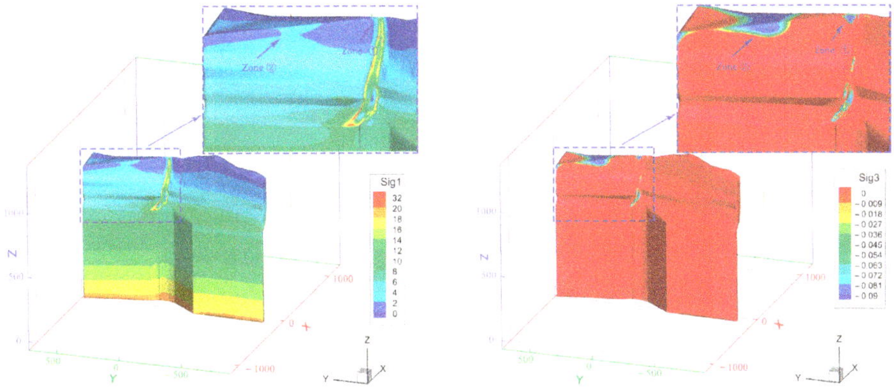

Figure 17. The stress field of feldspar porphyry.

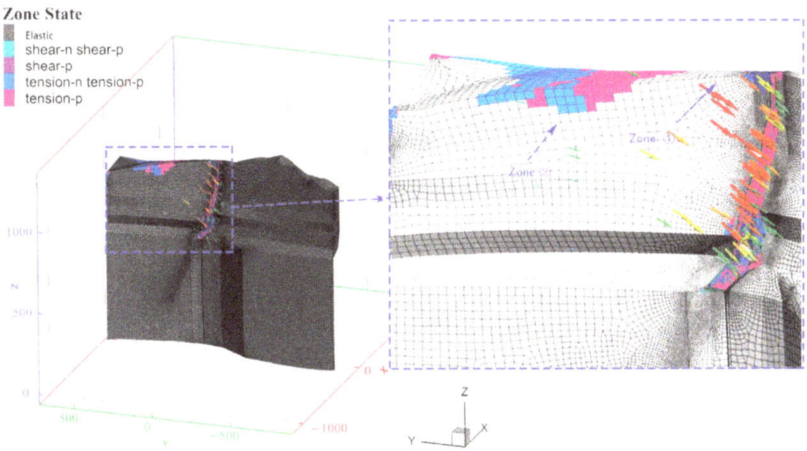

Figure 18. The plastic zone and the displacement of feldspar porphyry.

6.5.2. Analysis of Marble

As shown in Figure 19, there was obvious stress concentration at x = −240, z = 800 m (hangingwall) and x = −240, 1020 m (footwall) with the maximum principal stress of 30 MPa. The tensile stress concentration of the marble suspended part was clear, with the stress value close to the tensile strength of rock mass.

The nephogram of the plastic zone in Figure 20 showed that this part was mainly tensile plastic zone, and there were three-strip areas with the minimum principal stress close to 0 in the suspended part. Combined with Figure 21, the strip areas were due to non-uniform deformation, significant local tensile deformation, volume increase, and reduction of density during displacement. There was apparent maximum principal stress concentration in the non-suspended part of marble with a mixed plastic zone of shear and tensile plastic (Figures 19 and 20). According to the displacement vector diagram in Figure 22, the displacement form of the suspended part was mainly subsidence with torsion, and the maximum displacement reached 49 m. Marble was located in the middle of the orebody, and both the hanging and foot walls were orebody. After mining the orebody, the hanging and foot walls and the lower part had no supporting, and the upper outcrops were free. The marble (z = 900–1100 m) in the southwest wing of the orebody was suspended in the rock mass, to support unmined orebody, whose contact near the surface was bearing the weight of the whole marble mass, and the stress state was similar to the cantilever beam and in a very dangerous situation. At this time, the whole marble mass was in a tension stress field under the loading of its self-weight, and the tensile stress value was quite large, resulting in a seriously tensile stress concentration point at the contact of the surface and the surface subsidence at the southwest wing of the surface characterized by large range and depth.

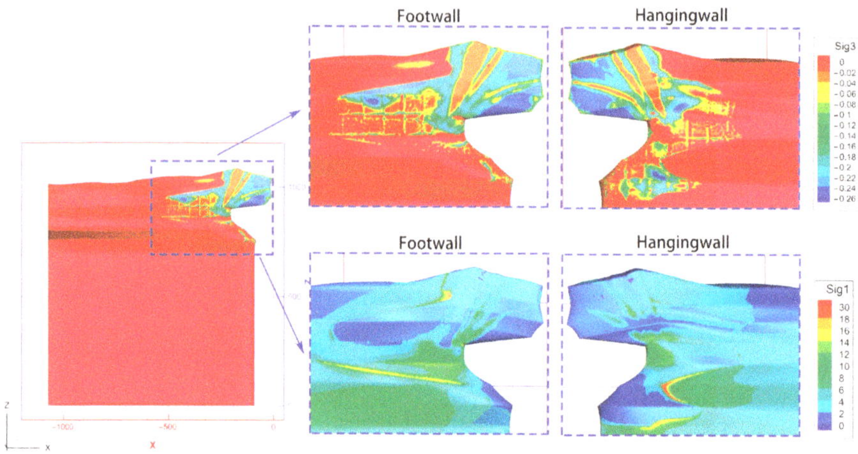

Figure 19. The stress field of marble (MPa).

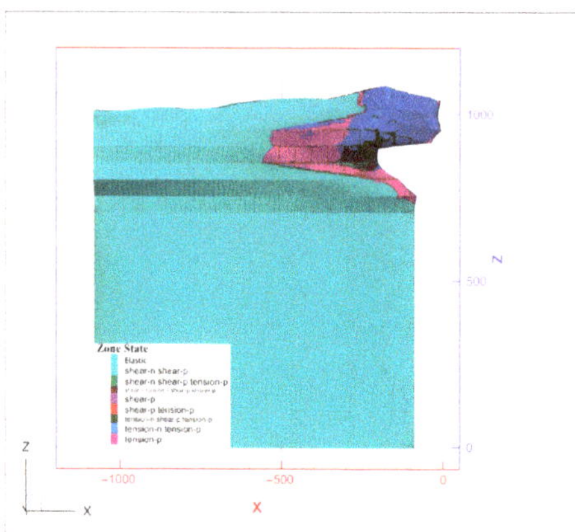

Figure 20. The plastic zone of marble.

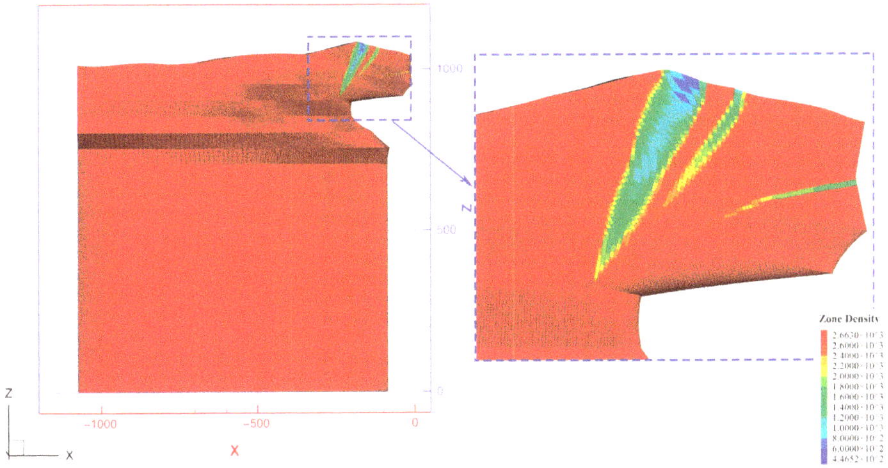

Figure 21. The density field of marble (kg/m^3).

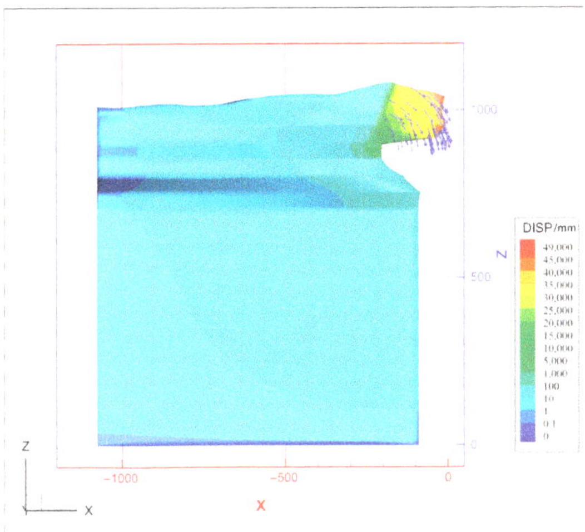

Figure 22. The displacement field and vectors of marble.

6.5.3. Analysis of Slate

As shown in Figure 23, there was a large tensile plastic zone of slate mass on the surface connected with the plastic zone of feldspar porphyry as a whole. In FLAC3D, tension-n and tension-p are generally recognised as tensile plasticity and tensile damage with plasticity. There would be a huge area of tensile fractures of the tensile plastic zone of hanging wall on the surface with a small range of coalescence to the plastic zone at the hanging wall of marble, which was presented as a collapse of rock mass and was consistent with the field observation. The depth and range of the tensile damage zone of the slate were illustrated in Figure 24. The feldspar porphyry occurred between the X-2 and X-3 sections, and its retention played a good role in supporting the surrounding rock of hanging and foot walls, with a relatively smaller tensile damage zone. There was a tensile damage coalescence from the hanging wall to surface around X-2 and X-3 sections with the greater stress value around the X-2 section close to the tensile strength of rock mass. At the X-2 section, there were two orebodies, which were in the direction of the hanging wall and footwall of marble. The collapse of the marble suspended part, and the mining of orebody directly led to the growing of mine-out areas, the increase of exposure face of hanging wall, the connection of tensile damage area of slate, and the risk of overall sliding and subsidence of surface.

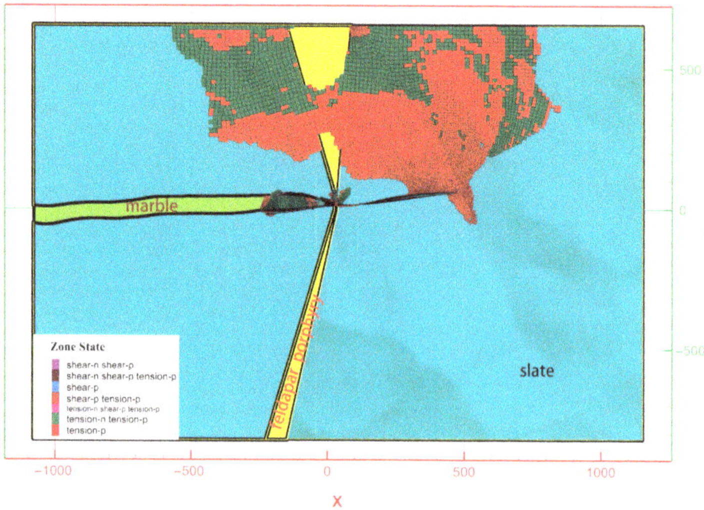

Figure 23. The plastic zone of mine surface.

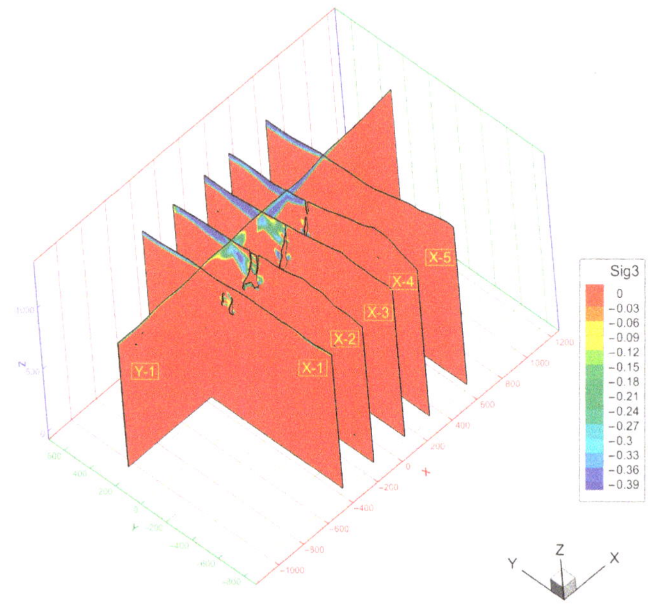

(**a**) section position

Figure 24. *Cont.*

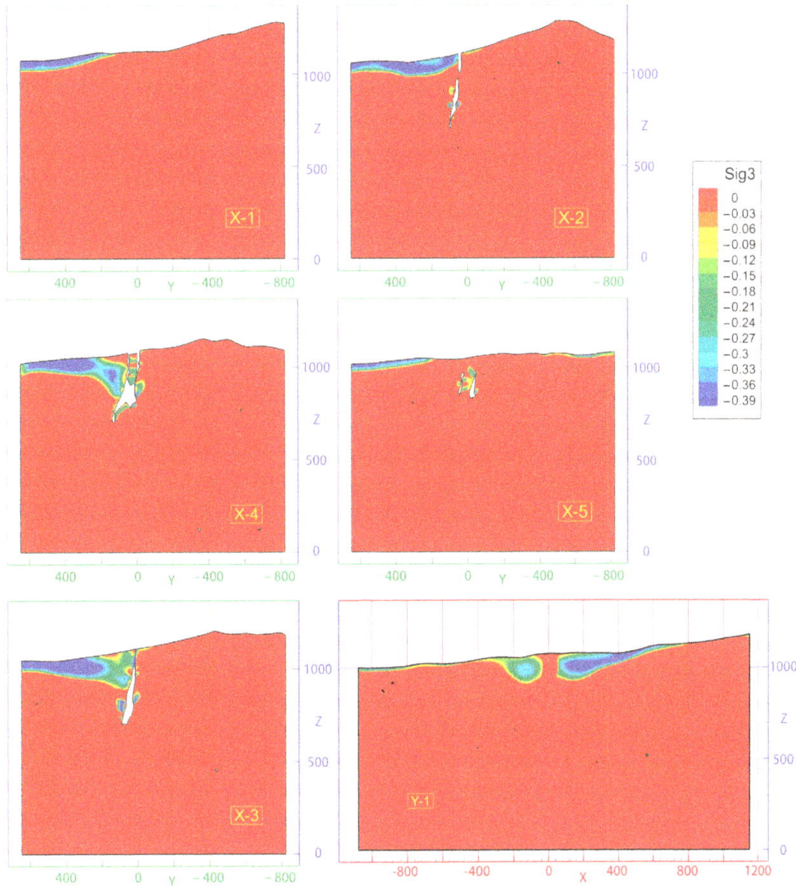

(b) The minimum principal stress section plan

Figure 24. The minimum principal stress field of sections of slate.

7. The Evolution of Surface Subsidence

The data collected by satellites (from 2009 to 2013) and unmanned aerial vehicles (from 2016 to 2019) are shown in Figure 25, Table 4 and Figure 26, respectively. Initially, there was a small subsidence area (A) in 2009, and with the regular excavation another two areas (B and C) were observed in 2012. Due to the collapse of marble mass caused by the extraction of ore rocks close to the marble mass, areas A and B were grown into a larger uniform area (AB) in 2016. In 2018, the development of the mine-out area increased the requirement of feldspar porphyry mass, which was located between area AB and area C to support the surrounding rock and eventually resulted in the subsidence connection of AB and C into ABC. There was a clear growth tendency in the total area of subsidence. The development of the subsidence area from 2009 to 2019 was in good agreement with the numerical simulation results.

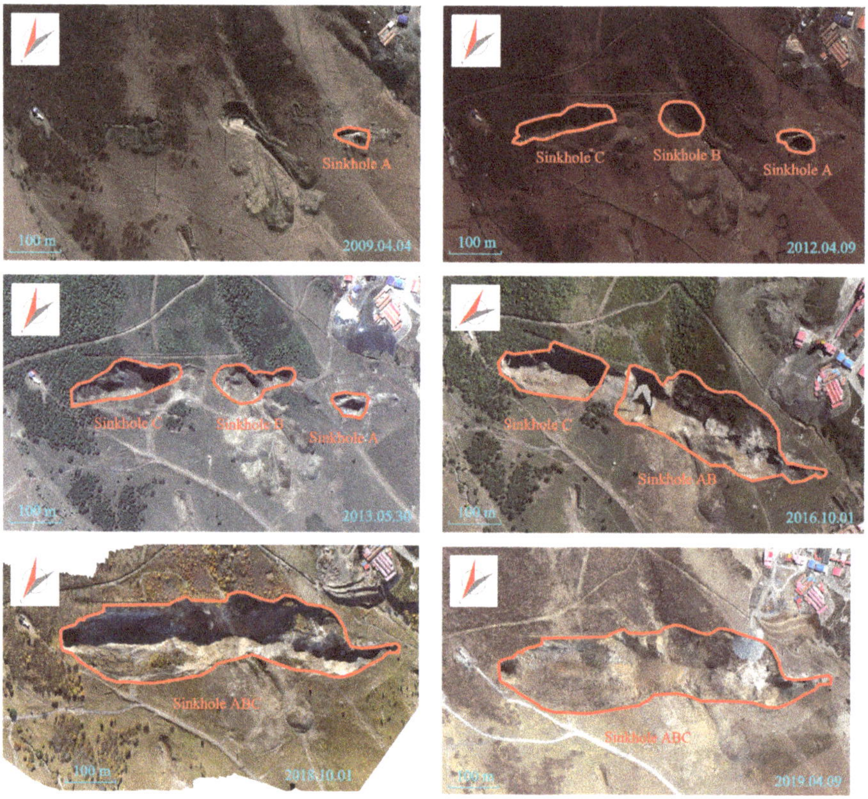

Figure 25. The evolution of surface subsidence.

Table 4. The area of surface subsidence.

Sinkhole \ Date	2009.04.04	2012.04.09	2013.05.30	2016.10.01	2018.10.01	2019.04.09
A	1704	2219	2488			
B		4370	7107			
C		7207	10,565	15,277		
AB				40,540		
ABC					72,407	74,238

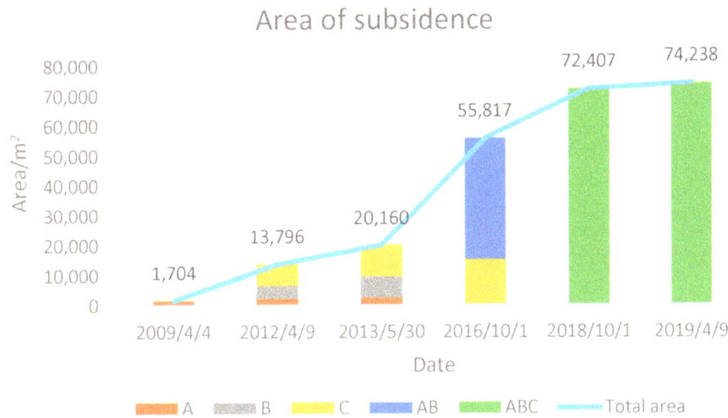

Figure 26. The area of surface subsidence.

8. Deformation Observation

December 2019, Twenty GPS deformation observation stations were installed, distributed around the collapse pit, which has been steadily monitored for more than 400 days. GPS2, GPS10, GPS11, GPS13, GPS14 and GPS16 are located on the hangingwall of the orebody (Figure 27). By comparing the numerical simulation results with the field monitoring results (Figures 28 and 29), it can be observed that the numerical simulation results are consistent with the field monitoring results, and this numerical simulation model can be used to simulate the real field situation. The numerical simulation calculates the displacement generated in the whole life cycle of the mine, and the monitoring data are the displacement of less than 500 days, so the monitored displacement is less than the result of the numerical simulation. The numerical simulation results and monitoring data show the same displacement distribution law and trend.

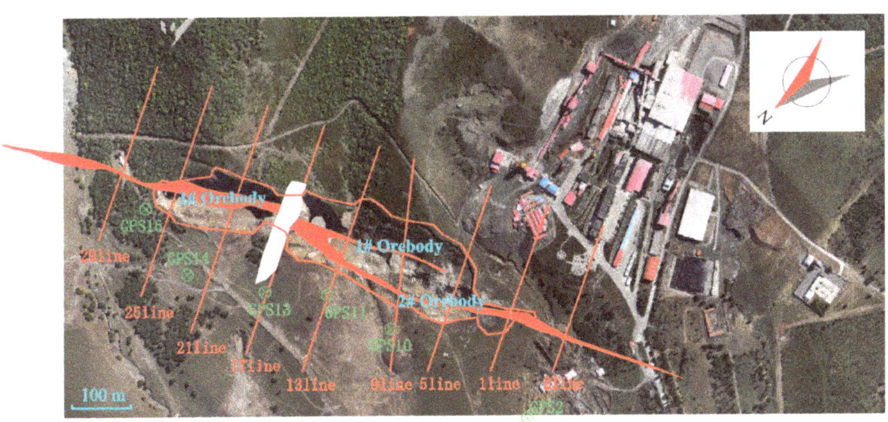

Figure 27. Position of GPS deformation observation stations.

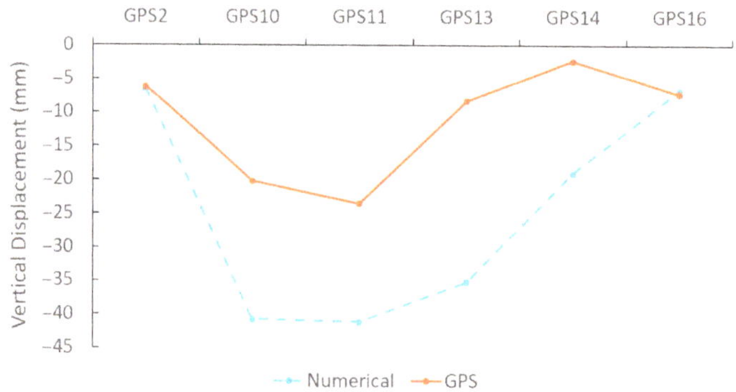

Figure 28. Comparison of the vertical displacement by GPS monitoring with the numerical results at the corresponding stations.

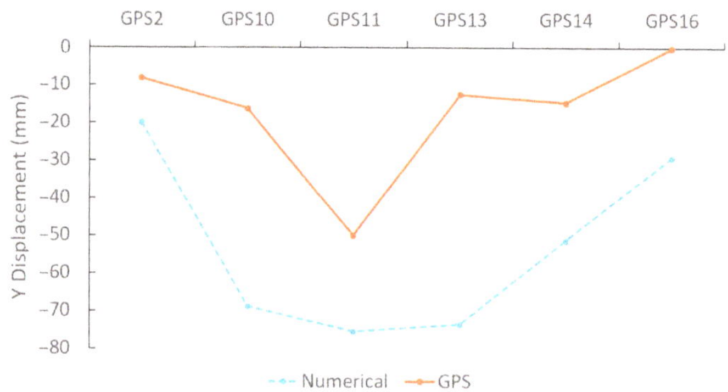

Figure 29. Comparison of the Y displacement by GPS monitoring with the numerical results at the corresponding stations.

9. Conclusions

This paper analysed the mechanism of rock mass displacement and surface subsidence induced by the excavation of near-surface steeply-inclined thick lead-zinc deposit by an integrated underground method of stoping and caving in Hongling Lead-Zinc Mine. A numerical model consisting of the orebody and surrounding rock, which were mainly combined by feldspar porphyry, marble and slate, was built from geological data by 3Dmine, MidasGTSNX and FLAC3D. The authenticity of modeling is strictly required when using this method for analysis, which can answer a lot of questions that cannot be explained by simplified models. UAV (Unmanned Aerial Vehicle) aerial survey was used to obtain the highly accurate surface topography, and combined with geological data such as mine borehole data, a real three-dimensional numerical calculation model of the mine was established. 2019 Li X using PFC2D surface subsidence for the red hill lead-zinc mine carried out some research [40], but because will be simplified into plane strain problem, it is not appropriate, cannot reveal the different parts in the form of the interaction in the process of rock mass in the mining and its damage, analysis of different rock mass destruction separately, have some limitations. The real three-dimensional numerical calculation model established in this manuscript can fully reflect the stress distribution, displacement change and failure of different rock masses in the process of mining during the simulation.

Relevant mechanical parameters were determined by laboratory tests or derivated by the Hoek-Brown criterion. Combined with the measured in-situ stress data, mining historical data and mining sequence, the development of collapses and subsidence caused was analyzed by numerical simulation. The results with high showed: (1) feldspar porphyry played an important role in supporting surrounding rock and surface of the orebody; due to continual stoping, the intersection of feldspar porphyry and orebody was completely plastic with a reduced support for the hanging wall; (2) the suspended marble caused by mine-out of hanging wall, footwall and its lower part had similar stress field with cantilever beam with the displacement of subsidence and torsion; the collapse of marble caused the horizontal and vertical extension of surface subsidence, the increase of exposed area of slate mass at hanging wall, and the collapse of the slate rock close to the hanging wall of ore body, which was consistent with the aerial image; (3) A large plastic zone ranging from the slate mass at hanging wall to the surface, connecting to the plastic zone of feldspar porphyry, brought the great potential to the appearance of large area of tension cracks on surface, which was uniform with the field observation. The simulation results had good correspondence with the evolution of the subsidence from 2009 to 2019. The current mining method integrating stoping and caving damaged the stability and safety of rock mass and surface, and the enlarging subsidence area put pressure on land acquisition, husbandry subsidy, water and soil conservation and mine reclamation. There was a risk to economic and safe operation of the mine and permanent destruction to the local ecological environment. It is suggested that the ground pressure control of the mine-out area and the management of surface subsidence should be carried out to achieve an ecological and economic all-win goal. Due to personal ability limitation, the cost of soil and water conservation and ecological reclamation is not calculated.

Author Contributions: Data curation, W.Y.; Funding acquisition, X.Z.; Writing—original draft, Y.Z.; Writing—review & editing, J.D. All authors have read and agreed to the published version of the manuscript.

Funding: This work was supported by the NSFC-Shandong Joint Fund (U1806208), State Key Laboratory of Nickel & Cobalt Resources Comprehensive Utilization, the Fundamental Research Funds for the Central Universities (N2001033).

Institutional Review Board Statement: Not applicable.

Informed Consent Statement: Not applicable.

Data Availability Statement: Data sharing not applicable.

Conflicts of Interest: The authors declare no conflict of interest.

References

1. Zhao, W.; Meng, Q. Researchon History and Development Trend of Domestic and Foreign Mining Subsidence. *J. Beijing Polytech. Coll.* **2010**, *9*, 12–15. [CrossRef]
2. Liu, B.; Liao, G. *Basic Law of Surface Movement in Coal Mine*; China Industrial Press: Beijing, China, 1965.
3. Jia, X.; Zhang, B.; Yang, N. Study on Classification of Destruction Degree of Mining Subsidence Land. *Coal Eng.* **2009**, *3*, 81–84. [CrossRef]
4. Yu, X.; Zhang, E. *Mining Damage Theory*, 2nd ed.; China Coal Industry Publishing House: Beijing, China, 2010.
5. He, G.; Yang, L.; Ling, G. *Mining Subsidence Theory*; China University of Mining & Technology Publisher: Xuzhou, China, 1991.
6. Ji, W. *Introduction to Collapse*; China City Press: Beijing, China, 1994.
7. Asaoka, A. Observational Procedure of Settlement Prediction. *Soils Found.* **1978**, *18*, 87–101. [CrossRef]
8. Whittaker, B.N. Surface subsidence aspects of room and pillar mining. *Nottm. Univ. Min. Dep. Mag.* **1985**, *1*, 59–67.
9. Bell, F.G.; Culshaw, M.G.; Cripps, J.C. *Engineering Geology of Underground Movement*; The Geological Society: London, UK, 1998.
10. Waltham, A.C. *Ground Subsidence*; Chapman and Hall: New York, NY, USA, 1989.
11. Zhang, Y. Prediction and control of surface subsidence in coal mine—Review and prospect of the turn of the century. In Proceedings of the 5th Youth Science and Technology Seminar of China Coal Society and the 3rd Youth Academic Annual Meeting of China Association for Science and Technology Satellite Conference, Taian, China, 5 November 1998; pp. 1–7.
12. Авершин, С.Г. СдвижениеГорныхПородПриПодземных Разработках; Coal Industry Press: Beijing, China, 1959.
13. Litwiniszyn, J. Fundamental principles of the mechanics stochastic medium. In Proceedings of the 3rd Conference Theoretical and Applied Mechanics, Bongalore, India, 24–27 December 1957.
14. Li, B.; Zhang, J.; Li, G. *Application of Stochastic Medium Theory in Mineral Engineering*; Hunan Science and Technology Press: Changsha, China, 2004.

15. Smolarski, A. *A Contribution to the Mechanics of a Horizontal-Homogeneous Stochastic Medium with Inclined Layers*; Bulletin de l'Academie Polonaise des Sciences cl: Warsaw, Poland, 1957; pp. 89–126.
16. Salamon, M.D.G. Elastic analysis of displacements and stresses induced by the mining of seam or reef deposits. *J. South. Afr. Inst. Min. Metall.* **1964**, *64*, 197–218.
17. Salamon, M.D.G. *Rock Mechanics of Underground Engineering*; Metallurgical Industry Press: Beijing, China, 1982.
18. Berry, D. An elastic treatment of ground movement due to mining—I. Isotropic ground. *J. Mech. Phys. Solids* **1960**, *8*, 280–292. [CrossRef]
19. Berry, D.; Sales, T. An elastic treatment of ground movement due to mining—II. Transversely isotropic ground. *J. Mech. Phys. Solids* **1961**, *9*, 52–62. [CrossRef]
20. Berry, D.; Sales, T. An elastic treatment of ground movement due to mining—III three dimensional problem, transversely isotropic ground. *J. Mech. Phys. Solids* **1962**, *10*, 73–83. [CrossRef]
21. Berry, D. An elastic treatment of ground movement due to mining—Corrigendum. *J. Mech. Phys. Solids* **1963**, *11*, 373–375. [CrossRef]
22. Siriwardane, H.J. A numerical procedure for prediction of subsidence caused by longwall mining. In *International Conference on Numerical Methods in Geomechanics*; August Aimé Balkema: Rotterdam, The Netherlands; Boston, MA, USA, 1985; pp. 1595–1602.
23. Alejano, L.R.; Ramírez-Oyanguren, P.; Taboada, J. FDM predictive methodology for subsidence due to flat and inclined coal seam mining. *Int. J. Rock Mech. Min.* **1999**, *36*, 475–491. [CrossRef]
24. Ambrožič, T.; Turk, G. Prediction of subsidence due to underground mining by artificial neural networks. *Comput. Geosci.* **2003**, *29*, 627–637. [CrossRef]
25. Xie, H.; Zhou, H.; Wang, J.; Li, L.; Kwasniew, S. Application and comparative analysis of FLAC in prediction of coal mining subsidence. *Chin. J. Rock Mech. Eng.* **1999**, *18*, 3–5. [CrossRef]
26. Huang, Y.; Fu, J.; Guo, Y. Research on Surface Strata Displacement Law for Mining the Steeply Inclined Thin Vein in the Ultra-deep well. *Min. Res. Dev.* **2015**, *35*, 59–62. [CrossRef]
27. Li, X.; Li, D.; Liu, Z.; Zhao, G.; Wang, W. Determination of the minimum thickness of crown pillar for safe exploitation of a subsea gold mine based on numerical modelling. *Int. J. Rock Mech. Min. Sci.* **2013**, *57*, 42–56. [CrossRef]
28. Zhou, K.; Du, X. Study on Stability of Residual Ore Recovery Based on Coupling of 3DMINE-MIDAS /GTS-FLAC3D. *China Saf. Sci. J.* **2011**, *21*, 17–22. [CrossRef]
29. Li, H.; Wang, W.; Xiao, Y.; Liu, H.; Liu, T. Stability analysis of actual goaf group based on Faro-3DMine and Midas-FLAC3D coupling technology. *Nonferrous Met. (Min. Sect.)* **2019**, *71*, 14–18. [CrossRef]
30. Bertuzzi, R.; Douglas, K.; Mostyn, G. Comparison of quantified and chart GSI for four rock masses. *Eng. Geol.* **2016**, *202*, 24–35. [CrossRef]
31. Hoek, E.; Carranza-Torres, C.; Corkum, B. Hoek-Brown failure criterion-2002 edition. *Proc. NARMS-Tac* **2002**, *1*, 267–273.
32. China, T.M.O.W. *GB50218-94 Standard for Engineering Classification of Rockmasses*; China Planning Press: Beijing, China, 2014.
33. Hao, X. Stability Analysis and Support Research on Rock Mass of +805 Level in Hongling Lead-Zinc Mine. Master's Thesis, Northeastern University, Shenyang, China, 2015.
34. Hoek, E.; Brown, E.T. Empirical strength criteria for rock masses. *J. Geotech. Eng. Div. Am. Soc. Civ. Eng.* **1980**, *106*, GT9.
35. Hoek, E.; Wood, D.; Shah, S. In A modified Hoek-Brown failure criterion for jointed rock masses. *ISRM Symp. Eurock* **1992**, *1*, 209–214. [CrossRef]
36. Hoek, E.; Diederichs, M. Empirical estimation of rock mass modulus. *Int. J. Rock Mech. Min. Sci.* **2006**, *43*, 203–215. [CrossRef]
37. Hoek, E.; Karzulovic, A. Rock mass properties for surface mines. In *Slope Stability in Surface Mining*; Hustrulid, W.A., McCarter, M.K., van Zyl, D.J.A., Eds.; Society for Mining, Metallurgical and Exploration (SME): Littleton, CO, USA, 2000; pp. 59–70.
38. Wang, K.; Diao, X.; Lai, J.; Huang, G. Engineering application comparison of strain softening model and Mohr-Columb model in FLAC3D. *China Sci.* **2015**, *10*, 55–59. [CrossRef]
39. Liu, H.; He, C.; Deng, K.; Bianca, Z.; Fan, H.; Lei, S.; Zhang, A. Analysis of forming mechanism of collapsing ground fissure caused by mining. *J. Min. Saf. Eng.* **2013**, *30*, 380–384.
40. Li, X.; Wang, D.; Li, C.; Liu, Z. Numerical Simulation of Surface Subsidence and Backfill Material Movement Induced by Underground Mining. *Adv. Civ. Eng.* **2019**, *2019*, 2724370. [CrossRef]

processes

Article

Modeling the Effects of Seasonal Weathering on Centrifuged Oil Sands Tailings

Umme Salma Rima [1,*], Nicholas Beier [2] and Ahlam Abdulnabi [3]

[1] Department of Civil and Environmental Engineering, CNRL/Markin Natural Resources Engineering Facility, University of Alberta, Edmonton, AB T6G 2W2, Canada
[2] Department of Civil and Environmental Engineering, Donadeo Innovation Centre for Engineering, University of Alberta, 9211-116 St NW, Edmonton, AB T6G 1H9, Canada; nabeier@ualberta.ca
[3] Thurber Engineering Ltd., 350-7330 Fisher St SE, Calgary, AB T2H 2H8, Canada; aabdulnabi@thurber.ca
* Correspondence: umme@ualberta.ca

Abstract: The oil sands industry employs different technologies at pilot and commercial demonstration scales in order to improve the dewatering rate of fluid fine tailings. Of these technologies, centrifugation has advanced to the commercial scale and is playing a major role in the fluid fine tailings management strategy. However, centrifuge technology on its own may not develop the required strength to ensure fine tailings can be incorporated into dry landform reclamation, which requires water contents close to their plastic limit. Hence, it is paramount to combine more than one technology to maximize post-depositional dewatering. Management of the tailings deposit to promote seasonal weathering (freeze–thaw, evaporation and self-weight consolidation) can promote further dewatering. Properly assessing the contributions of the seasonal weathering components is vital to optimizing this strategy. Using the geotechnical properties of centrifuged tailings, the effects of seasonal weathering on tailings were modeled under two different freezing temperature gradients. A coupled analysis combining FSConsol and Unsatcon was used to simulate the deposition scenario similar to the laboratory. The modeling results were found to match the laboratory response reasonably well, indicating the coupled approach proposed in this manuscript is valid and helps to predict the seasonal weathering effects on dewatering.

Keywords: fluid fine tailings; dewatering; modelling; seasonal weathering; freeze–thaw; evaporation

Citation: Rima, U.S.; Beier, N.; Abdulnabi, A. Modeling the Effects of Seasonal Weathering on Centrifuged Oil Sands Tailings. *Processes* **2021**, *9*, 1906. https://doi.org/10.3390/pr9111906

Academic Editor: Li Li

Received: 19 September 2021
Accepted: 22 October 2021
Published: 26 October 2021

1. Introduction

Alberta, Canada is considered to be the fourth-largest proven reserve of crude bitumen in the world [1]. Most of Alberta's oil is unconventional as it is trapped within oil sands and, hence, traditional drilling and pumping methods using the natural pressure differential cannot be employed here [2]. Instead, advanced extraction techniques such as oil sands mining and in-situ development are needed to extract the heavier oil/bitumen. With a combined estimated reserve of 1.8 trillion barrels of in-place reserves of in-situ crude bitumen, the Athabasca, cold Lake and Peace River deposits form a massive resource in Alberta. Of these regions, the Athabasca oil sand deposit (situated in northeastern Alberta in the Fort McMurray area) is the largest and only one to be shallow enough to allow for surface mining [3]. The extraction of bitumen from oil sands in a surface mining operation is a water-based process that generates large volumes of byproducts known as tailings. In general, tailings are a warm suspension of sand, fines (clays and silts), residual bitumen and process-affected water [4]. These are temporarily stored aboveground in dams referred to as tailings ponds, where the mixtures of coarse streams (primarily sand) form settled sand beaches near the deposition outlet and an aqueous slurry of fines and residual bitumen accumulates in the center of these ponds termed as thin fine tailings (TFT) [4,5]. When allowed to settle under quiescent conditions (self-weight consolidation under no further loading), TFT forms a material with a solids content (mass of solids divided by the total mass of tailings including bitumen) of around 30-40% by mass referred to as mature fine tailings (MFT) [6]. Dewatering of fluid fine tailings (the collective term for TFT and MFT) to recycle the released water and reduce the environmental footprint is very slow as these materials are highly dispersed, resistant to consolidation and can remain in a soft, fluid state for decades, thus creating a unique fundamental management problem for the oil sands industry [7]. Consequently, the inventory of tailings being stored in the ponds, covering a total area of 259 km^2, has been steadily increasing over time, and at present, the total volume of fluid fine tailings (FFT) stored in the pond already exceeds 1302 million m^3 [7,8]. In order to meet regulatory and closure requirements, FFT is needed to be dewatered so that these large volumes of FFT can be accommodated in the development of an environmentally acceptable reclamation plan.

In order to dewater and facilitate the reclamation efforts of the surface-mined FFT, different chemical, mechanical and environmental processes have been employed that could have the potential for effective tailings management in the oil sands industry [5]. One of these technologies, centrifugation, is currently being used by Syncrude and Canadian Natural Upgrading Ltd. (CNUL) as a key process technology to accelerate dewatering [6,7]. Centrifugation employs dredging FFT from the tailings pond and treating it with flocculant and/without coagulant prior to feeding to the centrifuge where solids are separated from the water via a centrifugal force [5]. The endproduct of centrifugation is known as centrifuged tailings. However, the achieved solids content from this technology is typically 50–55% [5], which is not sufficient (strength less than 1 kPa, as documented in [5,9]) to develop a trafficable surface (atleast 25 kPa for mobile equipment trafficability, as documented in [5]). Additionally, the oil sands industry is currently more focused on creating deep deposits (typically > 10–20 meters deep) to eliminate the large footprints typically required for thin-layered deposits [5,10]. However, deep deposits of these centrifuged tailings undergo settlement for centuries due to the extremely slow settlement times resulting from decreasing permeability as the deposit densifies [10].

Since environmental dewatering processes (freeze–thaw dewatering, evaporation, desiccation) are economical and cost-efficient, a combination of these processes can be considered as additional dewatering technologies contributing to creating a reclaimable deposit. Figure 1 shows a simplified diagram of the environmental processes contributing to dewatering from the tailings.

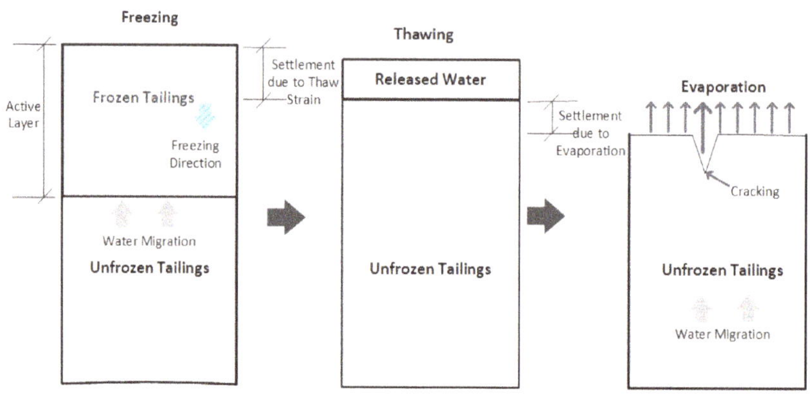

Figure 1. Simplified diagram of the seasonal weathering cycles.

The freezing process results in the formation of a three-dimensional reticulate ice network surrounding blocks of over-consolidated tailings. Upon thawing, dewatering occurs due to these structural changes within the frozen tailings, which in turn facilitates water removal from the underlying thawed tailings as ice melts [11]. Hence, tailings undergoing freeze–thaw dewatering develop thaw strain (ε) (the change in height prior to and after thawing divided by the total frozen height) that further allows for subsequent post-thaw consolidation through self-weight consolidation, thereby causing an increase in effective stress [11]. Consequently, the solids content increases and the void ratio decreases as the tailings material consolidates at a faster rate under self-weight [12]. When evaporation/drying is incorporated with freeze–thaw dewatering, the thawed tailings further dewater under the desiccation process, which is the process of drying and cracking [4]. As a result, further dewatering and higher undrained shear strength at the surface can be achieved, enabling reclamation and closure. Among all the environmental processes, the natural process of freeze–thaw dewatering has shown promise as a method to dewater, strengthen and reclaim FFT (as investigated by [11–14]). However, much of these works were focused on using thin-layered freeze–thaw dewatering but not so much on deep deposits. Hence, there is a need to develop a fundamental understanding of the effects of these environmental processes on the dewatering performances of deep-deposit tailings.

The objective of the research reported in this paper was to evaluate numerical approaches in order to simulate the dewatering of centrifuged tailings subjected to seasonal weathering under a controlled laboratory testing program. A coupled analysis methodology was developed here to validate two sequences of laboratory testing under two different freezing temperature gradients.

2. Tailings Material and Characterization

The centrifuged tailings samples studied in this research were received at the University of Alberta Laboratory in a 200L barrel from Syncrude Canada Ltd. The samples were homogenized thoroughly with a mixer and the mineralogy and geotechnical properties were determined upon delivery (Table 1).

Table 1. Geotechnical properties of the centrifuged tailings.

Property	Value
Water content, w (%)	89
Solids content, s (%)	53
Bitumen content (%)	5.7
Specific gravity, G_s	2.24
[1] Fines content (%)	87
[2] Clay content (Dispersed hydrometer) (%)	52
[3] Clay content (MBI) (%)	52
[4] D_{50} (μm)	1.5
Liquid limit (%)	57
Plastic limit (%)	26
Liquidity index	2

[1] Fines content = Material finer than 0.045 mm. [2] Clay content = Material finer than 0.002 mm. [3] Clay content by Methylene Blue Index (MBI). [4] Median particle diameter.

The initial water content (mass of water divided by the mass of dry solids including bitumen) of the sample was found to be 89% by mass, corresponding to a solids content of 53% by mass. The properties of the as-received centrifuged tailings show a fine-grained tailings material with a higher amount of clay content, high plasticity along with moderate water adsorption onto the clays. These values are influenced by the combined effects of the geologic origin, clay mineralogy, water chemistry and bitumen content [15].

3. Laboratory Setup

Figure 2 shows the schematic diagram of the laboratory freeze–thaw test setup. These small-scale freezing tests were carried out in cylindrical freezing cells (0.1 m dia × 0.22 m height) within a walk-in freezer where the samples were frozen from top-down (one dimensional) under two different temperature gradients of 0.083 and 0.028 °C/mm. These two temperature gradients were applied to the freezing cells through two temperature baths where the top boundary temperatures were set at −15 (to achieve 0.083 °C/mm) and −5 °C (to achieve

0.028 °C/mm) and the bottom boundary temperature was set at 0 °C in order to replicate the average temperature of Fort McMurray. The installation of the insulation wrap and the thermoelectric cooling plate were all applied in order to ensure one-dimensional freezing and to represent the freezing process that occurs in nature.

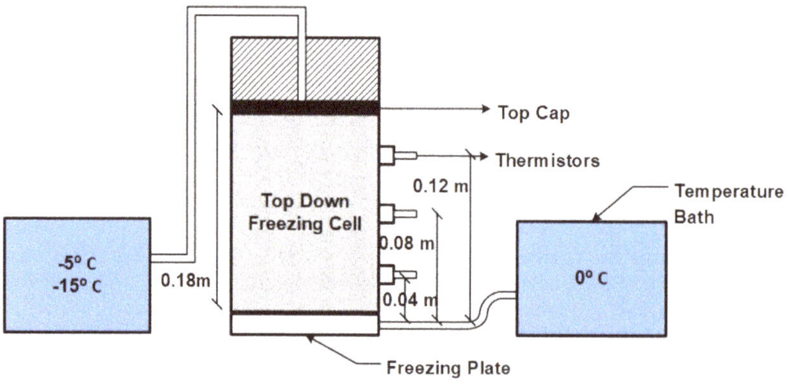

Figure 2. Schematic diagram of the freeze–thaw test setup (after [9,16]).

For the drying–wetting cycles, the drying tests were carried out in the freezing cells at room temperature (~20 °C) under which the weight loss due to evaporation was recorded daily along with the subsequent shear strength measurements. The volume of water that evaporated during each drying cycle was poured back into the cells to simulate the rainfall/wetting event.

Laboratory investigations were carried out in two phases: First-phase testing that is associated with five consecutive freeze–thaw cycles followed by a single cycle of drying–wetting–re-drying after wetting; and second-phase testing includes five alternate freeze–thaw and drying–wetting cycles to better represent the natural seasonal cycles. For both phases, each of the freeze–thaw cycles took seven days to complete, whereas the drying–wetting cycles varied. For the first phase of testing, the centrifuged tailings samples were subjected to a month-long (dried to a target actual evaporation/potential evaporation (AE/PE) ratio of 0.7 for each test) drying cycle followed by a single wetting event to simulate the rainfall. After the wetting event, another drying cycle was continued for around twenty days. Further, the second phase of testing included five drying–wetting cycles and hence, each of the drying cycles was run for a shorter duration (seven days) prior to a single wetting event. After the wetting event, the re-drying cycle was continued for another seven days. The detailed procedures of the first-phase and second-phase testing were documented in [9,16].

4. Coupled Modeling

4.1. Modeling Analysis Development

A one-dimensional coupled modeling approach was developed to simulate the effects of multiple freeze–thaw and drying–wetting cycles in the laboratory. The modeling exercises were conducted to simulate the laboratory results (water content) of centrifuged tailings samples subjected to seasonal weathering (multiple freeze–thaw consolidation and drying–wetting cycles) under two different temperature gradients (0.083 and 0.028 °C/mm), as mentioned in the above section. In this study, a coupled modeling analysis was conducted in two steps. First, the freeze–thaw analysis was coupled with the FSConsol model to incorporate the freeze–thaw process into consolidation modeling. Next, the coupled FSConsol model was coupled with the UNSATCON model to further incorporate the evaporation/drying cycles followed by the freeze–thaw cycles. FSConsol is a commercially available one-dimensional consolidation program that incorporates the large strain consolidation theory from Gibson et al. [17], whereas the UNSATCON program developed by Qi and Simms (documented in [18]) is a research code that simulates the tailings dewatering process induced by self-weight consolidation and evaporation while considering the stress/desiccation history and hydraulic hysteresis. Since none of these models incorporates the freeze–thaw consolidation directly, the change in water content due to thaw strain was calculated externally in an Excel sheet and applied to the active layer (the top layer of the ground that experiences above (thawing) and below 0 °C (freezing) during the year, also known as the frost depth (shown in Figure 1)) of the deposit (for the laboratory test, the active layer was equal to the total thickness of each of the samples) in FSConsol to account for the freeze–thaw consolidation.

Figure 3 shows the flowchart diagram of the coupling analysis. FSConsol was first run where the initial (height and void ratio) and boundary conditions similar to the laboratory were applied. A hydrostatic condition was assumed for the very first run. The boundary conditions at the top were specified as a constant water cap thickness of zero so that all the fluid will be drained off the top as the tailings consolidate. Similarly, the bottom boundary condition was specified as impermeable so that no fluid can exit through the bottom of the cell to simulate the laboratory condition. For numerical modeling of the oil sands tailings, the large strain consolidation theory is generally adopted [4] that requires one to input the compressibility (void ratio, e and effective stress) and saturated hydraulic conductivity (K_{sat}—void ratio) properties in FSConsol to be obtained from a large strain consolidation test. Using the large strain consolidation apparatus, the material properties of each of the samples were determined. After the consolidation analysis, the output results such as the void ratio (e), pore water pressure (pwp) and solids content (s) at different pre-sets depths were recorded. Next, solids contents were converted to the bulk density using the mass–volume relationship, and the thaw strain was applied to the analysis. The relationship between the thaw strain and bulk density for the centrifuged tailings samples under temperature gradients of 0.083 and 0.028 °C/mm is represented below using Equations (1) and (2), respectively:

$$\varepsilon = 0.0684 - 0.079 \ln\rho, \tag{1}$$

$$\varepsilon = 0.1887 - 0.276 \ln\rho, \tag{2}$$

where ε represents the thaw strain (unitless) and ρ represents the bulk density in gm/cm^3. The above two equations were obtained by fitting the laboratory testing data from the ε–ρ relationship. The constants of these equations are dependent on the boundary conditions and hence will be changed based on the available thaw strain data for the particular deposit. Based on the thaw strain, the void ratio and solids content profile from the initial FSConsol run were adjusted. The total height and the heights of all the observation points were adjusted accordingly due to the thaw settlements. Consequently, the input for the next FSConsol run allows for incorporating the freeze–thaw consolidation process (denoted as the F/T cycle to represent freeze–thaw cycles in the diagram). This was accomplished by changing the initial conditions to reflect the thawed void ratio (based on thaw strain) and pore water pressures (from the previous run) to be applied at the adjusted preset depths in the profile. All these steps prior to the coupling with UNSATCON were repeated during consecutive multiple freeze–thaw cycles (for example, the first-phase testing simulation where FSConsol incorporating the thaw strain analysis needs to be repeated for subsequent freeze–thaw cycles).

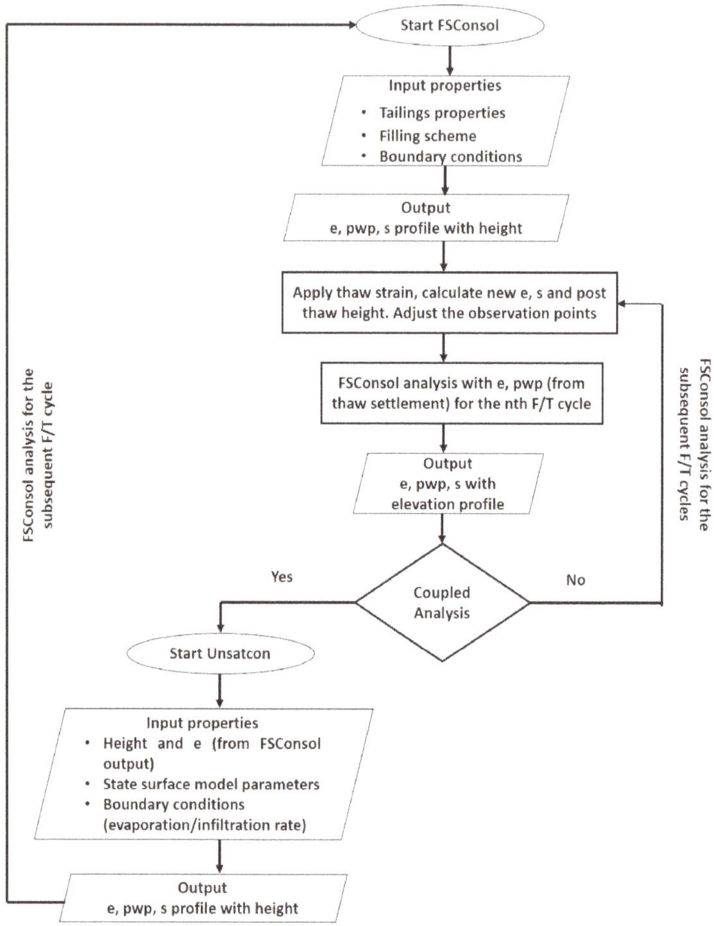

Figure 3. Flowchart diagram of coupling analysis.

Whenever a drying cycle was introduced, the coupled analysis was run by switching from FSConsol to the UNSATCON model (for example, the second-phase testing simulation where alternating analyses were required in each cycle). Hence, the output results (void ratio, elevation) from FSConsol were applied as an initial condition for the UNSATCON model. For modeling the unsaturated soil behaviour, the constitutive model (based on the state surface modeling approach) was selected, in which the void ratio and water content were expressed as the functions of the net normal and matric suction in the 3D space [19]. The parameters of this state surface model for the present study were obtained from the test conducted by Hurtado [20] on similar centrifuged tailings. The evaporation rate (in mm/day) was applied as a top boundary condition obtained from the laboratory tests, whereas no water flux through the bottom was allowed.

For the numerical modeling, a sequence similar to the laboratory was followed. Figure 4 shows the sequences of laboratory and modeling simulation of the first- and second-phase testing. However, due to the complexity of the coupling analysis in the second-phase testing (five alternate freeze–thaw and drying–wetting cycles), re-drying after the wetting event was not simulated in the numerical model.

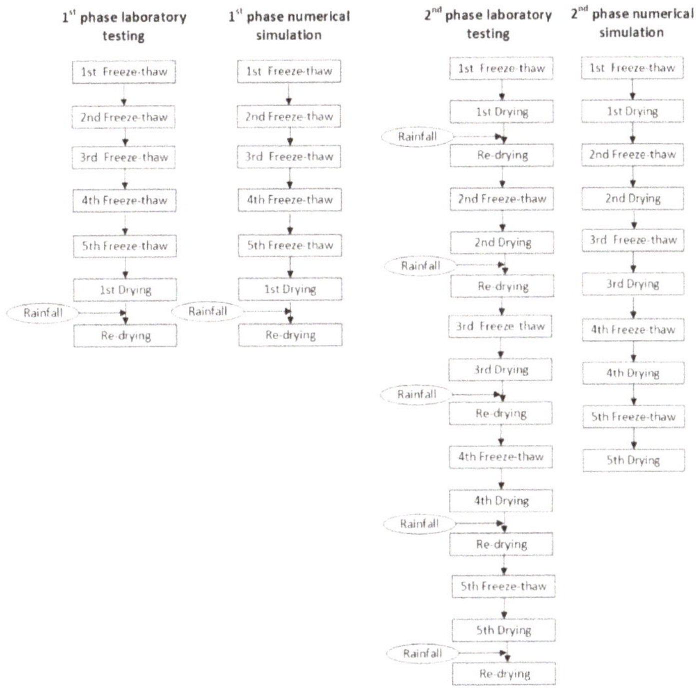

Figure 4. First- and second-phase laboratory testing and numerical simulation sequences.

4.2. Numerical Model Parameters

Tables 1–3 show the nessesary boundary conditions, intial conditions and summary of material and model parameters for the FSConsol and UNSATCON models, respectively. The initial thickness of each of the samples was 0.18 m, similar to the laboratory testing.

Table 2. Boundary conditions and summary of parameters for the FSConsol models.

Freeze–Thaw Cycle	Boundary Conditions	Material Properties				
		Compressibility ($e = A.\sigma'^B + M$) *			Permeability ($k = C.e^D$) **	
		A	B	M	C	D
0	Top: Constant water cap: Thickness 0 m Bottom: Impermeable	5.9548	−0.149	0	2×10^{-13}	15.832
1		3.7377	−0.123	0	2×10^{-12}	23.594
2		3.7956	−0.135	0	1×10^{-10}	11.465
3		3.516	−0.146	0	2×10^{-10}	10.723
4		3.516	−0.146	0	2×10^{-10}	10.723
5		3.516	−0.146	0	2×10^{-10}	10.723

* ($e = A.\sigma'^B + M$) Here, e = Void ratio and σ′= Effective stress in Pascal. ** ($k = C.e^D$) Here, k = Permeability in m/s.

To achieve numerical stability of the FSConsol model, a time step of 5 h was specified. The spatial discretization of this one-dimensional model was a total of 100 nodes, as per the recommendations provided by the manual of FSConsol. The sensitivity analysis was run with different timesteps including an hour and ten hours, and no significant change was observed.

Please note that the state surface model was developed by Qi [19], who solved a set of constitutive relationships incorporating volume change and water retention behaviour of unsaturated soils using finite difference techniques. All these in-depth formulations and algorithms were not studied in this paper. Only the parameters applied to similar centrifuged tailings were inputted in the models and these tests were performed by Hurtado [20].

Table 3. Boundary conditions and summary of parameters for the UNSATCON model.

Property			Value
Boundary conditions	Top (all cycles)		Desiccation is enabled. Evaporation rate data from laboratory (varied in each cycle)
	Bottom (all cycles)		No flux
State surface model parameters (mechanical: void ratio surface)	Plastic	a	2.4
		b	0.33
		c	0.015
		d	0.03
		f	6000
		g	5000
	Elastic	kappa	0.005
		kappa_s	0.001
State surface model parameters (hydraulic: water content surface)	Primary	C_d0	3
		C_w0	1.35
		lambda_se	0.15
		Lambda_sr	0.17
	Hysteresis	kappa_ss	0.04
Permeability	Multiplier	1st cycle	2×10^{-13}
		2nd cycle	2×10^{-12}
		3rd cycle	1×10^{-10}
		4th cycle	2×10^{-10}
		5th cycle	2×10^{-10}
	Power	1st cycle	23.594
		2nd cycle	11.465
		3rd cycle	10.723
		4th cycle	10.723
		5th cycle	10.723
	M (unsaturation effect)	All cycles	0.75
Numerical parameters	Number of nodes	All cycles	10
	Time step (s)		9

5. Modeling Results

5.1. Numerical Simulation of First-Phase Testing

The first-phase numerical modeling of laboratory testing was run for five consecutive freeze–thaw cycles under two freezing temperature gradients of 0.083 and 0.028 °C/mm followed by a single drying–wetting cycle. Two separate scenarios were modeled here using coupling analysis incorporating the material properties, volume change relationships and the laboratory testing sequences reported above. The model-predicted water contents along with the change in elevation after each cycle (denoted as F/T cycle for each freeze–thaw cycle and D-W cycle for the drying–wetting cycle in the figures) are displayed in Figures 5 and 6. Both figures show that the water contents were consistently decreasing with the freeze–thaw cycles, and by the fifth cycle, nearly half of the water had been lost (44 and 46% reduction in water content for the temperature gradients of 0.083 and 0.028 °C/mm, respectively). When drying/evaporation was incorporated in the numerical simulation, both these samples further dewatered to two-fold lower water content (Figure 5) for the higher-gradient sample and five-fold lower water content (Figure 6) for the lower-gradient sample, as compared to the initial value.

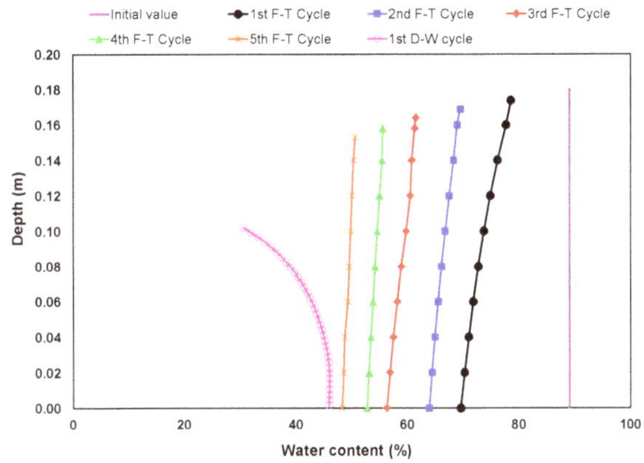

Figure 5. Water content profile simulation of centrifuged tailings at a temperature gradient of 0.083 °C/mm.

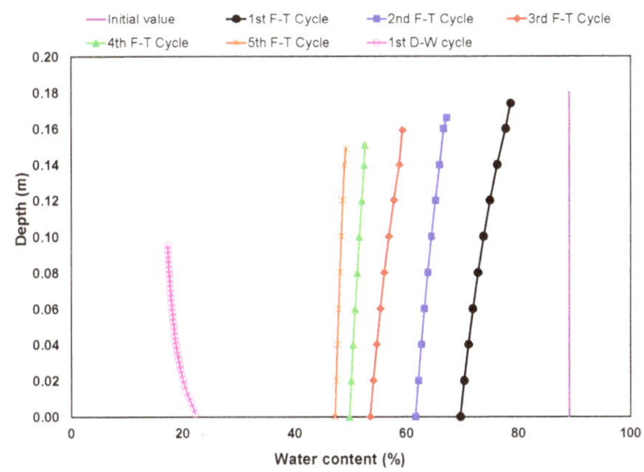

Figure 6. Water content profile simulation of centrifuged tailings at a temperature gradient of 0.028 °C/mm.

The centrifuged tailings samples at freezing temperature gradients of 0.083 and 0.028 °C/mm resulted in water contents of 48 and 47% at the surface, respectively, after five consecutive freeze–thaw cycles (the initial water content was 89%). Although the reduction in water contents for both the samples did not differ significantly from each other during the freeze–thaw analysis, the drying analysis from the UNSATCON model resulted in higher dewatering on the lower gradient (0.028 °C/mm) sample compared to the sample subjected to higher gradient. The lower gradient sample (as shown in Figure 6) experienced an average water content of 19% throughout the sample with the lowest one observed at the surface (17%). Conversely, the sample subjected to a higher freezing temperature gradient of 0.083 °C/mm resulted in an average water content of 42% (shown in Figure 5), with the lowest being observed at the surface was 31%. The final surface elevations for the higher- and lower-gradient samples were found to be 0.102 (43% reduction in thickness compared to the initial value) and 0.095 m (47% reduction in thickness compared to the initial value), respectively, at the end of the model run.

5.2. Numerical Simulation of Second-Phase Testing

The second-phase numerical simulations of the laboratory testing were run for five alternate freeze–thaw and drying–wetting cycles where each freeze–thaw cycle was followed by a drying–wetting–re-drying cycle, and this sequence was repeated five times per sample to represent five seasonal years in the field. The wetting event per cycle was introduced as a single event in the laboratory to simulate the rainfall where the volume of water that evaporated during atmospheric drying/evaporation for that particular cycle was poured back into the cell. As a result, the tailings surface was re-wetted, thereby allowing an increase in water content similar to the pre-drying phase. Upon wetting, the samples went through another drying cycle (with a similar time duration to the pre-wetting drying cycle) to investigate the tailings dewatering behaviour prior to and after the wetting event. The laboratory results suggest that the gain in solids content (or reduction in water content) achieved during the first seven days of the drying period was entirely depreciated by introducing the wetting event (the same amount of water was poured back into the cell that was evaporated) [9]. When the samples were re-dried for another seven days (at the end of fourteen days), upon wetting, an increase in solids content was observed similar to the pre-wetting cycle (within a difference in values of 0–1.4% by mass prior to and post-wetting drying). However, it was the last two cycles (fourth and fifth cycles) where the wetting event did not have any significant impact on increasing the water content of the tailings (particularly significant for the lower-gradient sample) because of the possible higher suction, thereby resulting in a further decrease in water content after fourteen days [9]. Nevertheless, incorporating all these components using an alternating simulation from FSConsol and UNSATCON in every cycle was very complex and, hence, the simulation was simplified by excluding the wetting and re-drying cycles.

Figures 7 and 8 show the model-predicted water content profiles for the two samples after each cycle (freeze–thaw and drying cycles were shown separately and denoted as F/T cycle for each freeze–thaw cycle and D cycle for drying cycle in the figures). The initial water content and elevation were found to be 89% and 0.18 m, respectively. The coupling analysis to simulate the second phase of testing shows that the centrifuged tailings samples subjected to temperature gradients of 0.083 and 0.028 °C/mm responded quite differently to seasonal weathering. After five alternate freeze–thaw and drying cycles, the samples with higher (0.083 °C/mm) and lower temperature gradients (0.028 °C/mm) were dewatered to nearly four-fold (surficial water content value was 25%, as shown in Figure 7) and six-fold (surficial water content value was 16%, as shown in Figure 8) lower water contents, respectively, at the surface when compared to the initial water content value. Both the samples dewatered similarly for the first two seasonal cycles (the first freeze–thaw–drying and second freeze–thaw cycles). It was during the second drying cycle that dewatering between the two samples started to differ. Evaporation from the lower temperature gradient sample reduced the water content by 29% at the surface compared to the other sample at the end of the second cycle. At the end of the third, fourth and fifth cycles, the sample subjected to a temperature gradient of 0.028 °C/mm resulted in a 28, 31 and 36% reduction in water contents at the surface, respectively, compared to the sample under a higher freezing temperature gradient (0.083 °C/mm). Both the higher- and lower-temperature-gradient samples dewatered considerably after five seasonal cycles with final elevations of 0.102 (43% reduction in thickness from the initial value) and 0.097 m (46% reduction in thickness from its initial value), respectively. The bottom few hundredths of a meter of the sample were not affected much by the seasonal weathering model run.

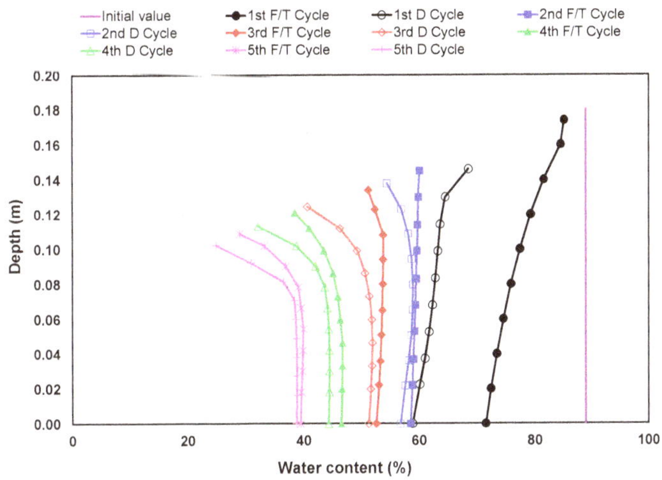

Figure 7. Water content profile simulation of centrifuged tailings at a temperature gradient of 0.083 °C/mm.

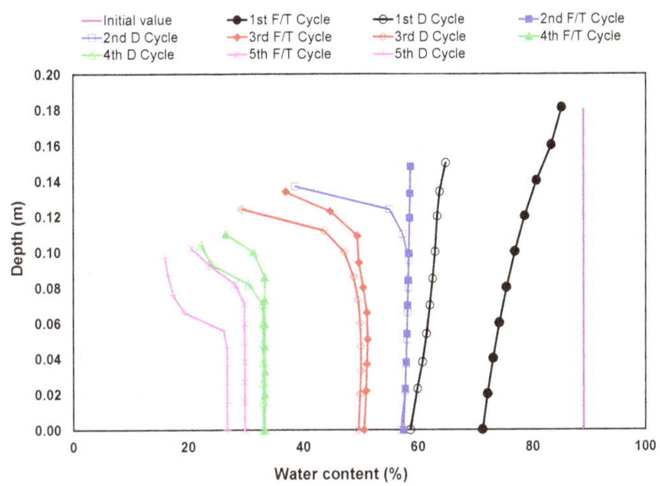

Figure 8. Water content profile simulation of centrifuged tailings at a temperature gradient of 0.028 °C/mm.

5.3. Comparison between the Model and Laboratory Results

Figures 9 and 10 illustrate the cumulative decrease in water content for the two centrifuged tailings samples under the temperature gradients of 0.083 and 0.028 °C/mm, respectively, after each seasonal cycle, where Figure 9 compares the first-phase laboratory testing and modeling results and Figure 10 compares the second-phase testing results between laboratory computation and model prediction. The water content obtained from the laboratory after each seasonal cycle was calculated on the basis of the thaw strain during freeze–thaw cycles and the changes in weight loss/gain during drying–wetting cycles. Hence, the water content per cycle obtained from the laboratory represents the average water content throughout the depth. In terms of numerical modeling, both the average and surface values were reported on the graph as the deposit surface is expected to dewater the most, given it is the most susceptible to the effects of seasonal weathering.

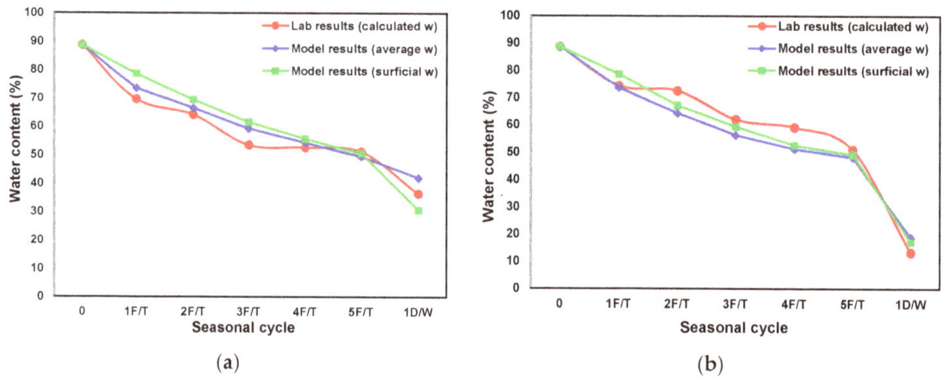

Figure 9. First-phase testing comparisons: Water content values per seasonal cycle for the centrifuged tailings samples under a temperature gradient of (**a**) 0.083 °C/mm and (**b**) 0.028 °C/mm.

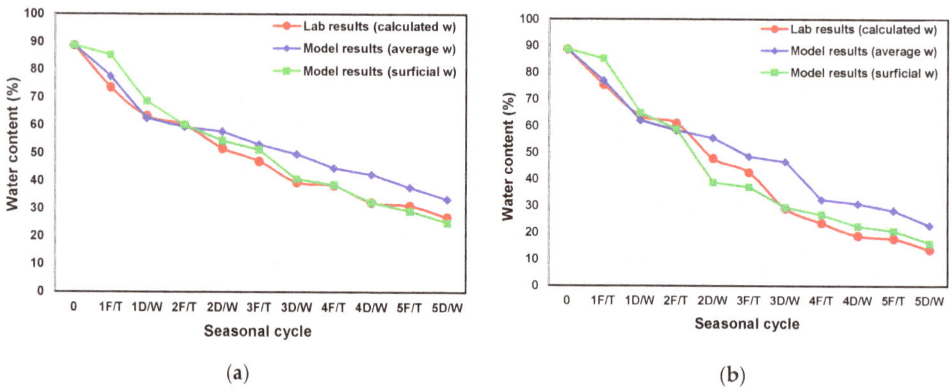

Figure 10. Second-phase testing comparisons: Water content values per seasonal cycle for the centrifuged tailings samples under a temperature gradient of (**a**) 0.083 °C/mm and (**b**) 0.028 °C/mm.

As evident in Figure 9, the differences between the model prediction and the laboratory computational results were marginal. However, the model overpredicted water content values compared to the laboratory results for the higher-temperature-gradient sample (Figure 9a), while it showed the reversed pattern for the lower-temperature-gradient sample (Figure 9b). The second-phase testing, as shown in Figure 10, shows a different trend. The laboratory results for both samples corroborated well with the model-predicted water contents at the surface.

Figure 11 shows the comparisons between the final (at the end of the tests) laboratory-measured and model-predicted water content profiles for the two centrifuged tailings samples subjected to temperature gradients of 0.083 and 0.028 °C/mm, respectively.

As shown in Figure 11a, first-phase testing results indicate that the simulation for the higher-temperature-gradient sample overall underpredicted water content values by 3%, as the average water content values along the depth were found to be 42 and 45%, respectively, from the modeling prediction and the laboratory measurement. Likewise, the lowest water content, observed at the surface, was found to be 31 and 39%, respectively, from the modeling and laboratory results (about an 8% underprediction of water content compared to the laboratory value). Conversely, the laboratory and model-predicted water content profiles correlated quite well for the lower gradient sample (Figure 11b). The model overall underpredicted water content values by about 1% throughout the depth and overpredicted this value by about 3% at the surface compared to the laboratory measurement. The model predicted an average water content of 19% throughout the depth with the lowest water content of 17% observed at the surface, whereas the laboratory-measured average value was found to be 20% along with the lowest water content of 14% observed at the surface. All these values from the lower-gradient sample suggest a stiff/solid consistency for the tailings, provided these values passed through the plastic limit of 26% (Figure 12).

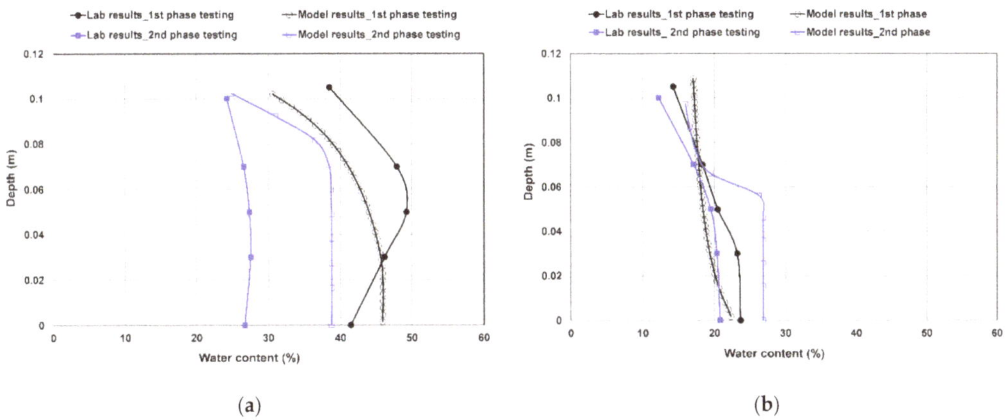

Figure 11. Laboratory versus. model water content profiles of centrifuged tailings at the end of the test under a temperature gradient of (**a**) 0.083 °C/mm and (**b**) 0.028 °C/mm.

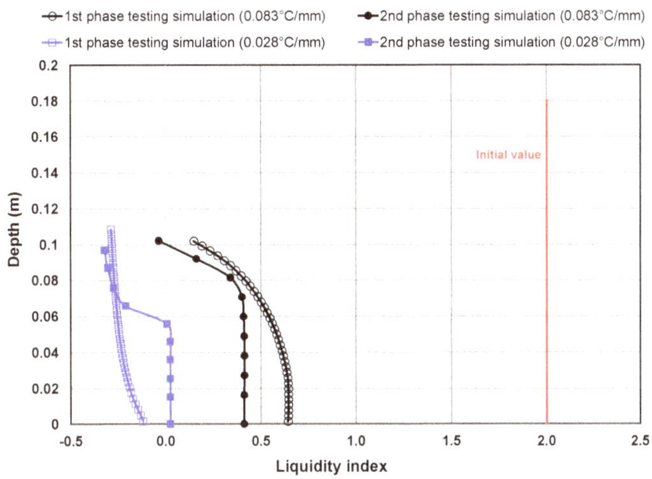

Figure 12. Liquidity index profiles of centrifuged tailings from coupled analysis.

As mentioned above, the numerical model to simulate the second-phase testing excluded the wetting event and re-drying cycle after wetting. Hence, the model predictions were expected to achieve higher water contents compared to the laboratory results. However, the laboratory results suggest that the water content reduction/solids content gain just prior to wetting (at the end of the seven-day drying cycle prior to the wetting event) and the post-wetting drying phase (at the end of another seven-day drying cycle post-wetting) was invariable for the higher-temperature-gradient sample (0.083 °C/mm) and, therefore, this sample is not supposed to be impacted much because of these wetting and re-drying event exclusions. It was the lower-gradient sample where the water content reduction/solids content gain prior to and after the wetting event was not similar at the higher cycles (fourth and fifth cycles). This is due to the higher suction values that prevented rainfall possibly causing any impact on the increasing water content upon wetting [9]. However, the modeling results that simulated the second-phase testing (as shown in Figure 11) suggest that the model overall overpredicted water content values by about 10 and 5% for the higher (Figure 11a) and lower (Figure 11b) temperature-gradient samples, respectively, as compared to the laboratory results. In contrast to that, the surficial water content values were comparable, given that the model overpredicted water content by almost 1 and 4% for the higher (0.083 °C/mm) and lower (0.028 °C/mm) temperature-gradient samples, respectively. The surface elevation of the higher-temperature-gradient sample was reduced to nearly half of its initial thickness (43 and 44% reduction of its initial thickness from the modeling simulation and laboratory results, respectively) at the end of the tests. Likewise, the surface elevation of the lower-gradient-temperature sample was reduced to 46 and 44% of the initial thickness from the numerical modeling and laboratory results, respectively.

Figure 12 shows the liquidity index profiles of the coupled analysis models. Similar to the first-phase testing, the second-phase testing also suggests that the surface of the lower-gradient sample was able to pass through the plastic limit. On the contrary, the higher-gradient sample was in a relatively softer consistency in the first-phase testing (could not reach the plastic limit), while the sample surface just reached the plastic limit in the second phase of testing. Overall, coupling analyses were able to predict the laboratory results reasonably well for the centrifuged tailings sample subjected to the lower temperature gradient (0.028 °C/mm) compared to the higher gradient (0.083 °C/mm).

6. Discussion

Both the laboratory results and the numerical model suggest that the centrifuged tailings samples subjected to seasonal weathering (multiple freeze–thaw and drying cycles) have the potential for improved dewatering compared to the as-received centrifuged tailings. Figures 5 and 6 show how multiple freeze–thaw cycles at two different freezing temperature gradients contributed to dewatering prior to drying/evaporation. The freeze–thaw process is known to alter the structure of the tailings/soil materials by redistributing moisture inside the tailings/soil particles that, in turn, improves dewatering upon thaw [11]. Consequently, higher dewatering/volume change can be expected during the very first freeze–thaw cycle followed by a gradual decrease in subsequent cycles due to the gradual decrease in available water/moisture inside the tailings materials [21,22]. Both of these two figures may look similar until the drying/wetting cycle is introduced. Fine-grained tailings subjected to freezing likely generate high suction/negative pore water pressure at the freezing front, thereby causing water migration upwards to the front (as shown in Figure 1) and a subsequent reduction in water content and shrinkage crack development in the tailings. The extent of these cracks is predominantly dependent on the temperature gradient, the number of freeze–thaw cycles and the physio-chemical interactions among the tailings particles and solutes [11,12,23]. Apart from the earlier findings (carried out by [12,23,24]), the visual observation from the laboratory testing also suggests that the lower freezing temperature gradient results in higher shrinkage and cracks, contributing to higher evaporation and solids content. Hence, the shrinkage and cracks developed during freeze–thaw cycles facilitated different evaporation rates for the centrifuged tailings under two different gradients, which contributed to the significantly different water content profiles in the drying–wetting cycle (as shown in Figures 5 and 6).

Figures 7 and 8 show how seasonal weathering in nature can contribute to dewatering of the centrifuged tailings deposit. At the start of the simulation, the tailings are first consolidated at the bottom. With subsequent cycles, the upward water flux due to consolidation becomes less than the applied evaporation at the surface and, hence, desiccation occurs. As a result, the tailings surface first becomes unsaturated after being exposed to the atmosphere, thereby reducing the water content considerably more than the bottom part. However, the void ratio profile (void ratio vs. elevation) from the previous freeze–thaw cycle cannot be specified within a single-layer deposit in the UNSATCON model and hence, an average void ratio was applied that can affect the drying cycle outputs. As shown in Figures 7 and 8, the water content profiles at the bottom few hundredths of a meter remain almost constant, which can be attributed to the input parameter specified as an average value (void ratio from the previous cycle).

The coupled analysis, as shown in Figure 9, correlates well with the first-phase laboratory testing results of the centrifuged tailings samples under temperature gradients of 0.083 and 0.028 °C/mm. The subsequent cycle has a lower impact on the volume changes due to the freeze–thaw cycles, thereby causing lower dewatering from the sample. Hence, the differences in water content values between the laboratory results and modeling predictions (both average and surficial water content values) are expected to be marginal. However, the results from the drying model, which has a significant impact on dewatering, corroborates well with the laboratory results and validates the efficacy of the coupled analysis model.

Similar to Figure 9, Figure 10 compares the second-phase laboratory testing with the modeling results of the centrifuged tailings under two different temperature gradients (0.083 and 0.028 °C/mm). In contrast to the Figure 9, the model-predicted water contents at the surface correlated well with the laboratory computation for both gradient scenarios of the investigated centrifuged tailings. The model-predicted average water content values deviated to an extent (up to 18% differences) from the laboratory-computed average values, and this can be attributed to the limitation of the drying (the UNSATCON model) model, where the input parameter (void ratio) has to be specified as an average value instead of a void ratio profile (varying with depth). The exclusion of wetting and re-drying after wetting events may also have an impact on the variations. However, given the fact that both the samples deviated in their average water content values compared to the laboratory results but correlated well in terms of surficial water content, the limitation of the drying model most likely predominantly contributed to these deviations.

Figure 11 shows the final water content profiles subjected to first- and second-phase testing for the two investigated centrifuged tailings under temperature gradients of 0.083 and 0.028 °C/mm, respectively. The laboratory results shown on the graph represent the oven-measured water content values at the end of the tests. Figure 11a shows that the higher-gradient centrifuged tailings sample predicted an overall lower water content profile than the laboratory measurement for the first-phase testing while the opposite trend was observed in the second-phase testing. The higher water content profile predicted in the model for the second-phase testing can be attributed to the preclusion of the wetting event and the re-drying cycle in the model. In addition to that, the limitation of the UNSATCON model to incorporate the void ratio/water content profile from the previous cycle results in the overall differences between the laboratory results and modeling prediction. However, the water content profiles developed to predict the laboratory testing (as shown in Figure 11b) supported the laboratory results considerably for the lower-gradient tailings sample. Although the drying model could predict the profile for the top few hundredths of a meter of the sample reasonably well following the evaporative flux incorporated in the top boundary, the lower part of the sample could not be predicted quite as well due to the limitation of defining input parameters. Figure 12 shows the change in consistency of tailings samples prior to and after seasonal weathering to illustrate the efficacy of these processes that can be applied in the field deposit.

Overall, this study presented the coupling of three components (incorporating the freeze–thaw process into the FSConsol, consolidation from FSConsol and atmospheric drying from UNSATCON) to develop a coupled model. Model-predicted dewatering behavior correlated well with the laboratory results that validated this coupling approach. However, the coupling analysis has a few limitations, such as:the incapability of running under a single program that can incorporate all the seasonal weathering processes (freeze–thaw, consolidation, desiccation, etc.) and the limitations of defining input parameters in the UNSATCON model, such as being unable (1) to define a void ratio profile (varying with depth) instead of an average void ratio per cycle; (2) to define data directly from the SWCC curve and compressibility curves instead of fitting data into a state surface model; and (3) to input separate files for the evaporation and infiltration under the same cycle. The numerical software available in the industry is yet to develop a program that can simulate the freeze–thaw, consolidation and desiccation of tailings under different boundary conditions and deposition scenarios. Switching between different modeling platforms to address all these seasonal weathering components is cumbersome and time-consuming. Nevertheless, the coupled analysis, in some ways, is able to validate the laboratory analysis based on which field behaviour can be predicted. It may not predict the long-term field behaviour given the complexity related to combining the multiple models for each cycle (each cycle is required to go through the flowchart each time as shown in Figure 1, which implies the flowchart needs to be repeated 100 times for a prediction of 100 years). However, these analyses can provide insight into predicting dewatering behavior of the future field deposits.

7. Conclusions

The coupling modeling methodology presented in this study was able to develop an approach that was validated by comparing it with the laboratory test results of centrifuged tailings under similar boundary conditions. This coupling methodology can assist in tailings management by predicting the short-term behaviour of the dewatering performance of the tailings deposits subjected to the natural seasonal weathering (freeze–thaw, consolidation and desiccation). Within limited scope, this coupled analysis demonstrated that freeze–thaw cycles can be considered as a part of the tailings management in regions where weather permits this. The freeze–thaw process may appear to achieve lower dewatering enhancement compared to atmospheric drying, but the shrinkage and/or cracks developed during the freeze–thaw cycles facilitate greater evaporation and desiccation during the subsequent drying cycle, thereby contributing to higher dewatering overall. The model also confirms the previous research and the present laboratory findings that suggest that a lower temperature gradient results in higher solid content/lower water content in post-thawed tailings. When the drying/evaporative component was incorporated, atmospheric drying was shown to significantly (half an order of magnitude lower) reduce the water content as compared to the multiple freeze–thaw cycles only. With the combined effects of alternate freeze–thaw and drying–wetting cycles, the tailings could even achieve a water content exceeding the plastic limit (as evident in the lower-temperature-gradient sample), enabling reclamation and closure. While time consuming, this approach can provide insights into predicting the dewatering behaviour of future field deposits so that it can be used for planning purposes.

Author Contributions: Conceptualization, U.S.R.; methodology, A.A. and U.S.R.; software, U.S.R.; validation, U.S.R. and N.B.; formal analysis, U.S.R.; investigation, U.S.R.; resources, U.S.R. and N.B.; data curation, U.S.R.; writing—original draft preparation, U.S.R.; writing—review and editing, U.S.R., N.B. and A.A.; visualization, U.S.R.; supervision, N.B.; project administration, N.B.; funding acquisition, N.B. All authors have read and agreed to the published version of the manuscript.

Funding: This research was funded by the Natural Science and Engineering Research Council of Canada (NSERC), Canada's Oil Sands Innovation Alliance (COSIA) and Alberta Innovates-Energy and Environment Solutions.

Data Availability Statement: Data are contained within the article.

Acknowledgments: The authors would like to thank the Natural Sciences and Engineering Research Council of Canada (NSERC), Canada's Oil Sands Innovation Alliance (COSIA) and Alberta Innovates—Energy and Environment Solutions for their financial support. The authors would also like to thank Louis Kabwe for providing the large strain consolidation parameters.

Conflicts of Interest: The authors declare no conflict of interest. The funders had no role in the design of the study; in the collection, analyses, or interpretation of data; in the writing of the manuscript, but have reviewed the results and approved to publish.

References

1. Government of Alberta. Oil Sands: Facts and Stats. 2021. Available online: https://www.alberta.ca/oil-sands-facts-and-statistics.aspx (accessed on 1 October 2021).
2. Cossey, H.L.; Batycky, A.E.; Kaminsky, H.; Ulrich, A.C. Geotechnical Stability of Oil Sands Tailings in Mine Closure Landforms. *Minerals* **2021**, *11*, 830. [CrossRef]
3. Alberta Energy Regulator (AER). *Alberta's Energy Reserves 2015 & Supply/Demand Outlook 2016–2025 (ST98-2016)*; Alberta Energy Regulator: Calgary, AB, Canada, 2016.
4. Jeeravipoolvarn, S. Geotechnical Behavior of In-Line Thickened Oil Sands Tailings. Ph.D. Thesis, University of Alberta, Edmonton, AB, Canada, 2010.
5. Oil Sands Tailings Consortium (OSTC) and Canada's Oil Sands Innovation Alliance (COSIA). *Technical Guide for Fluid Fine Tailings Management*; OSTC and COSIA: Calgary, AB, Canada, 2012. Available online: https://www.cosia.ca/uploads/documents/id7/TechGuideFluidTailingsMgmt_Aug2012.pdf (accessed on 31 July 2021).
6. Spence, J.; Bara, B.; Lorentz, J.; Mikula, R. Development of the Centrifuge Process for Fluid Fine Tailings Treatment at Syncrude Canada Ltd. In Proceedings of the World Heavy Oil Congress 2015, Edmonton, AB, Canada, 24–26 March 2015.
7. Alberta Energy Regulator. *State of Fluid Tailings Management for Mineable Oil Sands, 2019*; Alberta Energy Regulator: Calgary, AB, Canada, 2020. Available online: https://static.aer.ca/prd/2020-09/2019-State-Fluid-Tailings-Management-Mineable-OilSands.pdf (accessed on 31 July 2021).
8. Government of Alberta. Oil Sands Information Portal. Available online: http://osip.alberta.ca/map/ (accessed on 1 October 2021).
9. Rima, U.S.; Beier, N. Effects of seasonal weathering on dewatering and strength of an oil sands tailings deposit. *Can. Geotech. J.* **2021**, in press. [CrossRef]
10. Hyndman, A.; Sawatsky, L.; McKenna, G.; Vandenberg, J. Fluid Fine Tailings Processes: Disposal, Capping, and Closure Alternatives. In Proceedings of the 6th International Oil Sands Tailings Conference, Edmonton, AB, Canada, 9–12 December 2018; pp. 9–12.
11. Proskin, S.; Sego, D.; Alostaz, M. Oil Sands MFT Properties and Freeze-Thaw Effects. *J. Cold Reg. Eng. ASCE* **2012**, *26*, 29–54. [CrossRef]
12. Proskin, S.A. A Geotechnical Investigation of Freeze-Thaw Dewatering of Oil Sands Fine Tailings. Ph.D. Thesis, Department of Civil and Environmental Engineering, University of Alberta, Edmonton, AB, Canada, 1998.
13. Dawson, R.F.; Sego, D.C. Design Concepts for Thin Layered Freeze-Thaw Dewatering Systems. In Proceedings of the 46th Canadian Geotechnical Conference, Saskatoon, SK, Canada, 27–29 September 1993; pp. 283–288.
14. Johnson, R.L.; Pork, P.; Allen, E.A.D.; James, W.H.; Koverny, L. *Oil Sands Sludge Dewatering by Freeze-Thaw and Evaporation*; Report-RRTAC 93-8; Syncrude Canada and Alberta Conservation and Reclamation Council: Vegreville, AB, Canada, 1993. Available online: https://doi.org/10.7939/R33R0PT5T (accessed on 1 October 2021).
15. Scott, J.D.; Dusseault, M.B.; Carrier, W.D. Behaviour of the clay/bitumen/water sludge system from oil sands extraction plants. *Appl. Clay Sci.* **1985**, *1*, 207–218. [CrossRef]
16. Rima, U.S.; Beier, N. Effects of Multiple Freeze-Thaw Cycles on Oil Sands Tailings Behaviour. *Cold Reg. Sci. Technol.* **2021**, *192*, 103404. [CrossRef]
17. Gibson, R.; England, G.; Hussey, M. The theory of one-dimensional consolidation of saturated clays. *Geotechnique* **1967**, *17*, 261–273. [CrossRef]
18. Qi, S.; Simms, P.; Vanapalli, S. Piecewise-linear formulation of coupled large-strain consolidation and unsaturated flow. I: Model development and implementation. *J. Geotech. Geoenvironmental Eng.* **2017**, *143*, 04017018. [CrossRef]
19. Qi, S. Numerical Investigation for Slope Stability of Expansive Soils and Large Strain Consolidation of Soft Soils. Ph.D. Thesis, University of Ottawa, Ottawa, ON, Canada, 2017.
20. Hurtado, O. Desiccation and Consolidation in Centrifuge Cake Oil Sands Tailings. Master's Thesis, Carleton University, Ottawa, ON, Canada, 2018.
21. Eigenbrod, K.D. Effects of cyclic freezing and thawing on volume changes and permeabilities of soft fine-grained soils. *Can. Geotech. J.* **1996**, *33*, 529–537. [CrossRef]
22. Andersland, O.B.; Ladanyi, B. *Frozen Ground Engineering*, 2nd ed.; American Society of Civil Engineers & John Wiley & Sons Inc.: Hoboken, NJ, USA, 2004.
23. Pham, N.H.; Sego, D.C. Modeling Dewatering of Oil Sands Mature Fine Tailings using Freeze Thaw. In Proceedings of the International Oil Sands Tailings Conference, Lake Louise, AB, Canada, 7–10 December 2014.
24. Sego, D.C.; Dawson, R.F. *Influence of Freeze-Thaw on Behaviour of Oil Sand Fine Tails*; Alberta Oil Sands Technology and Research Authority: Edmonton, AB, Canada, 1992; p. 77.

processes

MDPI

Article

Influence of Yield Pillar Width on Coal Mine Roadway Stability in Western China: A Case Study

Qingwei Wang [1], Hao Feng [1], Peng Tang [1], Yuting Peng [1], Chunang Li [1], Lishuai Jiang [1,*] and Hani S. Mitri [2]

1 State Key Laboratory of Mining Disaster Prevention and Control, Shandong University of Science and Technology, Qingdao 266590, China; Lebron_wqw@163.com (Q.W.); fh1244950367@163.com (H.F.); tpsdustenactus@163.com (P.T.); pytjiaojiao86@163.com (Y.P.); mn1234560928@163.com (C.L.)

2 Department of Mining and Materials Engineering, McGill University, Montreal, QC H3A 0E8, Canada; hani.mitri@mcgill.ca

* Correspondence: lsjiang@sdust.edu.cn

Abstract: Roadway excavation technology in underground coal mines has an important impact on mining efficiency and production safety. High-efficiency and rapid excavation of underground roadways in coal mines are important means to improve the production efficiency of coal mines. To tackle the problems of instability of roadway and support difficulties, the tail entry of panel 3105 in Mataihao Mine was used as the case study. The methods of underground investigation, theoretical analysis, and FLAC3D numerical simulation were used to analyze the stability of the surrounding rock under different yield pillar widths. Through the stress field, displacement field, and plastic zone of roadway surrounding rock, the stability of the rock surrounding the roadway under different yield pillar widths (4 m, 6 m, and 8 m) was analyzed. The results show that, with the increase in the yield pillar width, the plastic zone failure and displacement of the roadway surrounding rock are mainly manifested in the narrow pillar rib, seam rib, roof, and floor. The plastic zone distribution changes slightly; the roadway displacement exhibits basic symmetry. The vertical stress and the displacement of the two sides increase with the increase in the yield pillar width, and the roof displacement and the ratio of tensile failure of the surrounding rock decrease with the increase in the yield pillar width. According to the dynamic evolution law of the rock surrounding the roadway along the goaf side, the effect of the yield pillar size is revealed, and a reasonable yield pillar width is determined. When the yield pillar width is 6 m, the plastic zone failure of the surrounding rock and the displacement of the two sides of the roof are the most balanced among the three schemes. This provides a reference for the selection of the narrow yield pillar size in coal mines under the same geological conditions.

Keywords: goaf-side entry; numerical simulation; stability of surrounding rock; pillar size optimization

Citation: Wang, Q.; Feng, H.; Tang, P.; Peng, Y.; Li, C.; Jiang, L.; Mitri, H.S. Influence of Yield Pillar Width on Coal Mine Roadway Stability in Western China: A Case Study. *Processes* **2022**, *10*, 251. https://doi.org/10.3390/pr10020251

Academic Editor: Li Li

Received: 1 December 2021
Accepted: 19 January 2022
Published: 27 January 2022

1. Introduction

Coal mining in western China has become the main production source of China in recent years thanks to the features of rich resource, shallow depth, and simple geology [1,2]. In the early coal exploration of western China, wide coal pillars (20–40 m in width) were commonly employed. However, in addition to the issue of recovery rate, the remaining wide pillars carry large amounts of overburden and stress, which lead to serious ground stability problems when mining the coal seams below the pillars [3]. Optimizing the pillar width will significantly improve the roadway stability by avoiding the unloading of concentrated pillar stress [4]. Therefore, studies on the goaf-side entry technique and other panel and pillar design with narrow pillar have attracted lots of attention by researchers and coal mining industries [5,6]. In this study, Mataihao Mine, a typical coal mine in western China, was taken as a case study to the investigate the optimization of the coal pillar size in the goaf-side roadway. Maleki [7,8] proposed that, in coal seams prone to collision and impact, the design of yielding coal pillars should consider the initial ground stress and mining stress, the size of the working face, and the characteristics of the rock above the coal seam. Xue et al. [9], through theoretical analysis, deduced an analytical expression for yield pillars width in goaf-side entry, and determined that the reasonable width of yield pillars is 4.8~6.9 m; Chen [10] analyzed the characteristics of impact appearance roadways along the goaf-side in the continuous panel of the mining area in the deep mining area of Shaanxi and Mongolia, and proposed the surrounding rock reinforcement method and the roadway driving method along the goaf; Zhang et al. [11], in view of the asymmetric displacement and failure characteristics of the roadway during the driving process, proposed the high-strength anchor beam net, asymmetric anchor beam truss structure, and prestressed anchor cable truss asymmetric support control plan, and successfully applied them to actual engineering; Wang et al. [12] proposed a new method for determining lateral support pressure. Through field monitoring and numerical simulation, the reasonable employing width of the section yield pillar was determined, and the conclusion was made that the yield pillar width should not be less than 4 m; Hou et al. [13] studied the evolution and distribution of the stress field, strain field, and plastic zone of yield pillars, which provided a theoretical basis for roadway support.

After scientific research and practice, the large-width yield pillars along the goaf have been changed to small-width yield pillars along the goaf. At present, many mines in the northwest region have implemented small-width yield pillars along the goaf. Basically, they weaken and eliminate the impact of rock pressure [14,15]. Once the test of goaf-side entry with yield pillars in Mataihao Mine is successful, this will greatly improve the safety, efficiency, and mining speed of the coal mine.

Although the surrounding rock stress is low and the supporting strength required for driving along the goaf-side completely is low, in engineering, harmful gases, water, falling gangue, and so on in the goaf area of the adjacent section can easily enter the roadway, which has a serious impact. The roadway is normally excavated, ventilated, and maintained. Therefore, the coal mines in our country generally adopt the roadway protection method of employing yield pillars and goaf-side entry [16,17].

Based on the engineering conditions of the large mining height working face, the three-dimensional numerical model of different yield pillar widths is established, and the stress state and mechanical characteristics of the surrounding rock of the large mining height goaf-side entry are studied through the dynamic simulation of the goaf-side entry and the impact of mining. The dynamic evolution law of surrounding rock displacement and stability during the period of driving and mining is studied, and the yield pillar size effect and influence mechanism are analyzed [18]. Based on numerical simulation analysis and field tests, the yield pillar width is optimized. In order to increase the recovery rate of coal resources, improve the control of the surrounding rock of the roadway, and reduce the cost of support, it is necessary to start from the actual situation on the site, that is, in-depth study of the evolution law of the mining stress field in the large mining height face, according to the dynamic evolution law of the surrounding rock of the goaf-side entry. The effect of yield pillar size and determining the reasonable yield pillar size was revealed.

2. Case Study

2.1. Geological and Geotechnical Overview of the Mataihao Mine

The 3-1 coal seam is being mined in Mataihao Mine. The buried depth of the coal seam is about 420 m, and the coal seam geological conditions are relatively simple. The working face adopts the full-height mining technology of large-cutting height and fully mechanized mining at the same time. The thickness of the top coal is 1 m, and that of the bottom coal is 1 m.

The panel of Mataihao Mine that has been mined is currently using wide yield pillars along the goaf side. The yield pillar size of the panel is 20 m along the goaf and the two lanes are driven together. Because the tail entry is affected by the two panels' advanced dynamic pressures, the support of the tail entry is difficult, which seriously affects safety and production. Starting from panel 3108, Mataihao Mine has increased the yield pillar size to 35 m. However, increasing the yield pillar size has caused considerable loss of resources, and if there is potential shock pressure, the 35 m yield pillar will remain the same. It is difficult to control the dynamic pressure load, and the dynamic pressure shock phenomenon may occur in severe cases. For these reasons, Mataihao Mine chose to use yield pillars to the goaf-side entry between the head entry of panel 3103 and the tail entry of panel 3105. To determine the yield pillar size, theoretical analysis and numerical simulation are carried out.

2.2. Theoretical Demonstration and Analysis of Yield Pillar Width in Goaf-Side Entry

A reasonable yield pillar width means that the yield pillar cannot be too large, within the peak of the mine stress in the direction of the coal seam, and not too small, because the yield pillar is broken and cannot be anchored with bolts [19]. If the yield pillar is too large, it is not safe and economical; if the yield pillar is too small, owing to the plastic fracture zone on both sides of the yield pillar, the stability and bearing capacity are low. The anchor bolt anchoring in the plastic fracture zone cannot play a role in controlling the surrounding rock. Therefore, the yield pillar size has a reasonable interval that meets the above requirements in the range of 0–20 m. The reasonable width of yield pillars can be obtained by theoretical calculation methods. According to the theory of elastoplastic, the yield pillar width is analyzed theoretically, and the theoretical calculation method of the yield pillar is determined [20–24].

Calculation formula for the reasonable yield pillar width:

$$\mathrm{X} < x_0 - \mathrm{x} \tag{1}$$

where X is a reasonable yield pillar width, x_0 is the width of the stress limit equilibrium zone, and x is the width of the roadway.

Calculation formula for the width of the stress limit equilibrium zone:

$$x_0 = \frac{hA}{2\tan\varphi_0} \ln\left(\frac{k_\gamma H + \frac{C_0}{\tan\varphi_0}}{\frac{C_0}{\tan\varphi^0} + \frac{P_x}{A}}\right) \tag{2}$$

$$A = \frac{\mu}{1-\mu} \tag{3}$$

where h is the height of the working surface, μ is the Poisson's ratio, A is the lateral pressure coefficient, φ is the coal internal friction angle, k is the stress concentration coefficient, γ is the rock bulk density, H is the buried depth of the roadway, C_0 is the cohesion, and P_x is the strength of roadway support.

According to the value of each parameter, the theoretical value of reasonable yield pillar width is obtained: x_0 = 14.5 m. Therefore, the reasonable yield pillar width is less than 8.5 m.

According to the engineering geological background of the Mataihao Coal Mine, theoretical calculations are carried out, and the reasonable theoretical value of yield pillar width is less than 8.5 m. To reasonably optimize the width of the yield pillar of the roadway, in the following chapters, the three-dimensional finite-difference software FLAC3D will be used to carry out the numerical simulation analysis of the case study.

3. Numerical Analysis of Ground Stability of Goaf-Side Entry

3.1. Establishment of Numerical Simulation Model for Goaf-Side Entry

3.1.1. Three-Dimensional Numerical Model

Goaf-side entry involves arranging and excavating the roadway in the relatively low stress area in the residual bearing pressure area on the side of the goaf after the overlying rock migration adjacent to the goaf is basically stable.

The main coal seam of panel 3103 and panel 3105 in Mataihao Mine is 3-1 seam, and the three-dimensional numerical models of 4 m, 6 m, and 8 m yield pillar widths are established based on the model boundary of the inclined midline of panel 3103 and panel 3105 [25–27].

The numerical model diagram is shown in Figure 1. When studying the mechanical properties of rock masses in the collapse zone through numerical simulation, scholars all over the world generally recognize the double-yield model [28]. The mechanical properties of goaf are shown in Table 1. The type of analysis is elastoplastic. As the surrounding rock in this mine does not present notable strain-softening behavior, the Mohr–Coulomb model is employed to improve the simulation efficiency [29–31]. The model is 300 m long along the strike of the panel, of which the strike length of the work face is 220 m, with a boundary of 40 m at the front and rear; the length of the panel is 210, 212, and 214 m; and the height is 120 m. The in situ stress regime was first created in the model using the initial stress feature with gravity loading and a horizontal-to-vertical stress ratio of 1.5. The experimental results from previous rock mechanical testing [32] illustrated a linear yield or failure envelop, well described by the Mohr–Coulomb criterion. The rock mass properties, as listed in Table 1, are estimated from the intact rock properties using the generalized Hoek–Brown failure criterion [33].

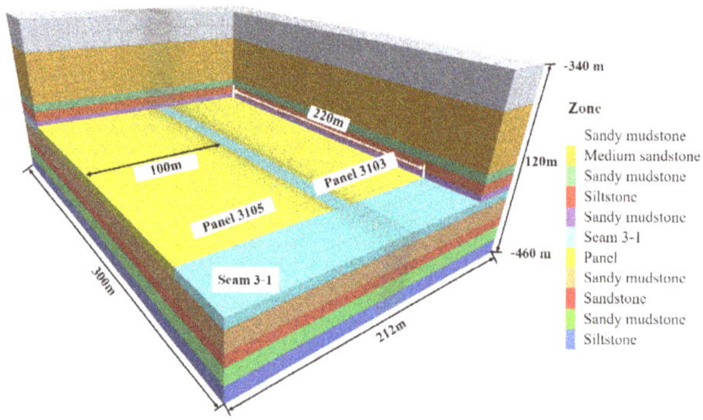

Figure 1. FLAC3D model diagram.

Table 1. Rock mechanical properties of modeled materials.

Strata	Lithology	K (GPa)	G (GPa)	ρ (kg/m3)	φ (°)	C (MPa)	σ_t (MPa)
Roof	Sandy mudstone	14.7	2.14	2350	31	2.4	1.7
	Medium sandstone	20.8	1.26	2580	30	2.8	1.2
	Sandy mudstone	13.1	2.46	2420	31	3.0	2.0
	Siltstone	6.7	4.96	2720	36	3.89	2.93
	Sandy mudstone	7.4	1.08	2380	32	2.0	1.12
Coal seam	Coal	3.16	2.46	1600	31	3.0	1.08
Floor	Sandy mudstone	6.2	1.8	2300	30	2.1	1.28
	sandstone	12.2	2.4	2500	28	2.5	2.1
	Sandy mudstone	7.8	1.4	2350	31	2.1	0.8
	Siltstone	9.6	7.13	2720	37	4.13	2.85
Goaf	Goaf	19.9	1	1700	30	0.001	0

K is bulk modulus, G is shear modulus, ρ is density, φ is friction angle, C is cohesion, and σ t is tensile strength.

The entry is driven along the goaf-side of 3103 with a 6 m wide yield pillar, as illustrated in Figure 2. The numerical simulation plan is shown in Figure 3. A typical geological column based on core logging was carried out in an adjacent panel 11,050, as shown in Figure 4. First, panel 3103 is mined. Because, after the completion of panel 3103, its head entry is destroyed, and tail entry 3105 of the research object needs to be re-excavated, the head entry 3103 part of the model is directly replaced by the double-yield model of panel 3103. After the mining of panel 3103 is completed, tail entry 3105 is excavated, and bolt and anchor cable support is carried out. After the roadway is excavated and supported, panel 3105 is mined.

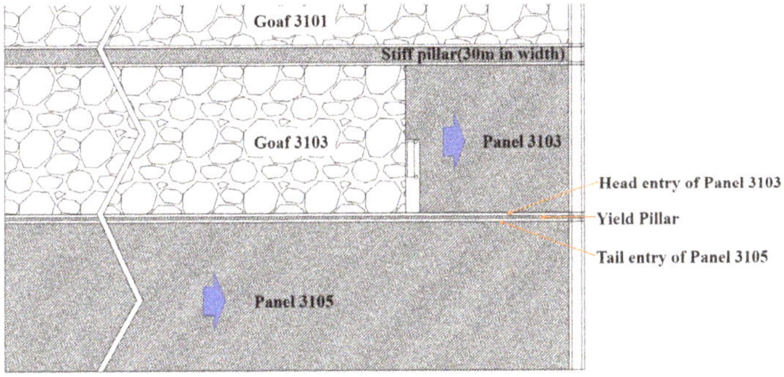

Figure 2. Plan view of local panel layout.

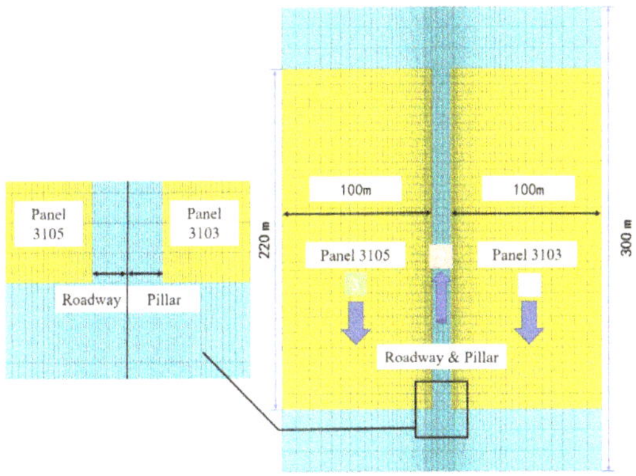

Figure 3. Plan view of numerical simulation of yield pillars goaf-side entry.

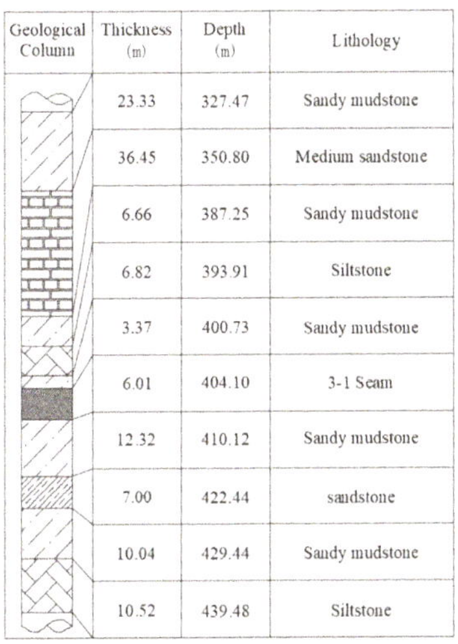

Geological Column	Thickness (m)	Depth (m)	Lithology
	23.33	327.47	Sandy mudstone
	36.45	350.80	Medium sandstone
	6.66	387.25	Sandy mudstone
	6.82	393.91	Siltstone
	3.37	400.73	Sandy mudstone
	6.01	404.10	3-1 Seam
	12.32	410.12	Sandy mudstone
	7.00	422.44	sandstone
	10.04	429.44	Sandy mudstone
	10.52	439.48	Siltstone

Figure 4. Typical geological column.

3.1.2. Sensitivity Analyses

- Base boundary

To verify the boundary sensitivity, the same model as in Figure 2 was established and the displacement (The displacement direction of the roof in the text is vertical downward, and the displacement direction of the two ribs is the horizontal direction.) at 120 m of the roadway excavation was monitored. Figures 5 and 6 illustrate that the distance between the yield pillar is 20 m, 40 m, 60 m, and 80 m, the roadway displacement and stress distribution are different. The stress values in the roof, floor, and two ribs of one roadway section are marked in Figure 6. As can be seen, when the boundary increases from 20 m to 40 m, the stress distribution in surrounding rocks changed; however, the change becomes negligible with the continuous increase in boundary size. Therefore, to reduce the calculation volume and the influence of the boundary effect on the model, the model with the boundary of 40 m is chosen.

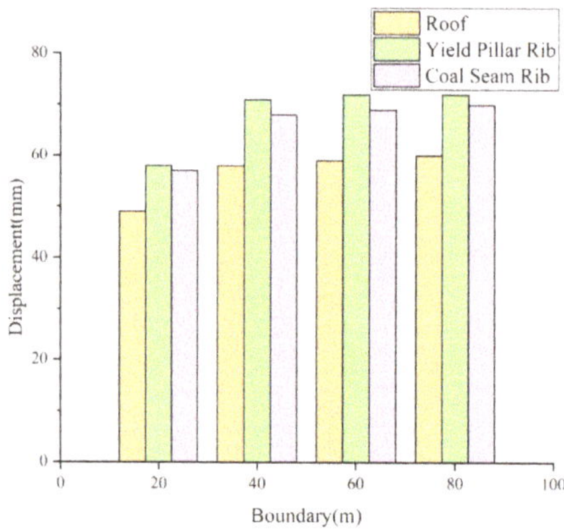

Figure 5. The influence of boundary effects on the model.

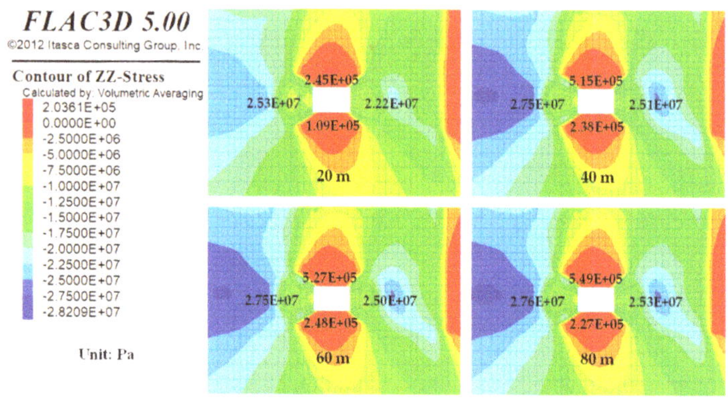

Figure 6. The influence of boundary effects on the stress of the model.

- Mesh dependency

The mesh density of the numerical model, especially for rock mass near roadway, might also be sensitive to the simulation results. Figure 7 illustrates the simulation results of roadway displacement at the last mining step with respect to different mesh density in its surrounding rock, and the number of zones in the roadway surrounding the area of model 1, 2, and 3 is 18,000, 36,000, and 54,000, respectively. As can be seen, the mesh sensitivity of displacement varies for roof and rib. Owing to the weaker properties, the ribs tend to fail more in tension, which leads to the effect of tension weakening being more significant than roof. As a result, displacements of rib are more sensitive to mesh density. As there is no notable difference between model 2, which is employed to carry out the simulations in the present study, and model 3, which has a denser mesh, its mesh density and simulation results are validated. This mesh dependency study indicates that the mesh density of the numerical model, especially for the areas where notable tensile failure will take place, should be calibrated.

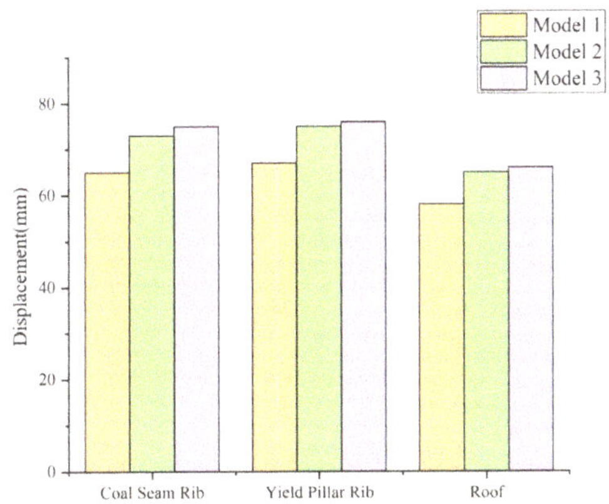

Figure 7. The effect of mesh dependency on the model.

3.1.3. Simulation Scheme and Roadway Support

After panel 3103 is mined, the rock mass compaction, overlying rock migration, and stress adjustment in the caving zone form the surrounding rock conditions and stress environment before goaf-side entry. After the yield pillars with widths of 4 m, 6 m, and 8 m are employed, respectively, the tail entry of panel 3105 was excavated, and the roadway excavation width × height was 6 m × 4 m. The roadway support follows the excavation, and the roadway support design is shown in Figure 8.

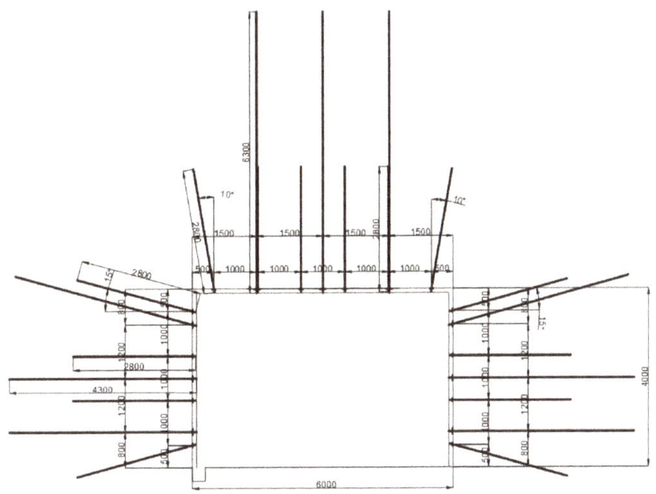

Figure 8. Roadway support design.

The roadway support is simulated using structural elements in FLAC3D, and the support design used in the simulation is the same as the on-site application support scheme [34]. For the convenience of presentation, Figure 9 shows the support of a 2 m section of the tail entry of panel 3105. All rebar bolts and cable bolts are partially grouted with resin cartridges; the parameters are listed in Table 2. To comprehensively evaluate the surrounding rock and stress environment for goaf-side entry and optimize the reasonable layout of the goaf-side entry, the study is carried out from three aspects: the stress distribution of the surrounding rock, the distribution of the plastic failure zone, and the displacement and evolution characteristics of the surrounding rock during the roadway driving. The influence of yield pillar width on the behavior of underground pressure and the stability of surrounding rock during goaf-side entry is summarized.

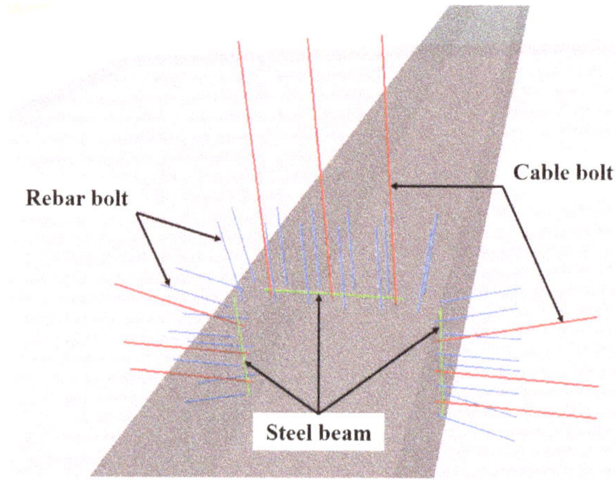

Figure 9. Roadway support design simulation.

Table 2. Parameters of the rebar and each cable bolt.

Type of Bolt	Bolt Length (mm)	Grout Length (mm)	Diameter (mm)	Tensile Strength (KN)
Rebar bolt	2800	1400	22	335
Roof cable bolt	6300	3000	21.8	510
Rib cable bolt	4300	2000	21.8	510

3.2. Stress State and Mechanical Characteristics of Rock Surrounding the Roadway Goaf-Side Entry with Different Yield Pillar Widths

3.2.1. Boundary Stress Distribution Coal Seam Goaf before Goaf-Side Entry

The goaf-side entry is arranged in the area where the supporting pressure of the coal body is relatively low under the large lateral structure of the goaf. The disturbance to the overlying rock layer during the roadway excavation generally does not affect the stability of the large structure.

After the completion of mining panel 3103, the stress is re-adjusted and distributed in the adjacent surrounding rock to form the mining stress field in the upper section, which forms the surrounding rock and stress environment during the excavating of panel 3105 head goaf-side entry. Figure 10 shows the vertical stress distribution in the yield pillar in the monitoring section and the driving position of the tail entry of panel 3105 with different yield pillars. By observing the stress state of the surrounding rock before excavating with yield pillars of different widths, the larger the yield pillars' width in isolated goaf, the higher the vertical stress of surrounding rock with goaf-side entry. Because the stress reduction area in the yield pillar causes the coal and rock mass to yield and fail owing to the impact of mining, it is formed by loosening and decompression. Therefore, the smaller the yield pillar width, the worse the conditions of the surrounding rock during roadway driving. In the process of redistribution, it is sensitive to stress changes, volatile failure, poor bearing capacity, and other mechanical characteristics. Owing to the large span of the roadway (6 m), the mechanical properties and stress state of surrounding rock on both sides of roadway roof and floor as well as the two sides of coal rock mass are different.

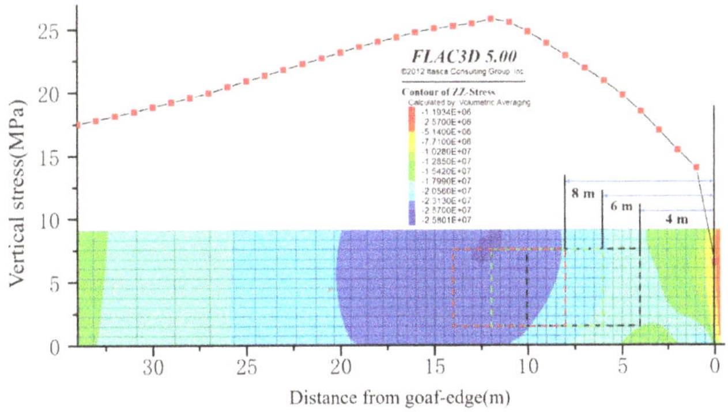

Figure 10. Vertical stress distribution in the yield pillar; the position of roadway is shown in dotted lines for different yield pillar widths.

3.2.2. Stress State of Surrounding Rock of Roadway Driving along Goaf under Different Yield Pillar Widths

Figure 11 shows the vertical stress distribution in the surrounding rock after goaf-side entry with different yield pillar widths. When the yield pillar width is 4 m, the maximum vertical stress in the yield pillar is only 14.5 MPa. When the pillar width is 6 m, the maximum stress in the pillar is 16.9 MPa. When the pillar width is 8 m, the maximum stress in the pillar is 18.8 MPa. With the increase in pillar width, the vertical stress in the pillar increases. When the yield pillar width is 8 m, the maximum vertical stress in the yield pillar is more than twice of the original rock stress, and there is obvious stress concentration.

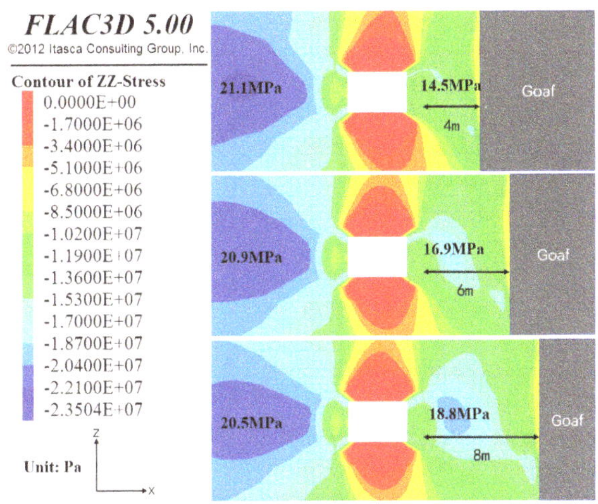

Figure 11. Vertical stress distribution in surrounding rock of yield pillar widths of 4 m, 6 m, and 8 m along the goaf side.

After the surrounding rock stress is redistributed, the concentrated stress shifts to the depth of the coal seam. The yield pillar width has almost no effect on the width of the stress limit equilibrium zone, but has a significant effect on the peak value of the vertical stress. When the yield pillar width is 4 m, the peak value of vertical stress in a deep coal seam is 21.1 MPa, which is 2.41 times the original rock stress. When the yield pillar width is 6 m, the peak value of vertical stress in a deep coal seam is 20.9 MPa, which is 2.38 times that of the original rock stress. When the yield pillar width is 8 m, the peak value of vertical stress in a deep coal seam increases to 20.5 MPa, which is 2.34 times that of the original rock stress. When the yield pillar width is 6 m, the vertical stress distribution of the yield pillar side and coal seam side is more uniform. When the pillar width is 4 m, the vertical stress is concentrated on the side of the coal seam. When the pillar width is 8 m, the vertical stress is concentrated on the pillar side. The internal stress distribution in the shallow surrounding rock (less than 2 m away from the surrounding rock surface) is almost not affected by the size of the yield pillar.

3.2.3. Development Characteristics of Tensile Failure of Surrounding Rock of Goaf-Side Entry

After roadway excavation, the original triaxial stress state is broken. Because the tensile strength of coal rock is far lower than the compressive strength, and the joints and fissures and other weak structural planes have almost no tensile strength, the surface and shallow surrounding rock of roadway are prone to tensile fracture and convergent displacement in roadway space. Therefore, the extension range of the shallow tensile failure zone in the plastic zone of roadway surrounding rock can be used as an effective index to measure the stability of roadway surrounding rock and analyze the failure mechanism of surrounding rock. At the same time, the plastic failure zone caused by the influence of mining in the upper section before roadway excavation can be distinguished from the plastic failure zone caused by the influence of goaf-side excavation.

Through the FLAC3D built-in FISH language programming, the tensile failure elements in the shallow surrounding rock of goaf-side excavation were identified and counted, and the distribution of tensile failure zones along axial and vertical sections of a certain section of goaf-side excavation with different yield pillar widths is shown in Figure 12. By comparing the distribution of the tensile failure zone of the surrounding rock of goaf excavation with different yield pillar widths in Figure 12 under the condition of different yield pillar widths, the common feature of surrounding rock failure is that the tensile failure zone is more widely distributed and extends deeper on the roof and floor of the roadway. This is because the tail entry of panel 3105 is excavated in the coal seam, and the roof strata, two sides, and floor are all weak coal bodies, which can easy lead to displacement and failure.

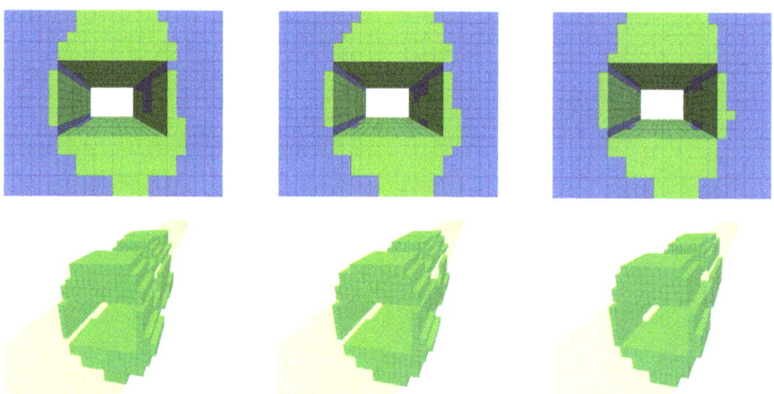

(a) (b) (c)

Figure 12. Distribution of tensile failure zones in goaf excavation with different yield pillar widths: (**a**) model—4 m; (**b**) model—6 m; (**c**) model—8 m.

The ratio of the number of tensile failure elements calculated by the FISH language program to the total number of surrounding rock elements in the shallow part of the roadway was obtained, and the tensile failure ratio of the shallow surrounding rock is shown in Table 3.The failure characteristics of surrounding rock in different yield pillar widths were analyzed in combination with Figure 12. When the yield pillar width is 4 m, closest to the mines' yield pillar, the most dramatic effect by mining of coal and rock roadway roof and floor yield pillar side caused severe tensile fracture, because, before the roadway, the surrounding rock mechanics properties are poor, and the depth of the scope of tensile fracture at the side of the pillar to the roof and floor is 32.05% of the whole roadway surrounding the rock of shallow tensile failure. When the yield pillar width is 6 m, the mechanical properties of rock mass before excavation are improved, the tensile failure in the roof is slightly smaller than that of the 4 m yield pillar, and the failure is relatively symmetrical. Moreover, 29.73% of the shallow surrounding rock in the whole roadway is subjected to tensile failure. When the yield pillar width is 8 m, the mechanical properties of surrounding rock before excavation are good, but the stress of the surrounding rock is high. Tensile failure occurs in 28.69% of the shallow surrounding rock in the whole roadway.

Table 3. Tensile failure ratio of surrounding rock in goaf excavation with different yield pillar widths.

Pillar Width (m)	Number of Zones within 2 m from Entry Surface	Number of Tensile Failure Zones	Ratio of Tensile Failure (%)
4	1248	400	32.05
6	1248	371	29.73
8	1248	358	28.69

3.3. *Evolution of Surrounding Rock Displacement of Goaf-Side Entry under Different Yield Pillar Widths and the Effect of Yield Pillar Size*

In order to comprehensively study the surrounding rock displacement in the whole section of the roadway during excavating and panel mining, eight displacement monitoring nodes are arranged in the monitoring section, as shown in Figure 13. Note that the roof displacement is considered as the vertical displacement towards the floor, the floor displacement is considered as the vertical displacement towards the roof, and the converging displacement of the two ribs is horizontal displacement towards the other rib.

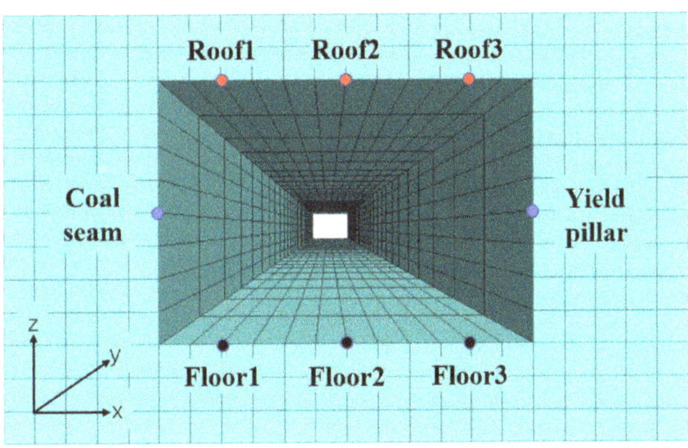

Figure 13. Layout of measuring points of roadway displacement.

Along the direction of roadway excavation, the right side is the yield pillar left, and the left side is the coal seam to be recovered. Displacement measuring points are arranged on the left side of the roof of the roadway roof, the center of the roof, and the right side of the roof to monitor the amount of roof subsidence and displacement, which are marked in red as Roof 1, Roof 2, and Roof 3, respectively, arranged on the coal seam and yield pillar; the displacement measuring points are moved closer to each other horizontally and marked in blue; and the displacement measuring points are arranged on the left, center, and right of the bottom of the roadway floor to monitor the displacement of the bottom drum, marked in black as Floor 1, Floor 2, and Floor 3.

3.3.1. Evolution of Roof Displacement during Goaf-Side Entry

Under the condition of keeping yield pillars of different widths, the displacement evolution characteristics of the roof in the monitoring section of goaf excavation during excavation are shown in Figure 14. For convenience, the absolute value of roof subsidence displacement is adopted in the figure. Monitoring begins after excavation of the roadway in the monitoring section and ends after excavation of the roadway. It can be seen from Figure 14 that the continuous displacement time of surrounding rock is long, the stability speed of surrounding rock is slow, and the creep characteristics are obvious after the roadway is excavated. Specifically, during the process of 20 m~60 m in the advance monitoring section of the driving face, the roof subsidence increases by 7 mm~8 mm under the condition of each yield pillar width. During the process of 60 m~100 m in the advance monitoring section of the driving face, the roof displacement tends to be stable and the subsidence amount increases by about 4 mm.Different widths of yield pillars along the goaf-side mean that roadways are driven in different loose and broken surrounding rock environments and different stress concentrations in stress environments. Comparing Figure 14a–c, the influence of the yield pillar width on the roof displacement of roadway driving along the goaf is not only reflected in the amount of subsidence, but also has a significant influence on the shape of the roof displacement.

The monitoring positions of the roof of the roadway deformed rapidly after the roadway was excavated, and then the displacement rate slowed down with the continuous advancement of the driving face, but a certain displacement rate was still maintained. For different yield pillar widths, the average roof displacement of the monitoring section at different stages of roadway excavation (the distance between the excavation work surface and the monitoring section) and the ratio of the total displacement during the roadway excavation period are shown in Table 4. First, the yield pillar width has a significant impact on the total roof displacement during the goaf-side entry. The roof subsidence is the largest when the pillar width is 4 m, the second largest when the pillar width is 6 m, and the roof displacement is the smallest when the pillar width is 8 m. Secondly, the ratio of existing displacement to the total displacement at different stages during roadway excavation (hereinafter referred to as the displacement ratio) is also affected by the yield pillar width. When the yield pillar width is 4 m, the roof displacement speed is slow after the roadway is excavated, and the displacement is small. The roof displacement after the excavating work surface passes the monitoring point accounts for 73.51% of the total displacement during the excavating period. The roof displacement ratio shows an upward trend with the increase in the yield pillar width. When the yield pillar width is 8 m, the roof displacement ratio at the same stage is 79.46%, and the displacement ratio difference is 5.95%. When the driving surface is advanced to 60 m from the monitoring section, the roof displacement and displacement ratio of the roadway under each yield pillar condition increase. It can be seen that the excavating displacement mainly occurs from the time the section is excavated to the lagging excavating surface of 0–60 m, that is, the excavating influence period. The difference between the roof displacement ratio of the 4 m yield pillar and the 8 m yield pillar is 2.95%, and the evolution trend of roof displacement remains unchanged. When the driving surface is further advanced to 100 m from the monitoring section, the roof displacement ratio difference of different yield pillar widths is significantly reduced, and the roadway roof displacement ratio difference between the 4 m yield pillar and the 8 m yield pillar is only 0.1%. At this stage, the roof displacement was 4.35 mm and 2.29 mm, respectively, and the roof displacement ratio was 98.56% and 98.66%, respectively.

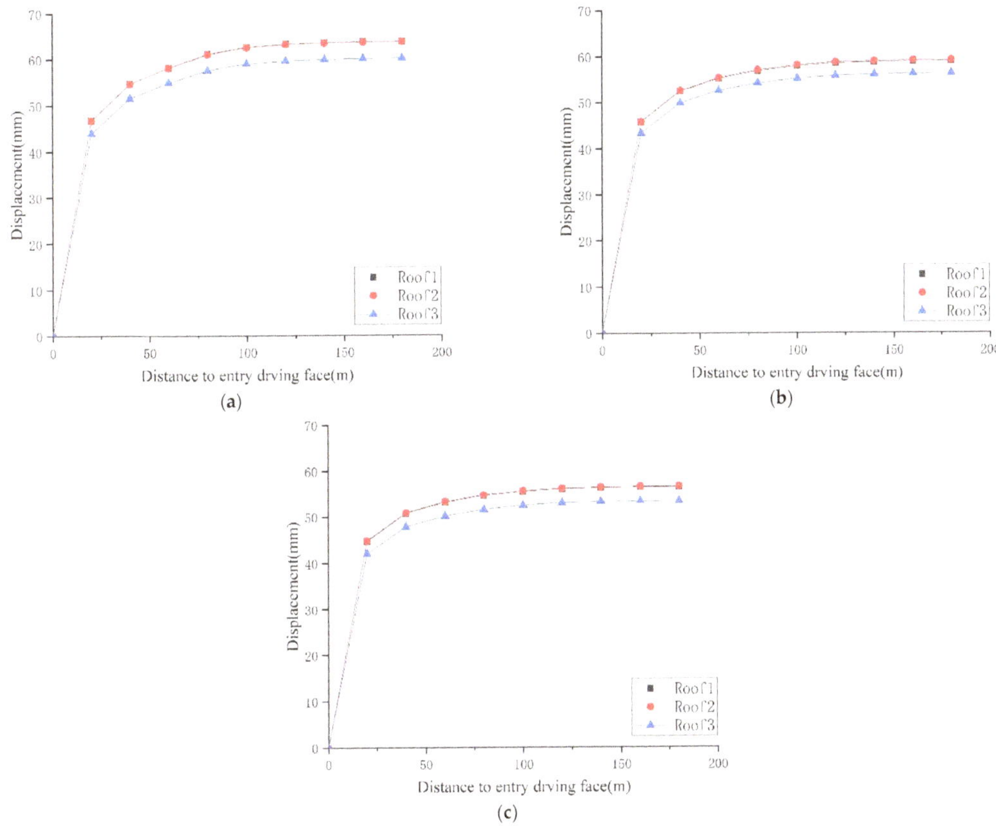

Figure 14. Roof displacement during roadway driving with different yield pillar widths: (**a**) model 4 m; (**b**) model 6 m; (**c**) model 8 m.

Table 4. Evolution of average roof subsidence during roadway driving with different yield pillar widths.

Pillar Width (m)	Average Roof Sinking during Excavating	The Average Subsidence and Displacement Ratio When the Distance of the Lagging Excavating Working Surface Is L_e		
		L_e = 20 m	L_e = 60 m	L_e = 100 m
4 m	62.30 mm	45.80 mm 73.51%	57.05 mm 91.57%	61.40 mm 98.56%
6 m	57.95 mm	45.05 mm 77.74%	54.47 mm 93.99%	57.11 mm 98.55%
8 m	55.26 mm	43.91 mm 79.46%	52.23 mm 94.52%	54.52 mm 98.66%

The above-mentioned roadway roof displacement and displacement rate evolution characteristics are affected by the yield pillar width and the difference is caused by the different surrounding rock and stress environment of the road excavation position with different yield pillar widths [35]. The smaller the yield pillar width, the worse the mechanical properties of the surrounding rock under the influence of mining in the upper section of the goaf-side entry, and the lower the stress of the surrounding rock. Therefore, the weak and broken surrounding rock rapidly deforms after the roadway is driven, and the displacement rate is high at the initial stage of roadway driving. After entering the stable stage after excavation, because the surrounding rock stress is low, the stress environment is good, and the creep amount in the later stage is small. The greater the yield pillar width, the better the mechanical properties of the surrounding rock during roadway excavation, but the higher the stress concentration at the roadway location. After the roadway is excavated, the more complete surrounding rock will gradually deform and fail during the stress adjustment process, and the initial displacement rate of the roadway will be slightly lower, but it is affected by a high-stress environment in the later stage, and still maintains a high creep value [36].

Under the conditions of different yield pillar widths, the vertical roof displacement distribution cloud diagram of the goaf-side entry monitoring section after the roadway excavation is completed is shown in Figure 15. Through comparison, it can be seen that the yield pillar width also has a significant effect on the roof displacement of the goaf-side entry. Figure 16 respectively lists the displacement evolution law of three different monitoring nodes on the roof during the roadway excavation. Under the influence of different yield pillar widths, the relative displacement and displacement rate of each position of the roof are significantly different, and the displacement and displacement of the roof subsidence after the completion of the roadway are also significantly different, as shown in Figure 16. It can be seen from Figures 14–16 that, during the goaf-side entry, the overall displacement of the roadway roof is large when the yield pillar is 4 m, and significant asymmetric large displacement occurs on the side of the yield pillar; when the yield pillar width is 6 m, the roadway displacement of the roof is slightly larger than the 8 m yield pillar, but the difference is not much.

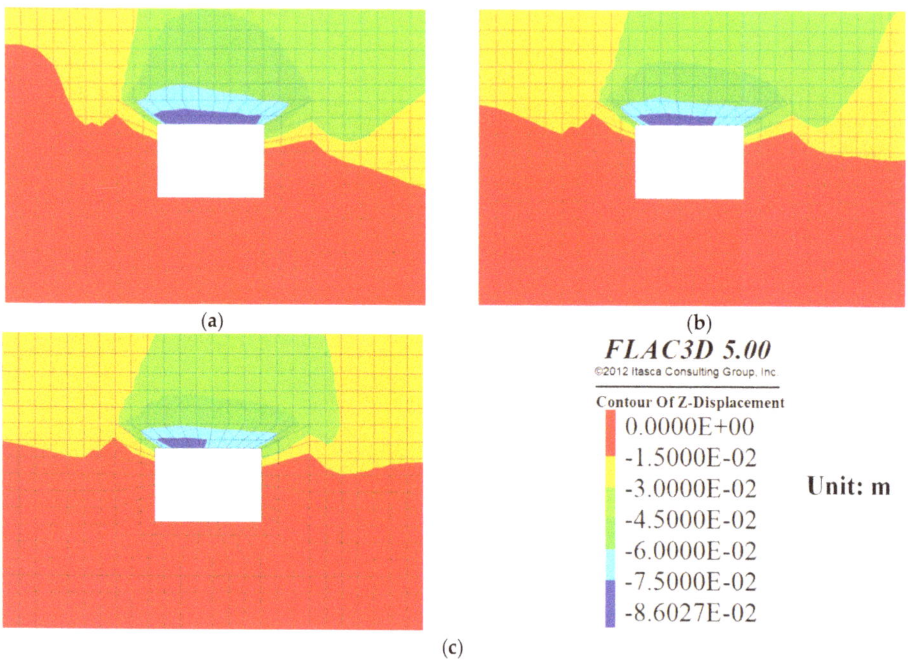

Figure 15. Vertical displacement distribution of the roof of the goaf-side entry with different yield pillar widths: (**a**) model 4 m; (**b**) model 6 m; (**c**) model 8 m.

3.3.2. Displacement Evolution of Two Sides during Roadway Driving along the Goaf-Side

With the conditions of different yield pillar widths, the displacement evolution characteristics of the two sides of the goaf-side entry during the excavating stage are shown in

Figure 17. For ease of presentation, the absolute value of the displacement of the two sides is used in the figure.

Like the roof displacement characteristics, both sides are rapidly deformed after driving along the goaf-side, and then the displacement rate slows down with the continuous advancement of the driving face, but a certain displacement rate is still maintained. The moving of the two sides shows the creep characteristics of the sustained displacement time is shorter than the roof sinking, and the surrounding rock has a faster stable speed. This feature is significantly affected by the yield pillar widths. The approaching amount and approaching displacement of the two sides of the roadway monitoring section at different stages of driving influence are shown in Tables 5 and 6. When the yield pillar width is 8 m, the displacement speed of the two sides of the roadway in the monitoring section is fast and the amount of displacement is large, especially the sudden displacement of the yield pillar. When the lagging excavating working surface is 20 m, the displacement ratio of the two sides approaching reaches 87.71%. The displacement ratio of the two sides basically shows a downward trend as the yield pillar width decreases. When the yield pillar width is 4 m, the displacement ratio of the two sides at the same stage is 86.45%. When the driving surface is further advanced to 60 m from the monitoring section, the two coal bodies move closer and converge in the roadway under the influence of the driving. The displacement ratio of the 4 m yield pillar of the roadway and the coal seam reaches 96.97% and 98.09%, respectively, while the displacement ratio increases with the increase of the yield pillar width in the coal seam and there is no obvious law for the yield pillar. When the yield pillar width is 8 m, the ratios are 96.84% and 98.75%, respectively. From the excavating work surface to the monitoring section, it is 60 m from the monitoring section until the roadway excavating is completed, that is, the excavating stable period.

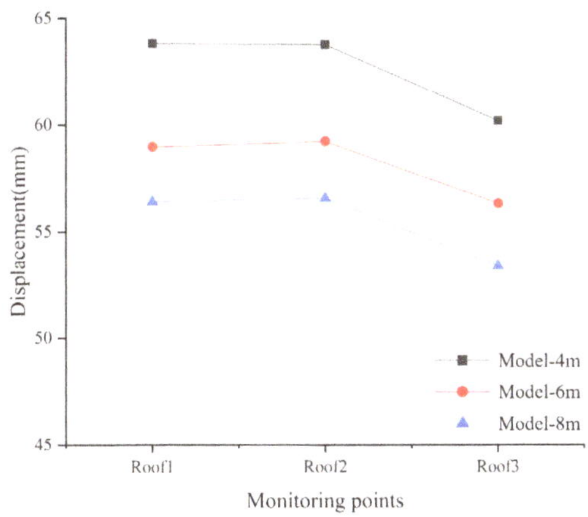

Figure 16. Roof subsidence shape diagram during the excavation of the different yield pillar widths.

Table 5. Evolution of coal seam displacement during roadway driving with different yield pillar widths.

Pillar Width (m)	Cumulative Displacement of Coal Seam	Displacement and Displacement Ratio of the Coal Seam When the Distance of the Lagging Excavating Working Surface Is L_e		
		L_e = 20 m	L_e = 60 m	L_e = 100 m
4 m	61.21 mm	55.69 mm 90.98%	60.04 mm 98.09%	60.92 mm 99.53%
6 m	67.55 mm	62.54 mm 92.58%	66.57 mm 98.55%	67.31 mm 99.64%
8 m	71.33 mm	66.71 mm 93.52%	70.44 mm 98.75%	71.14 mm 99.73%

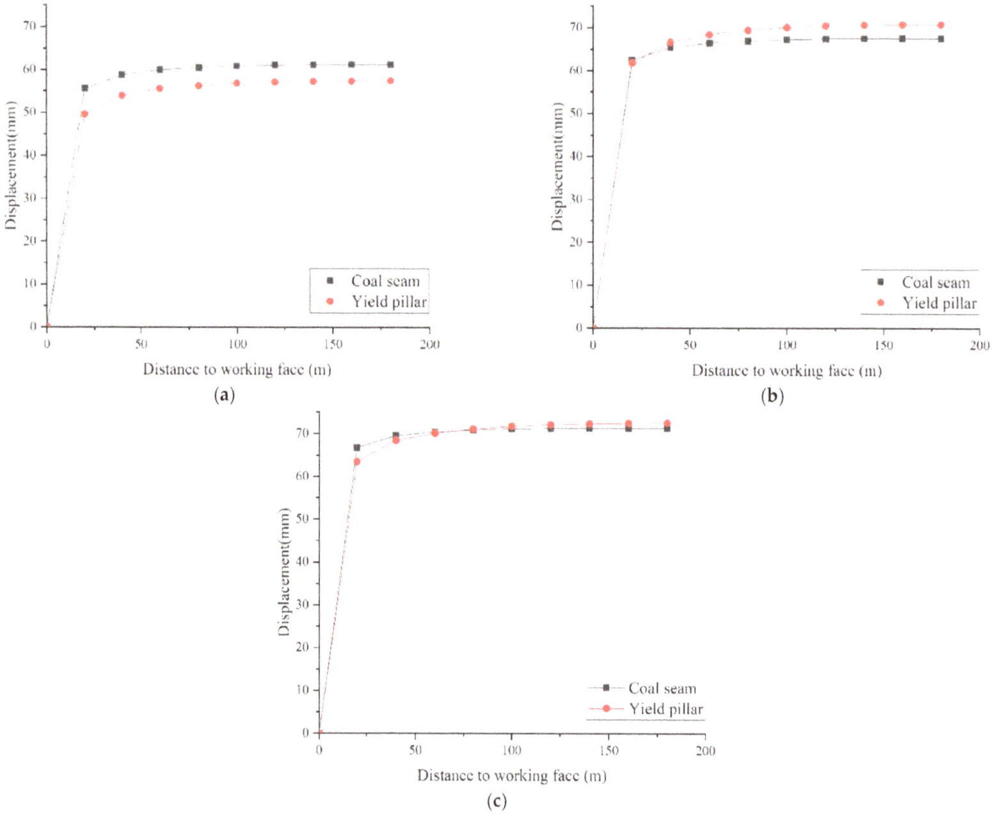

Figure 17. Two sides of roadways' displacement with different yield pillar widths during driving: (**a**) model 4 m; (**b**) model 6 m; (**c**) model 8 m.

Table 6. Evolution of yield pillar displacement during roadway driving with different yield pillar widths.

Pillar Width (m)	Cumulative Displacement of Yield Pillar	Displacement and Displacement Ratio of the Yield Pillar When the Distance of the Lagging Excavating Working Surface Is L_e		
		L_e = 20 m	L_e = 60 m	L_e = 100 m
4 m	57.36 mm	49.59 mm 86.45%	55.62 mm 96.97%	56.90 mm 99.20%
6 m	70.73 mm	61.83 mm 87.41%	68.47 mm 96.80%	70.15 mm 99.18%
8 m	72.41 mm	63.51 mm 87.71%	70.12 mm 96.84%	71.81 mm 99.17%

Under the conditions of different yield pillar widths, the horizontal displacement distribution cloud diagrams of the two sides of the goaf-side entry monitoring section after the roadway excavation is completed are shown in Figure 18. Through comparison, it can be seen that the yield pillar widths have two effects on the goaf-side entry. The deformed shape of the rib also has a significant impact. It can be seen from Figures 18 and 19 that, when the yield pillar width is 6 m and 8 m, the coal body of the pillar side is significantly affected by the mining of the upper section. The difference between the displacement of the two sides is significantly affected by the yield pillar width. When the yield pillar width is 4 m, the displacement of the yield pillar ledge is less than the displacement of the coal seam after the roadway driving is completed, the displacement of the yield pillar ledge reaches 56.90 mm, and the displacement of the coal ledge is 60.92 mm between the two sides The difference in displacement is 3.98 mm, and the two sides show weak asymmetric displacement. When the yield pillar width is 6 m and 8 m, the displacement difference between the two sides is 2.84 mm and 0.67 mm, respectively. The displacement difference between the two sides is small, and the weak asymmetric displacement is negligible.

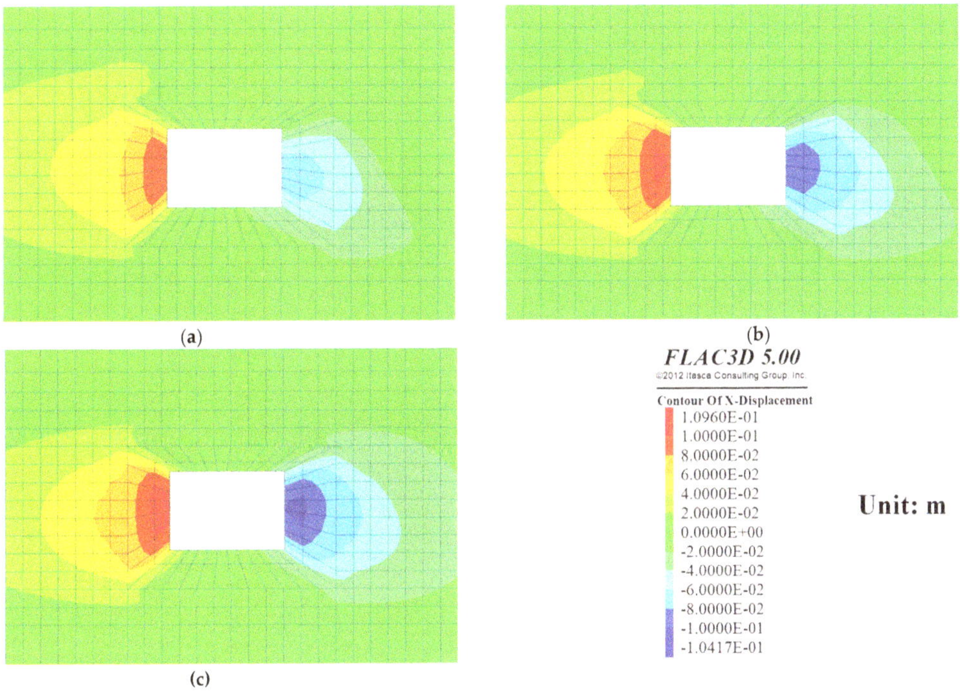

Figure 18. The horizontal displacement distribution of the two sides of the different yield pillar widths during the excavation: (**a**) model 4 m; (**b**) model 6 m; (**c**) model 8 m.

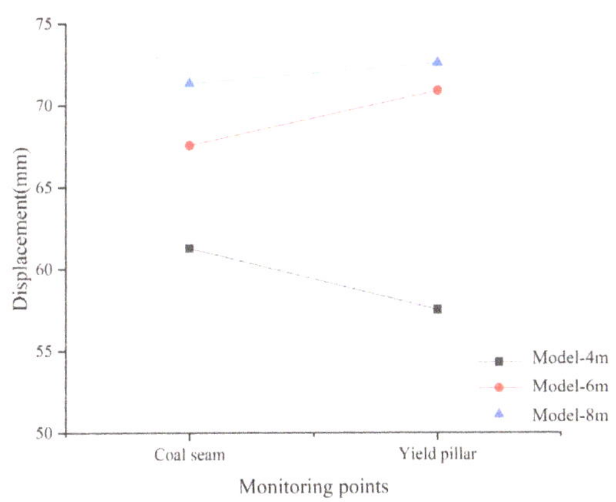

Figure 19. Displacement of the two sides of the different yield pillar widths during the excavation.

3.3.3. The Influence of Different Yield Pillar Width on the Displacement of Surrounding Rock during Roadway Driving

From the displacement and evolution of surrounding rock, the influence of yield pillar width on the appearance of pressure in the mine during the goaf-side entry is analyzed. Under the current geological occurrence and engineering technical conditions, when the yield pillar width is 4 m, the roof subsidence is the largest; when the yield pillar width is 8 m, the displacement of the two sides of the roadway is the largest among the three schemes, and the position of the roadway before the excavation of the roadway is a vertical stress concentration area, which is prone to impact accidents; when the yield pillar width is 6 m, the displacement of the roof, the two sides, and the floor is small. This law confirms the theoretical correctness and design superiority of employing yield pillars to protect the roadway when driving in completely along the goaf-side. Because the roadway is located in the coal seam, the physical and mechanical properties of the roof rock layer are the same as those of the two ribs and the floor coal. The approaching displacement is greatly affected by the width of the yield pillar, and when the yield pillar width is 6 m, more coal is saved than when the yield pillar width is 8 m. Based on the above considerations, when the yield pillar width is 6 m, the plan is the best. Therefore, when the wide yield pillars are changed to yield pillars in Mataihao Mine, the scheme with a yield pillar width of 6 m is preferred.

3.4. Field Monitoring and Observation

The tail entry 3105 is a test goaf-side entry with small yield pillars, with a yield pillar width of 6 m. Before mining in panel 3105, the tail entry is only affected by the excavation, and the displacement of the surrounding rock of the roadway is relatively small. In order to deeply study the mechanism of displacement and failure of surrounding rock of goaf-side entry, and to provide guidance and basis for the layout and support design of roadway under similar conditions, the monitoring of the surrounding rock displacement of the tail entry 3105 was carried out, and the failure mechanism and the control strategy research provide detailed and reliable actual data. Set up monitoring points when driving to 200 m, record data every day, and obtain on-site monitoring data, as shown in Figure 20.

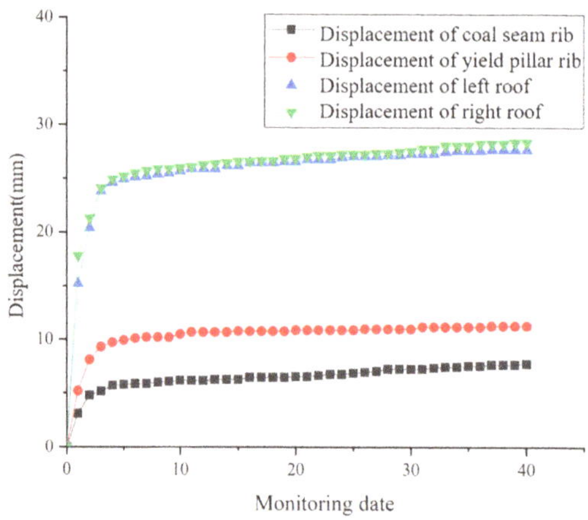

Figure 20. Yield monitoring of roof and two-side displacement data analysis.

In field practice, the amount of surrounding rock displacement is a direct indicator of engineering research to measure the stability of the surrounding rock of the roadway and the reliability of support [37,38]. Through the field data analysis, the detection starts when the roadway excavation reaches the detection point. With the passage of the working face, the roadway surrounding rock is relatively stable, the displacement is small, and there is no obvious asymmetry in the roadway roof. The actual displacement state of the surrounding rock of the roadway is shown in Figure 21. The displacement of the yield pillar is larger than that of the coal seam, but it is also basically stable. To validate the employed numerical approach and suggested design, a field test was carried out at the head entry of panel 3105. The field monitoring results indicate the success of the surrounding rock of roadway control by employing a yield pillar with a 6 m width and serve to validate the numerical approach.

Figure 21. The actual state of displacement of surrounding rock in tail entry 3105.

4. Discussion

Through the study of displacements, plastic failure evolution of surrounding rock, and ground pressure behavior during the excavation of the entry along the goaf, it is shown that after the overburden structure of the mined-out area in the upper section is stabilized, and the goaf-side entry mainly plays a role of isolation. The yield pillar width has a greater impact on the surrounding rock conditions and the mechanical environment where the roadway is located.

The displacement evolution characteristics of the two sides of the goaf-side entry under the influence of the yield pillar width are consistent with the laws and mechanisms of the roof displacement evolution characteristics, and they are all caused by the different surrounding rock and stress environment of the roadway location under different yield pillar widths. When the yield pillar width is small, it is necessary to ensure the timeliness and effectiveness of temporary support and one-time support during the excavation to avoid problems such as roof collapse caused by the rapid displacement of broken surrounding rock after the roadway is excavated. When the yield pillar width is large, it is necessary to carry out regular and timely monitoring of surrounding rock displacement behind the working surface of the roadway and prevent slow creep behaviour from causing support failures and large displacements.

5. Conclusions

In this study, it is shown that adopting reasonable yield pillar is crucial for the efficiency and safety of the goaf side roadway. A reasonable pillar width means that it cannot be too large, otherwise it will be stiff enough to attract high stress in the direction of the coal seam. Moreover, it cannot be too small, otherwise it will fail, and it will not be possible to anchor broken rock. According to the theoretical calculations, with the existing geological conditions in seam 3-1, the reasonable yield pillar width is 4 to 6 m. According to field data measurement, when the yield pillar width is less than 8.5 m, the roadway maintains a stable condition.

Considering comprehensively the pressure and support of the mine along the goaf with high mining height, the isolation and ventilation safety of the goaf, the recovery rate of coal resources in the mining area, and the social and economic benefits, a 6 m yield pillar is employed along the goaf-side, which is more conducive to the roadway. During excavation and mining, the surrounding rock is stable, and the yield pillar can accommodate the depth of anchoring support, which can reduce the amount of roadway maintenance. At the same time, the yield pillar has a better isolation effect; it can further improve the coal mining rate as well as the economic and social benefits.

Author Contributions: Conceptualization, Q.W.; methodology, L.J.; software, Q.W.; validation, H.S.M., H.F. and P.T.; data curation, Q.W.; writing—original draft preparation, Q.W.; writing—review and editing, C.L. and Y.P.; funding acquisition, L.J. All authors have read and agreed to the published version of the manuscript.

Funding: This research was funded by the National Natural Science Foundation of China, grant number 52074166; the China Postdoctoral Science Foundation, grant number 2020T130385; the Natural Science Foundation of Shandong, grant number ZR2021YQ38; and the Climbing Project of Taishan Scholar in Shandong Province, grant number tspd20210313.

Institutional Review Board Statement: Not applicable.

Informed Consent Statement: Not applicable.

Data Availability Statement: Not applicable.

Conflicts of Interest: The authors declare no conflict of interest.

References

1. Fan, L.; Liu, S. Respirable nano-particulate generations and their pathogenesis in mining workplaces: A review. *Int. J. Coal Sci. Technol.* **2021**, *8*, 179–198. [CrossRef]
2. Zhang, B.; Wang, P.; Cui, S.; Fan, M.; Qiu, Y. Mechanism and surrounding rock control of roadway driving along gob in shallow-buried, large mining height and small coal pillars by roof cutting. *J. China Coal Soc.* **2021**, *46*, 2254–2267.
3. Colwell, M.; Hill, D.; Frith, R. ALTS II, A longwall gateroad design methodology for Australian collieries. In Proceedings of the 1st Australian Conference on Ground Control in Mining, Sydney, Australia, 10–13 November 2003; pp. 123–135.
4. Qi, Q.; Li, Y.; Zhao, S.; Zhang, N.; Zheng, W.; Li, H.; Li, H. Seventy years development of coal mine rock burst in China: Establishment and consideration of theory and technology system. *Coal Sci. Technol.* **2019**, *47*, 1–40.
5. Du, J.; Meng, X. *Mining Science*; China University of Mining and Technology Press: Xuzhou, China, 2009; pp. 122–130.
6. Hou, C.; Li, X. Stability principle of large and small structures of surrounding rock in gob-side driving roadway of fully mechanized caving. *J. China Coal Soc.* **2001**, *26*, 1–7.
7. Maleki, H.N. Ground response to longwall mining: A Case study of two-entry yield pillar evolution in weak rock. *Colo. Sch. Mines Q.* **1988**, *83*, 51.
8. Maleki, H.N. An analysis of violent failures in US coal mines—Case studies. In *Proceedings of Mechanics and Mitigation of Violent Failure in Coal and Hard-Rock Mines*; Maleki, H., Wopat, P.F., Repsher, R.C., Tuchman, R.J., Eds.; SP 01-95; US Bureau of Mines: Washington, DC, USA, 1995; pp. 5–26.
9. Xue, K.; Fu, B. Research and application of reasonable width of coal pillars in roadway excavation. *J. Undergr. Space Eng.* **2018**, *14* (Suppl. 1), 403–408.
10. Chen, W. Study on control technology of rockburst in coalmines of Western Erdos. *Coal Sci. Technol.* **2018**, *46*, 99–104.
11. Zhang, G.; He, F. Asymmetric failure and control measures of large cross-section entry roof with strong mining disturbance and fully-mechanized caving mining. *Chin. J. Rock Mech. Eng.* **2016**, *35*, 806–818.
12. Wang, D.; Li, S.; Wang, Q.; Li, W.; Wang, F.; Wang, H.; Peng, P.; Ren, G. Experimental study of reasonable coal pillar width in fully mechanized top coal caving face of deep thick coal seam. *Chin. J. Rock Mech. Eng.* **2014**, *33*, 539–548.
13. Hou, C.; Mines, C.O. Key technologies for surrounding rock control in deep roadway. *J. China Univ. Min. Technol.* **2017**, *46*, 970–978.
14. Wang, W.; Hou, C. Study of mechanical principle of floor heave of roadway driving along next goaf in fully mechanized sub-level caving face. *J. Coal Sci. Eng.* **2001**, *7*, 13–17.
15. Wang, Y.; He, M.; Yang, J.; Fu, Q.; Gao, Y.; University, T. The structure characteristics and deformation of "short cantilever beam" using a non-pillar mining method with gob-side entry formed automatically. *J. China Univ. Min. Technol.* **2019**, *48*, 718–726.
16. Mark, C.; Mclinda, G.M.; Dolinar, D.R. Analysis of rock bolting systems. In Proceedings of the 20th International Conference on Ground Control in Mining, Morgantown, WV, USA, 7–9 August 2001; pp. 123–125.
17. Mark, C.; Dolimar, D.R. Development and application of the coal mine roof rating (CMRR)—A decade of experience. *Int. J. Coal Geol.* **2005**, *46*, 85–103. [CrossRef]
18. Esterhuizen, G.S.; Gearhart, D.F.; Klemetti, T.; Dougherty, H.; van Dyke, M. Analysis of gateroad stability at two longwall mines based on field monitoring results and numerical model analysis. *Int. J. Min. Sci. Technol.* **2019**, *29*, 35–43. [CrossRef]

19. Shabanimashcool, M.; Li, C.C. Numerical modelling of longwall mining and stability analysis of the gates in a coal mine. *Int. J. Rock Mech. Min. Sci.* **2012**, *51*, 24–34. [CrossRef]
20. Bai, J.; Hou, C.; Huang, H. Numerical simulation study of stability of narrow coal pillar of roadway driven along the goaf. *Chin. J. Rock Mech. Eng.* **2004**, *20*, 3475–3479.
21. Qian, M.; Shi, P.; Xu, J. *Mining Pressure and Strata Control*; China University of Mining and Technology Press: Xuzhou, China, 2010. (In Chinese)
22. Tu, S.; Bai, Q.; Tu, H. Pillar size determination and panel layout optimization for fully mechanized faces in shallow seams. *J. Min. Saf. Eng.* **2011**, *28*, 505–510.
23. Jiang, L.; Liu, H.; Lian, X.; Zhang, W. Research on rational width of coal-pillar in shallow-buried medium-thick coal-seam. *Coal Min. Technol.* **2012**, *7*, 105–108.
24. Yavuz, H. An estimation method for cover pressure re-establishment distance and pressure distribution in the goaf of longwall coal mines. *Int. J. Rock Mech. Min. Sci.* **2004**, *41*, 193–205. [CrossRef]
25. Esterhuizen, G.S.; Mark, C.; Murphy, M.M. Numerical model calibration for simulating coal pillars, gob and overburden response. In Proceedings of the 29th International Conference on Ground Control in Mining, Morgantown, WV, USA, 27–29 July 2010; pp. 1–12.
26. Esterhuizen, G.S. A stability factor for supported mine entries based on numerical model analysis. In Proceedings of the 31st International Conference on Ground Control in Mining, Morgantown, WV, USA, 27–29 July 2012; pp. 1–9.
27. Badr, S.A. Numerical Analysis of Coal Slender Pillars at Deep Longwall Mines. Ph.D. Thesis, Colorado School of Mines, Golden, CO, USA, 2004; pp. 23–56.
28. Zhang, Z.; Bai, J.; Chen, Y.; Yan, S. An innovative approach for gob-side entry retaining in highly gassy fully-mechanized longwall top-coal caving. *Int. J. Rock Mech. Min.* **2015**, *80*, 1–11. [CrossRef]
29. Li, W.; Bai, J.; Peng, S.; Wang, X.; Xu, Y. Numerical modeling for yield pillar design: A case study. *Rock Mech. Rock Eng.* **2013**, *48*, 305–318. [CrossRef]
30. Piotr, M.; Zbigniew, N.; Tafida, B. A statistical analysis of geomechanical data and its effect on rock mass numerical modeling: A case study. *Int. J. Coal Sci. Technol.* **2021**, *8*, 312–323.
31. Yan, S.; Bai, J.; Wang, X.; Huo, L. An innovative approach for gateroad layout in highly gassy longwall top coal caving. *Int. J. Rock Mech. Min. Sci.* **2013**, *59*, 33–41. [CrossRef]
32. Yang, X. Research and Application of Surrounding Rock Stability Control Technology in Soft Rock Roadway Influenced by Multiple Disturbances. Ph.D. Thesis, Shan Dong University of Science and Technology, Qingdao, China, 2020.
33. Hoek, E.; Carranza-Torres, C.; Corkum, B. Hoek-Brown failure criterion—2002 edition. In Proceedings of the NARMS-Tac 2002, Toronto, ON, Canada, 7–10 July 2002; Volume 1, pp. 267–273.
34. Jiang, L.; Zhang, P.; Chen, L.; Hao, Z.; Sainoki, A.; Mitri, H.S.; Wang, Q. Numerical approach for goaf-side entry layout and yield pillar design in fractured ground conditions. *Rock Mech. Rock Eng.* **2017**, *50*, 3049–3071. [CrossRef]
35. Wang, M.; Bai, J. Failure mechanism and control of deep gob-side entry. *Arab. J. Geosci.* **2015**, *8*, 9117–9131. [CrossRef]
36. Peng, S.S. *Coal Mine Ground Control*, 3rd ed.; Peng SS Publisher: Morgantown, WV, USA, 2008; pp. 229–267.
37. Singh, R.; Mandal, P.K.; Singh, A.K.; Kumar, R.; Sinha, A. Optimal underground extraction of coal at shallow cover beneath surface/subsurface objects: Indian practices. *Rock Mech. Rock Eng.* **2008**, *41*, 421–444. [CrossRef]
38. Hou, C. *Ground Control of Roadways*; China University of Mining & Technology Press: Xuzhou, China, 2013.

processes

Article

Numerical Investigation on the Impact of Tailings Slurry on Catch Dams Built at the Downstream of a Breached Tailings Pond

Shitong Zhou [1,2] and Li Li [2,*]

1 School of Resources and Safety Engineering, Central South University, No.932 South Lushan Road, Changsha 410083, China; zhoushitong@csu.edu.cn
2 Research Institute on Mines and Environment, Department of Civil, Geological and Mining Engineering, École Polytechnique de Montréal, C.P. 6079 Succursale Centre-Ville, Montréal, QC H3C 3A7, Canada
* Correspondence: li.li@polymtl.ca

Abstract: Tailings storage facilities (TSFs) are known as a time-bomb. The numerous failures of TSFs and the heavy catastrophic consequences associated with each failure of TSFs indicate that preventing measures are necessary for existing TSFs. One of the preventing measures is to construct catch dams along the downstream near TSFs. The design of catch dams requires a good understanding of the dynamic interaction between the tailings slurry flow and the catch dams. There are, however, very few studies on this aspect. In this study, a numerical code, named LS-DYNA, that is based on a combination of smoothed particle hydrodynamics and a finite element method, was used. The numerical modeling shows that the tailings slurry flow can generally be divided into four stages. In terms of stability analysis, a catch dam should be built either very close to or very far from the TSF. When the catch dam with an upstream slope of a very small inclination angle is too close to the tailings pond, it can be necessary to build a very high catch dam or a secondary catch dam. As the impacting force can increase and decrease with the fluctuations back-and-forth of the tailings slurry flow, the ideal inclination angle of the upstream slope of the catch dam is between 30° and 37.5°, while the construction of a catch dam with a vertical upstream slope should be avoided. However, a catch dam with steeper upstream slopes seems to be more efficient in intercepting tailings flow and allowing the people downstream to have more time for evacuation. All these aspects need to be considered to optimize the design of catch dams.

Keywords: tailings dam; impacting force; kinetic energy; numerical simulation; smoothed particle hydrodynamics (SPH)

Citation: Zhou, S.; Li, L. Numerical Investigation on the Impact of Tailings Slurry on Catch Dams Built at the Downstream of a Breached Tailings Pond. *Processes* **2022**, *10*, 898. https://doi.org/10.3390/pr10050898

Academic Editor: Farhad Ein-Mozaffari

Received: 13 February 2022
Accepted: 29 April 2022
Published: 2 May 2022

1. Introduction

Mines produce large volumes of mine waste each year in terms of tailings and waste rocks [1]. A small portion of these materials can be sent back to underground voids as backfill [2–18]. The recycling and reuse of these materials for in- and out-mine sites are also increasingly seen [19]. All these practices can reduce the amount of mine waste to be disposed of on the ground surface. However, due to the swelling of rocks by blasting and grinding, the large amount of tailings has to be sent by pipes to a place on the ground surface contained by dams, called a tailings storage facility (TSF), which then becomes a new perpetuity infrastructure in the environment [20–25].

Tailings are a byproduct of mill processing, containing a large portion of fine particles smaller than 80 μm and water [26]. Upon disposition in a TSF, tailings remain saturated for a long time. This is particularly true in the regions where the water balance is positive. When a TSF fails, a large volume of tailings can be instantaneously released. Their flow usually transforms into a mud flood, causing significant damage to the environment and infrastructures, and even loss of lives [27]. In 1965 in Chile, the town of El Cobre was destroyed and more than 200 people were killed due to the failure of an upstream TSF [28]. Since then, a lot of work has been done on the analysis of failure mechanisms of TSFs [29–33]. Despite considerable progress in the studies, failures of TSF continue to take place worldwide. In February 1994, in South Africa, the village of Merriespruit was destroyed and 17 people were killed due to the failure of an upstream TSF. On that occasion, the TSF was, for the first time, qualified by a judge as a time-bomb [34]. In China, the failure of a tailings dam at Shiqiaozi, Haicheng, Liaoning Province, on 25 November 2007 resulted in 10 people being killed, 3 reported missing, and 17 injured. Another case of a TSF failure at Taoshi, Linfen City, Xiangfen County, Shanxi Province, on 8 September 2008 involved 277 people being killed and 33 injured [28]. In 2014, in Canada, the failure of a TSF of the Mount Polley mine resulted in the release of 23.6 million cubic meters of tailings into the environment [35,36]. In November 2015, at the Samarco mine in Brazil, the failure of a TSF resulted in the death of 17 people and a flow of tailings sludge of about 800 km up to the Atlantic coast [28,37,38]. In January 2019, near Brumadinho in Brazil, 248 people were killed and 22 were reported missing due to the failure of a TSF [28,39]. These endless failures of TSFs worldwide indicate that big efforts are still needed to prevent more casualties issued from TSF failures.

The instability of a TSF can be related to a number of influencing factors, varying from technical and engineering to maintenance and management [27,29,30,34,38–41]. The technical and engineering aspects affecting the stability of TSFs include the design and construction of TSFs and their foundations. At current stage, it is impossible to entirely eliminate any risk of TSF failure [33,42]. Rather, the probability of TSF failures can even increase with time due to the persistence of existing TSFs, the ceaseless increase of the number of mines and TSFs, and the more and more pronounced climate changes, which usually have a negative impact on the stability of TSFs. Efforts should be continuously made to improve the stability of TSFs [43], install monitoring systems [44], and even forecast the scale of destructions associated with the failure of a TSF [45].

Currently, thickened or filtered tailings are widely considered as the solution of solving the physical instability of TSFs. It is, however, problematic with chemically reactive mine waste because the unsaturated tailings favor the oxidation and generation of acid mine drainage [46–48]. In addition, unsaturated tailings can also be prone to liquefaction, as long as the degree of saturation approaches full saturation [49], and this is probably the case in the regions where the water balance is positive. Considering this as a sure and certain solution to avoid a TSF failure could be irresponsible [21]. Moreover, work is needed to secure existing conventional TSFs. In all cases, risk analysis on the breach of TSFs is necessary [50–65]. In many cases, preventing measures can be necessary to avoid catastrophic consequences in case of a TSF failure. One of the preventing measures is to build catch dams near the downstream of a TSF. Upon failure of the TSF, the catch dams must remain stable to hold at least part of the released tailings slurry, delaying and reducing the flood of tailings. If the catch dam is not properly designed and constructed, it can fail upon the dynamic impact of a tailings slurry flood. This was the case after the breach of a TSF in March 2020 in Yichun, Heilongjiang, China. Ten catch dams had to be built, one after another, along the downstream to stop the tailings flood [66].

The purpose of this study is to obtain a good understanding of the characteristics of the tailings slurry flow after a breach of a TSF, and on the dynamic interaction between the flooding tailings and catch dams. This can ultimately help to obtain a safe and economic design of catch dams to avoid catastrophic consequences after the failure of a TSF.

Over the years, a number of works have been published on the dynamic interaction between water and downstream obstacles [67–72]. One sees also some publications on the dynamic interaction between debris flow and check dams [73–76], or even between granular material flows and obstacles [77]. These works provide a useful insight into the dynamic interaction of water, debris, and granular material with obstacles. However, very few works have been made on the dynamic interaction between tailings slurry and downstream obstacles.

Wang et al. [78] conducted numerical modeling for a specific project to study the flow characteristics of tailings upon the failure of a tailings dam. The effects of downstream dams were briefly presented without any detailed quantitative studies. Zeng et al. [79] evaluated the effect of check dam height against the debris flow of tailings slurry. A similar work has been published by Wang et al. [80], who analyzed the risk associated with the possible failure of tailings dams and the effect of a preventing dam height, built 110 m downstream from the tailings pond. All these works provide some insights on the effects of check dams on the flow of the tailings slurry. However, they were mostly for some specific projects with site-specific conditions. Much more work is needed to obtain a better understanding of the effect of downstream catch dams on the flow of a tailings slurry.

In this study, the dynamic interaction between a tailings slurry flow and a downstream infrastructure is investigated through numerical modeling, using a hybrid numerical approach with the smoothed particle hydrodynamics-finite element method (SPH-FEM). The kinetic energy of the tailings flow and the striking forces on the downstream infrastructure are evaluated. The influences of the position and upstream slope angle of the catch dam on the impact of the catch dam are, for the first time, studied. The results provide some guidelines on the layout and design of catch dams.

2. Numerical Code

Studying the characteristics of tailings slurry flow needs a numerical code that can simulate the large deformation and displacement with the time of fluid-like materials, while the fluid–structure interaction behavior can be evaluated by the conventional finite-element or finite difference method.

In this study, a numerical code, called LS-DYNA [81], based on a combination of smoothed particle hydrodynamics (SPH) and the finite element method (FEM) was used. SPH is a meshless particle method based on the Lagrangian formulation for fluid dynamics' simulation. It was originally developed by Lucy [82], Gingold, and Monaghan [83] for three-dimensional astrophysical analysis. It is based on the hypothesis that the problem domain can be divided into a set of discrete elements, which possess or can be attributed to a certain material property. With the SPH method, large deformations can be handled without element distortions. It can thus be used to simulate dynamic problems involving large deformations, such as fluid problems with a free surface [84].

The application of the SPH method involves two approximations to represent the simulated space: the kernel and particle [85]. Figure 1 schematically shows a presentation of the SPH kernel function, W. On the figure, r is the distance between the two particles; λh denotes the influence area of the kernel function; λ is determined by the kernel function and space dimensions to satisfy the normalization condition.

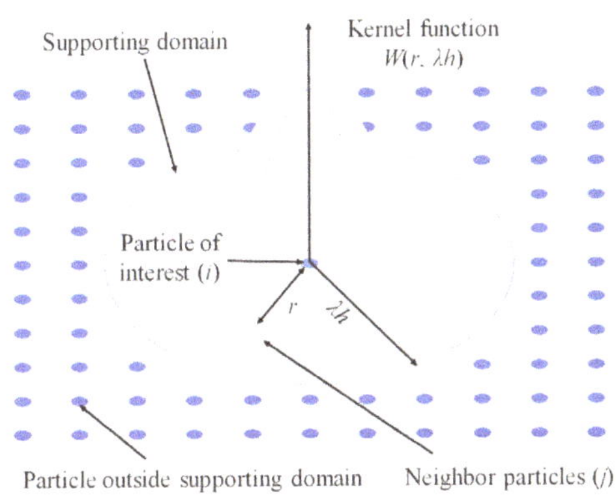

Figure 1. A schematic presentation of the SPH kernel function W (reproduced with minor changes from Atif et al. [86], with permission from Copyright Clearance Center, Inc. (CCC) on behalf of Springer-Verlag, Dordrecht, The Netherlands).

In the approximation of the kernel, the spatial distance between the particles is covered by a smoothing length, over which the properties are smoothed by a smoothing kernel function. In particle approximation, the state of a system is represented by a set of particles, which possess individual material properties and move according to the conservation equations. A given particle, *i*, with its given velocity, density, mass, and pressure, interacts with particle *j* within the smoothing length. Considering the conservations of mass, momentum, and energy leads to the following expressions [87]:

$$\frac{d\rho_i}{dt} = \sum_{j=1}^{N} m_j v_{ij}^{\beta} \frac{\partial W_{ij}}{\partial x_i^{\beta}} \tag{1}$$

$$\frac{dv_i^{\alpha}}{dt} = \sum_{j=1}^{N} m_j \left(\frac{\sigma_i^{\alpha\beta}}{\rho_i^2} + \frac{\sigma_j^{\alpha\beta}}{\rho_j^2} + \prod{}_{ij} \right) \frac{\partial W_{ij}}{\partial x_i^{\beta}} \tag{2}$$

$$\frac{de_i}{dt} = \frac{1}{2} \sum_{j=1}^{N} m_j \left(\frac{\sigma_i^{\alpha\beta}}{\rho_i{}^2} + \frac{\sigma_j^{\alpha\beta}}{\rho_j^2} + \prod{}_{ij} \right) v_{ij}^{\alpha} \frac{\partial W_{ij}}{\partial x_i^{\beta}} \tag{3}$$

where ρ_i and ρ_j denote the densities of particles *i* and *j*; *N* denotes the number of particles; α and β denote the space vectors; x_i^{β} denotes the coordinate component of *x* along β; W_{ij} denotes the smoothing function between particles *i* and *j*; v_{ij}^{α} and v_{ij}^{β} represent the velocities of particle *i* relative to particle *j* along α and β; m_j denotes the mass of particle *j*; v_i^{α} and v_j^{α} denote the velocity components of particles *i* and *j* along α; $\sigma_i^{\alpha\beta}$ and $\sigma_j^{\alpha\beta}$ denote the total stress tensors of particles *i* and *j*; $\prod_{ij}$ denotes the artificial viscosity; e_i represents the internal energy of the particle *i*; and *t* represents the time.

While the SPH method is advantageous in dealing with fluids and large deformation problems, this dynamic meshless technique is much less efficient in dealing with static and small deformation problems of solid bodies than FEM [2,88]. The combination of SPH and FEM methods is thus a natural choice to deal with the dynamic interaction between fluid and infrastructures. This explains the wide application of the SPH and FEM hybrid algorithm in the simulation of the fluid–solid interaction [89].

Figure 2 shows a possible interaction between fluid particles and structure elements. The interaction between fluid and a structure is achieved by a penalty-type 'node-to-surface' contact algorithm. The FEM elements and SPH particles are considered as master and slave parts, respectively. The reaction force between a SPH particle and the contact surface of the structure is then proportional to the depth of the slave part on the surface of the master part. The forces exerted by the SPH particles on the FEM part are then calculated by summing up the forces exerted by each SPH particle on the FEM part.

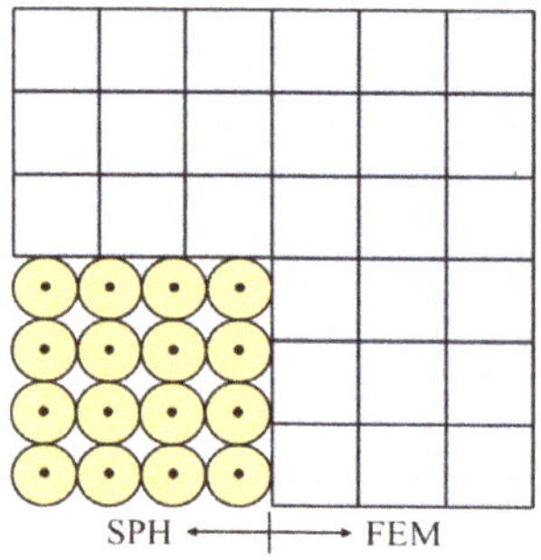

Figure 2. The coupling of SPH and FEM (reproduced with minor changes from Koneshwaran et al. [90], with permission from Copyright Clearance Center, Inc. (CCC) on behalf of Elsevier Science & Technology Journals, Amsterdam, The Netherlands).

In this study, the explicit dynamic solver of LS-DYNA® version R10.0 was employed to perform the numerical modeling based on SPH theories [91]. The applicability (usually called "validation") of the numerical code SPH was tested against the experimental results of a slump test (see details given in Appendix A). The results indicate that the numerical code is capable of simulating the dynamic interaction between a tailings slurry and a structure.

3. Numerical Simulations of Dynamic Interaction between Tailings Flow and A Catch Dam

3.1. Numerical Models

Figure 3 shows a numerical model to simulate a largely reduced scale tailings pond. The tailings pond is 0.5 m long, 0.5 m wide, and 0.5 m high. The tailings slurry is the same as that used to test the applicability of the numerical code, presented in Appendix A. It is characterized by a density of 1826.6 kg/m^3 and a dynamic viscosity of 0.43 Pa·s to simulate the tailings slurry having a solids content by mass of 70% of Gao [92]. The catch dam is simulated by a rigid triangular prism 0.2 m high and 0.5 m wide with an upstream slope angle α varying from 0° to 90°. The distance between the tailings pond and the catch dam is 0.8 m. The movement of the tailings particles is allowed in all directions, but only inside the open box defined by the back wall of the tailings pond, a horizontal floor and two vertical longitudinal walls. The floor and the three walls are rigid and their movements are not allowed in any direction. The tailings slurry is simulated by particles with SPH, while the rigid catch dam is built with FEM shell elements.

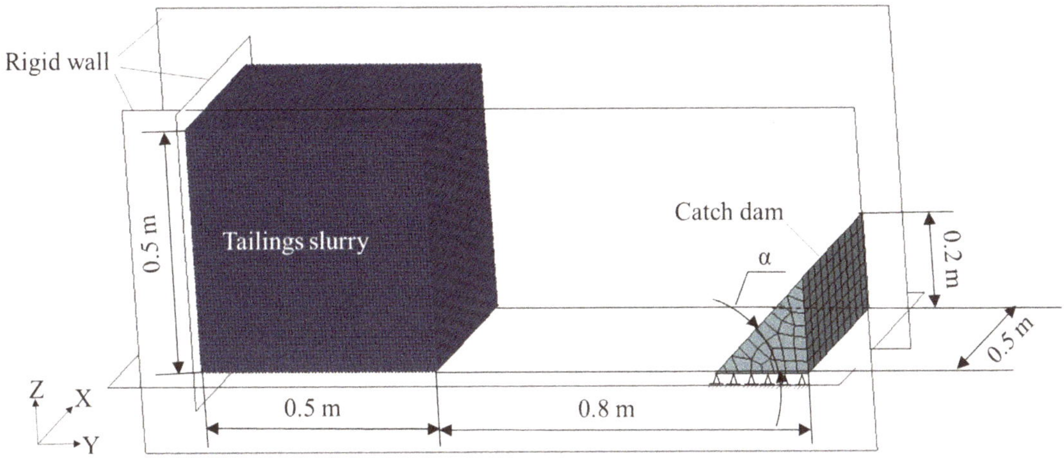

Figure 3. The numerical model and boundary conditions of tailings pond break model.

In this study, the flow behavior of the tailings slurry and the impact of the tailings slurry on the catch dam are analyzed by a contact algorithm with the tailings slurry and the catch dam defined as slave and master parts, respectively. The gravitational acceleration is applied along the vertical downward direction. The three vertical walls and the horizontal floor are fixed along their outside surfaces. The catch dam is fixed along its bottom. The simulations were done by considering a duration of 2.5 s, starting from the breach of the tailings pond.

For each numerical model, sensitivity analyses were made to make sure that the distinct elements were dense enough for the tailings and voids simulated by SPH (see Appendix B). These analyses showed that the tailings can be modeled with 357,911 SPH particles and a particle-to-particle spacing of 7 mm.

3.2. Numerical Results

Before the breach of the tailings pond, the tailings slurry had the highest gravitational potential energy. After the breach of the tailings pond, the tailings slurry flowed. The gravitational potential energy and kinetic energy are interconverted. It is noted that the kinetic energy K and gravitational potential energy P designate the total kinetic energy and total gravitational potential energy of the whole tailings slurry. At a given time, they can be obtained as follows:

$$K = \sum_{i=1}^{i=N} \frac{m_i v_i^2}{2} \tag{4}$$

$$P = \sum_{i=1}^{i=N} (m_i g h_i) \tag{5}$$

where m_i, v_i, and h_i are the mass, velocity, and elevation of SPH particle i, respectively; N is the total number of SPH particles of the tailings slurry; and g is the gravitational acceleration.

Figure 4 shows the flow states of the tailings particles at different times (T) starting from the breach of the tailings dam, with the catch dam having an upstream slope angle of $\alpha = 30°$. At $T = 0$ s, when the tailings dam instantaneously collapses, the potential energy of tailings slurry is the highest and ready to be released. No particle moves and the kinetic energy is zero (Figure 4a). At $T = 0.24$ s, the upper front part of the tailings falls, moves forward and reaches the catch dam with a maximum velocity of 4.40 m/s (Figure 4b). At $T = 0.5$ s, all the upper part of the tailings starts moving. The front line of the tailings flood rushes away from the catch dam at a velocity of 3.31 m/s. Due to this high velocity and the slope inclination of the catch dam's upstream, the movement of the front line is upward, almost parallel to the surface of the upstream slope (Figure 4c). At $T = 0.76$ s, more tailings at the top start to move and the front line of the tailings flood continues rushing away from the catch dam, but falling toward the floor. The maximum velocity of the front line is 3.92 m/s (Figure 4d). At $T = 1.2$ s, all the particles of tailings that can move have moved. It is interesting to note that the tailings slurry near the upstream of the catch dam is not only higher than the catch dam, but also higher than the remnants of the tailings pond. This is probably due to the drag effects by the friction of the upper mobile particles and the suction of front particles. In addition, it should be noted that a back flow of the tailings slurry takes place at the toe of the catch dam's downstream (Figure 4e). This back flow may have a negative impact on the stability of the catch dam. At $T = 2.5$ s, the crest of the tailings flood moves back toward the tailings pond (Figure 4f). This phenomenon is well-known as sloshing [93,94]. Finally, one notes that the tailings particles near the rigid back walls of the tailings pond always show delayed movement, probably due to the friction and viscosity along the tailings–wall interfaces (see Figure 4b,c,f).

Figure 5 shows the variation of the tailings' kinetic energy (Figure 5a) and impacting force (Figure 5b) exerted by the tailings flood on the catch dam's upstream slope surface as a function of time since the breach onset of the tailings dam. One sees that the flow characteristics of the tailings slurry can generally be divided into four stages:

- Stage 1: Initiation. This stage starts at the breach onset of the tailings dam at $T = 0$ s and ends at $T = 0.38$ s when the kinetic energy and impacting forces reach their first peak values. Upon the breach of the tailings dam at $T = 0$ s, the front part of the tailings tends to flow, driven by the potential energy. Once started, the movement of the tailings particles accelerates upon their residual potential energy and the pushing forces exercised by the upper and back particles. Subsequently, the kinetic energy continuously increases with time. At $T = 0.24$ s, the front line of the tailings flood reaches the toe of the catch dam. The upstream slope of the catch dam starts to receive an impacting force. After then, the tailings slurry flows upward over the upstream slope of the catch dam. A part of the kinetic energy has to be transferred into the potential energy required to allow the rising of the tailings particles. This leads to a decreased pace in the increase of kinetic energy until a peak value at $T = 0.38$ s when the impacting force also reaches its first peak value.

- Stage 2: Peak stage. This stage starts at $T = 0.38$ s when the kinetic energy and impacting forces reach their first peak values, and ends at $T = 0.76$ s when the front line of the tailings flow touches the floor. In this stage, the kinetic energy and impacting force first exhibit a decrease from their peak values and then an increase until a second peak value. At $T = 0.50$ s, the front of the tailings flow rises to the maximum height of 37 cm. At $T = 0.76$ s, the front line of the tailings flow touches the floor and the kinetic energy starts to decrease. Despite their changes with time, the magnitudes of kinetic energy and impacting force remain at relatively high levels.
- Stage 3: Ebb. This stage starts at $T = 0.76$ s when the kinetic energy starts to decrease, and ends at $T = 1.2$ s when the crest of the tailings flood starts to retreat back toward the tailings pond. During this stage, the kinetic energy continuously decreases while the impacting force keeps almost constant.
- Stage 4: Sloshing. This stage starts from $T = 1.2$ s when the crest of the tailings flood starts to retreat toward the tailings pond. In this stage, the crest of the tailings slurry can first move back toward the tailings pond, and after, it can then move forward toward the catch dam. This back-and-forth can take several rounds until an equilibrium is reached. Despite these back-and-forth movements, the kinetic energy and impacting forces decrease with time.

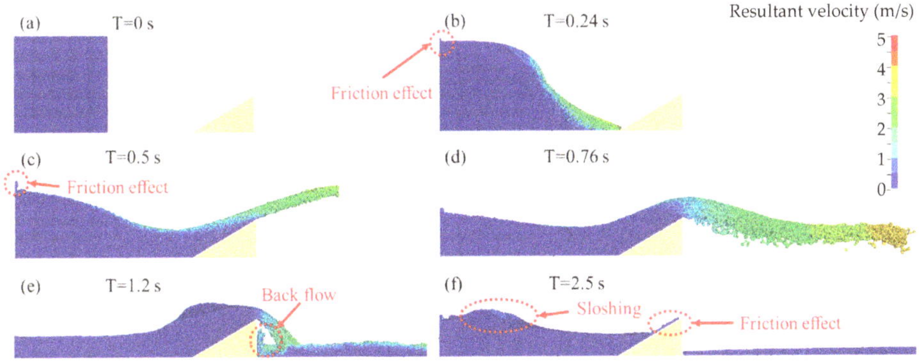

Figure 4. Flow states of the tailings particles at (**a**) $T = 0$ s, (**b**) $T = 24$ s, (**c**) $T = 0.5$ s, (**d**) $T = 0.76$ s, (**e**) $T = 1.2$ s, and (**f**) $T = 2.5$ s starting from the breach of the tailings dam; simulation with a catch dam having an upstream slope angle of 30°.

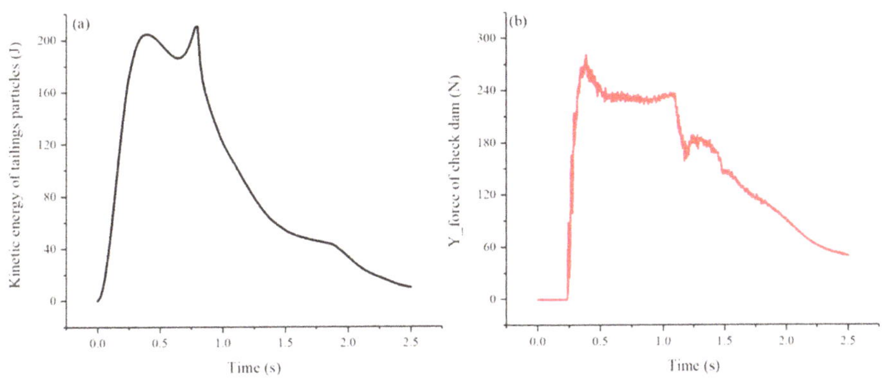

Figure 5. Evolution of (**a**) kinetic energy of tailings slurry and (**b**) impacting force in Y direction on the catch dam.

Additional simulations have been done to analyze the effects of the upstream slope inclination angle of the catch dam on the kinetic energy and impacting forces. Figure 6 shows the flow states of the tailings particles when the catch dam's upstream slope starts to be more or less fully covered by the tailings slurry, with the inclination angle α varying from 15° to 90°. Firstly, one sees that the tailings flow rises smoothly along the slope when the catch dam's upstream slope has a small inclination angle. Backflow of the tailings front and splashing are observed when the slope inclination becomes abrupt. Secondly, as the distance between the front toe of the tailings and the catch dam's downstream slope toe is kept constant at 0.8 m (see Figure 3), the distance between the front toe of the tailings and the catch dam's upstream slope toe decreases as the inclination angle α decreases. The tailings flow thus meets the catch dam and covers the upstream slope earlier with a smaller inclination angle α. The catch dam can play its role of resistance earlier, largely reducing the maximum value of velocity. From this point of view, it seems that the catch dam should be constructed as close as possible to the tailings dam. However, when the catch dam with an upstream slope of a very small inclination angle is very close to the tailings pond, a greater quantity of tailings slurry may flow out over the catch dam. Another catch dam along the downstream of the catch dam may be necessary. These results indicate that an optimization process is necessary by considering the position, height, and upstream slope inclination angle of the catch dam, as well as the volume of tailings slurry to be confined.

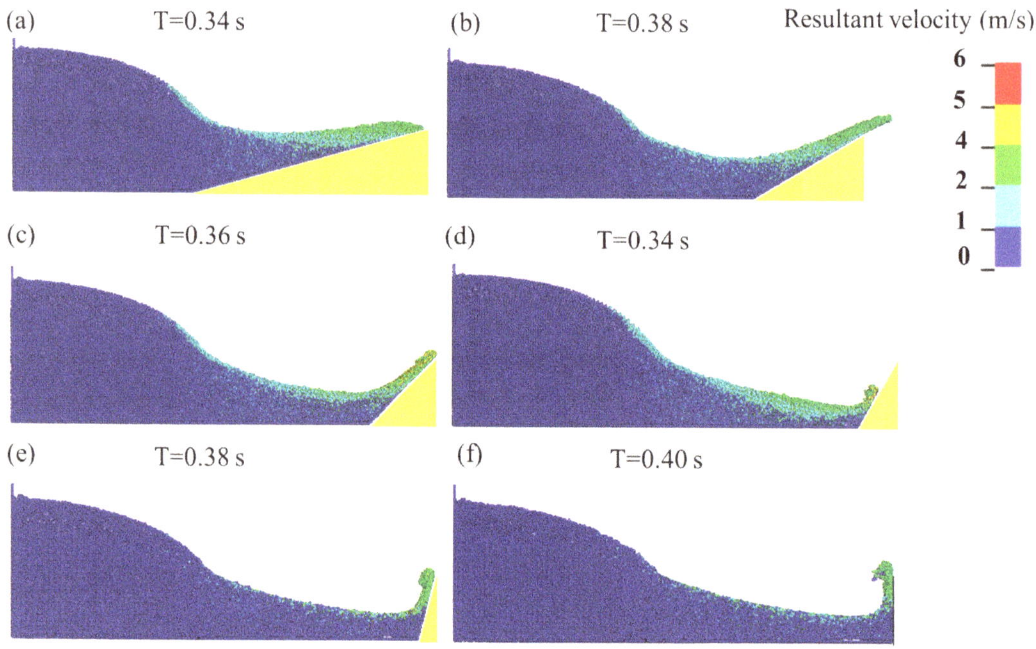

Figure 6. Flow states of the tailings particles when the catch dam's upstream slope starts to be more or less fully covered by tailings slurry, with the inclination angles α varying from 15° to 90°: (**a**) 15°; (**b**) 30°; (**c**) 45°; (**d**) 60°; (**e**) 75°; (**f**) 90°.

Figure 7a shows the impacting forces as a function of time for different values of α, while Figure 7b presents the peak and trough values of the impacting forces as a function of the upstream slope inclination angle α of the catch dam. One sees that the trends of impacting force evolution for different inclination angle values are quite similar (Figure 7a). For a given inclination angle α, the impacting force increases to a peak value rapidly after the tailings flow reaches the catch dam. After then, the impacting force decreases to a minimum value. After a short stage of fluctuations accompanied with slight increases, the impacting force dramatically decreases to a trough value (almost −400 N for the case of $\alpha = 90°$) and sharply increases to almost the same level before falling. After then, another wave of fluctuations can be observed, but the impacting force generally decreases toward zero. It is interesting to note that these results have a certain similarity with those obtained by the numerical modeling performed by Calvetti et al. [95] on the interaction between the flow of dry granular material and rigid barriers. In the latter case, the impacting force also increases with time until a peak value. After then, the impacting force decreases almost monotonously with time, showing neither a second peak, nor a negative value. This last behavior is a particularity belonging to the flow of tailings slurry associated with the sloshing and possible suction.

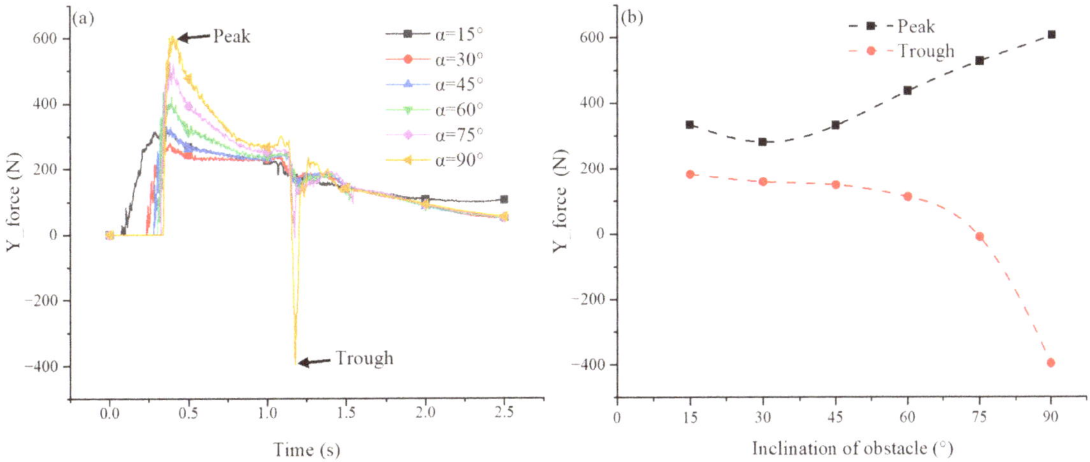

Figure 7. The Y force exerted on the obstacle along the Y direction: (**a**) time-history of Y force under different inclination angles of the obstacle; (**b**) peak and trough values of Y force as a function of upstream slope inclination angle of catch dam.

By comparing the flow state of tailings and the impacting force evolution, one notes that the impacting forces reach their maximum peak value when the catch dam's upstream slope is almost fully covered by the tailings flow, while the minimum trough value corresponds to the moment of backflow, which tends to pull the upstream slope by suction toward the tailings pond. These results are important for the stability analyses of catch dams. On the one hand, the impacting force should be kept as small as possible in order to avoid pushing the catch dam away, but on the other hand, negative values should be avoided because the upstream slope of the catch dam can be disintegrated by the pulling forces if the catch dam is built with an uncemented or lowly cemented, cohesionless material. The difference between the peak and trough values should also be minimized in order to prevent a cycle loading on the catch dam's upstream slope. From Figure 7b, one sees that the smallest peak value of the impacting force and the smallest difference between the peak and trough values of the impacting force are obtained when the inclination angle α is 30°. The largest peak value of the impacting force and the largest difference are observed when the inclination angle α is 90°. These results tend to indicate that the optimum inclination angle α is 30°, while the construction of the catch dam with a vertical upstream slope represents the highest risk and should be avoided in terms of stability for the catch dam.

Figure 8a shows the evolution of kinetic energy for different upstream slope inclination angles α. When the catch dam is absent (i.e., $\alpha = 0°$), the kinetic energy of the tailings flow increases continuously and reaches a peak value of 256.5 J at $T = 0.58$ s, followed by a gradual decrease due to the frictional resistances along the side walls and floor. These results are very important because they indicate that a catch dam should be built either very close to or very far from the tailings pond. The worst design is to construct the catch dam at a position where the kinetic energy reaches the maximum value and a very strong catch dam is necessary.

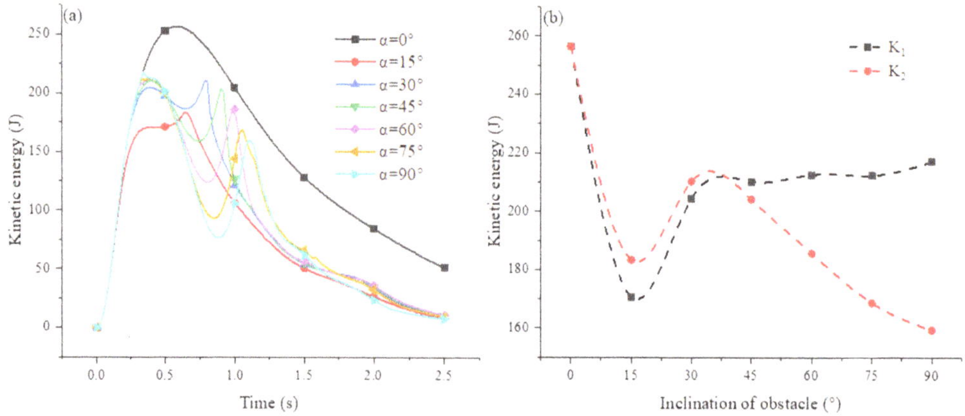

Figure 8. Kinetic energy of tailings flow under different inclinations of obstacle. (**a**) Time-history of kinetic energy; (**b**) first and second peak of kinetic energy.

From Figure 8a, one sees that two peak values in the kinetic energy can generally be observed when a catch dam is constructed at the downstream of the tailings pond. They can be designated as K_1 for the first one and K_2 for the second one. The first peak value K_1 for the inclination angle $\alpha = 15°$ is smaller and arrives earlier than the other cases, probably due to the shorter distance between the tailings downstream slope toe and the catch dam's upstream slope toe compared to the other cases. When the inclination angle α is larger than 45°, the magnitude and arrival time of the first peak value K_1 become insensitive to further variation in the inclination angle α. However, the second peak value K_2 arrives later and becomes smaller as the inclination angle α increases from 45° to 90°.

Figure 8b shows the variation of the two peak values (K_1 and K_2) as a function of the inclination angle α. When $\alpha = 15°$, both K_1 and K_2 are small with a relatively large difference between them. At $\alpha = 90°$, the value of K_1 becomes the largest while the value of K_2 becomes the smallest (except in the case without a catch dam). This case should be avoided in terms of the stability of the catch dam; this is consistent with the previous analysis of the impacting force. At an inclination angle of $\alpha = 37.5°$, the magnitudes of K_1 and K_2 become almost the same even though their values are relatively high. This result is consistent again with the previous analysis of impacting forces as a function of the inclination angle α. It seems that the optimal inclination angle α, in terms of the stability of the catch dam, ranges from 30° to 37.5°. It is interesting to note that waste rocks typically have a repose angle of 37° [96]. This study tends to indicate that waste rocks can be the ideal building materials to construct catch dams. However, detailed analyses should be made in the design of catch dams because waste rocks and other cohesionless materials cannot submit any negative (pulling) forces.

Figure 9 shows the state of the tailings flow at the arrival of the second peak value K_2. One sees that the front of the tailings flow is closer to the catch dam with a higher value of the inclination angle α. Disastrous consequences can be expected if any infrastructures, lives, and equipment are inside the range of the front line. From this figure, one also sees that the delay of the arrival for K_2 increases as the catch dam's upstream slope becomes steeper. This is further illustrated in Figure 10, which shows the variation of the arrival time for the second peak value K_2 as a function of the inclination angle α. When a tailings dam fails, the most important thing is an early warning to the people located along the downstream of the tailings pond. From this perspective, a larger α seems to be better to intercept the tailings flow and allows the people downstream to have more time for evacuation. However, a steep upstream slope is not the favorite choice for the stability of catch dams. These results show again the complexity in designing a catch dam. Optimization must be done by considering the several influencing factors and possible consequences.

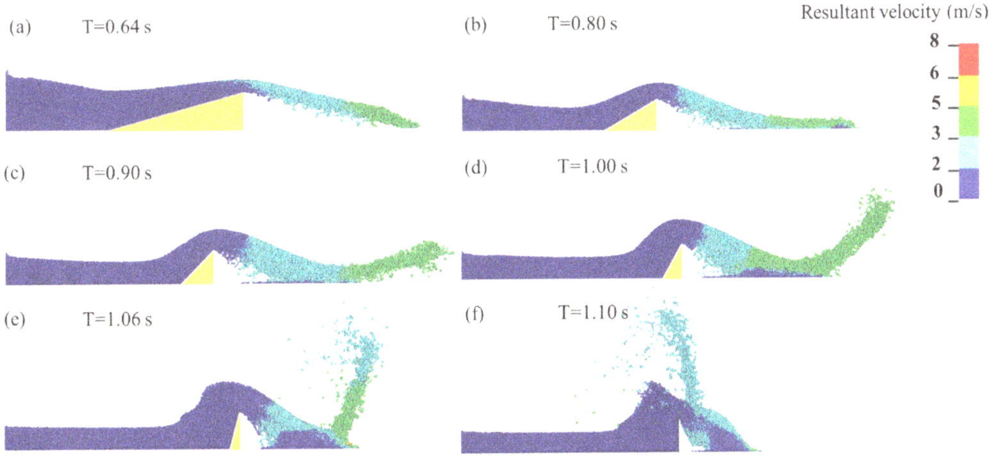

Figure 9. Tailings flow forms corresponding to the moment of K_2 appearing: (**a**) 15°; (**b**) 30°; (**c**) 45°; (**d**) 60°; (**e**) 75°; (**f**) 90°.

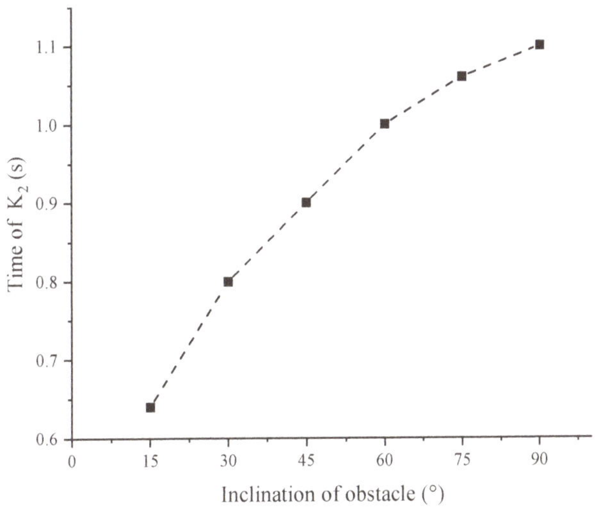

Figure 10. Arrival time of K_2 as a function of slope inclination angle α.

4. Discussion

In practice, a breach of the tailings dam is usually accompanied with a release of a large quantity of tailings slurry, which transforms into a mud flood and produces extremely destructive consequences on the downstream personnel, infrastructures, and environment. To prevent such undesirable consequences, catch dams can be constructed along the downstream of tailings ponds. The catch dams need to be properly designed. This requires a good understanding on the rheological behavior of the tailings slurry and on the mechanical responses of the catch dam upon the flooding of tailings slurry. To this end, a series of numerical simulations were performed with LS-DYNA. The numerical results provide some interesting insight on the interaction behavior between the tailings flow and catch dams. The tailings flow can generally be characterized into four stages. Ideal distances between the tailings pond and catch dam should exist. Optimal upstream slope angles of the catch dam have been identified. These numerical results are thus very useful and helpful for the design and construction of catch dams. However, one has to keep in mind that these numerical results were obtained based on some specific case. The study involves several limitations and should be considered as preliminary. Much more work is necessary. For instance, the numerical model used in this study is very small due to the limitation of available computing resources. More numerical simulations with large-dimension models are necessary in the future to analyze the rheological behavior of tailings slurry and the mechanical behavior of catch dams on a large scale.

In this study, the numerical simulations were realized by considering a tailings slurry having a solids content by mass of 70%, a density of 1826.6 kg/m^3, and a dynamic viscosity of 0.43 Pa·s [92]. It is well-known that the viscosity and flow characteristics of a tailings slurry depend largely on the solids content, as shown by Chen et al. [65]. More work is necessary to analyze the influence of the solids content on the flow of the tailings slurry and the impact forces on the downstream catch dams.

In practice, the quantity of overflow of the tailings slurry through the catch dam and the maximum distance of the tailings flow, as well as the height of the tailing slurry flood along the two side walls can be of great interest. These are closely related to the height and volume of the tailings pond, the distance between the tailings pond and catch dam, the geometry of the catch dam, and even the downstream topography (side wall inclinations, floor inclination). More work is necessary to better understand the effects of these influencing factors on the tailings flow's characteristics and impact on catch dams in the future.

Another limitation is associated with the simulations of a single catch dam. In practice, the construction of one catch dam may be insufficient to stop the tailings flood, especially when a tailings flood has already started and one has to construct several catch dams somewhere along the downstream far from the tailings ponds [66]. More work is necessary to consider several catch dams positioned at different distances.

Finally, one has to mention that the numerical models have been partly validated against slump tests. Field or laboratory experimental work with tailings ponds and catch dams is necessary to further validate or calibrate the numerical models.

5. Conclusions

In this work, the flow characteristics of a tailings slurry and its impact force on a catch dam built downstream near the tailings pond after the breach of the tailings dam were analyzed through numerical modeling with a SPH-FEM code, named LS-DYNA. The applicability of the numerical code has been shown through a numerical reproduction of a tailings slurry slump test reported in the literature. Despite the several limitations of the numerical models, several conclusions can be drawn from the numerical modeling performed in this study.

The flow characteristics of the tailings slurry can generally be divided into four stages:

- Stage 1, called "initiation", tailings particles start to move and accelerate.
- Stage 2, called the "peak stage", the kinetic energy and impacting force first exhibit a decrease and then an increase until second peak values.
- Stage 3, called the "ebb", the kinetic energy decreases and the crest of the tailings flood retreats back toward the tailings pond.
- Stage 4, called "sloshing", the crest of the tailings slurry can first move back toward the tailings pond, and then forward toward the catch dam.

The first analysis shows that the catch dam should be constructed as close as possible to the tailings pond to reduce the maximum velocity of the tailings flow. However, when the catch dam with an upstream slope of a very small inclination angle is too close to the tailings pond, a greater quantity of tailings slurry may flow out over the catch dam. The catch dam needs to be higher or the construction of a secondary catch dam is necessary. Further analysis indicates that a catch dam should be built either very close to or very far from the tailings pond. The worst design is to construct the catch dam at a position where the kinetic energy reaches the maximum value where the construction of a very strong catch dam is necessary.

The numerical results show that the impacting force can increase and decrease with the fluctuations and back-and-forth motion of the tailings slurry flow. A positive and large impacting force may have the effect to push the catch dam away, while a negative force may pull particle material away on the upstream slope of the catch dam. A cycle loading on the catch dam's upstream slope is possible due to the difference between the peak and trough values. From this point of view, the ideal inclination angle α is between 30° and 37.5°, while the construction of a catch dam with a vertical upstream slope should be avoided. Waste rocks can be an ideal building material to construct catch dams. However, detailed analyses should be made because waste rocks are cohesionless and cannot submit to any negative (pulling) forces.

While the previous analyses focusing on the stability of catch dams indicate that the abrupt upstream slope of catch dams should be avoided, the arrival time of peak kinetic energy values tends to indicate that a larger value of α seems to be better to intercept the tailings flow and allow the downstream people to have more time for evacuation. Optimization must be done by considering the stability of the catch dam and evacuation time.

Author Contributions: S.Z.: conceptualization, numerical modeling, analysis, literature, and writing and editing of the original draft. L.L.: project administration, methodology, supervision, and editing of the original draft. All authors have read and agreed to the published version of the manuscript.

Funding: This work was financially supported by the China Scholarship Council (202006370203), the Natural Sciences and Engineering Research Council of Canada (RGPIN-2018-06902, ALLRP-566888-21), and industrial partners of the Research Institute on Mines and the Environment (RIME UQAT-Polytechnique; http://rime-irme.ca/). The authors are grateful for their support.

Institutional Review Board Statement: Not applicable.

Informed Consent Statement: Not applicable.

Data Availability Statement: Data are contained within the article.

Acknowledgments: The authors gratefully acknowledge the financial support from the NSERC and the industrial partners of the Research Institute on Mining and Environment (RIME). The first author expresses her gratitude for the financial supports from the Central South University (CSU) and the China Scholarship Council (CSC).

Conflicts of Interest: The authors declare no conflict of interest.

Appendix A. Applicability of LS-DYNA to Simulate the Rheological Behavior of Tailings Slurry

The slump test is widely used to assess the workability or fluidity of slurry or concrete [97,98]. Here, the slump test results of Gao [92] were used to test the applicability of the numerical code LS-DYNA in simulating the rheological behavior of tailings slurry. The slump tests were conducted by following ASTM C-143 with an Abraham cone 300 mm high, 200 mm in the base diameter, and 100 mm in the top diameter. The tailings slurry has a solids content by mass of 70%, a density of 1826.6 kg/m^3, and a dynamic viscosity of 0.43 Pa·s.

Figure A1 shows the numerical model to simulate the slump test of Gao [92]. The tailings slurry is treated as an incompressible fluid flow and represented by the material model MAT_NULL and the equation of state *EOS_MURNAGHAN in LS-DYNA [81]. On Figure 3, one sees a meshless infinite rigid wall at the bottom to represent the ground surface to prevent vertical movement of the fluid passing through it. The friction coefficient between the tailings slurry and the rigid wall is automatically set to be 0.466 [99]. As the numerical model of SPH involves several model controlling parameters [100,101], sensitivity analyses of these model controlling parameters are necessary to ensure stable numerical results. In this study, the sensitivity analyses of model controlling parameters led to a particle spacing of 4 mm (85,980 particles), a smoothing length factor of 1.20, a time step scaling factor of 0.45, and an approximation theory of 15, respectively.

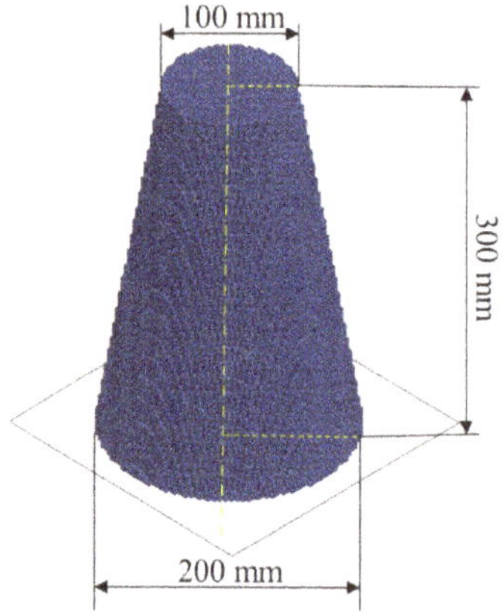

Figure A1. The numerical model of the slump test.

Figure A2 shows the variations of the shape and slump height of the tailings slurry with time once the Abraham cone was removed by a lift, obtained by the numerical modeling. As the tailings slurry has a good fluidity, the main slump occurs very quickly within 0.42 s and reaches 273 mm. After then, a further slump becomes extremely slow, with a final slump height of 276 mm. This value is quite close to the test results with a slump height of 286 mm (with a relative error of 3.50%).

Figure A3 shows a comparison between the final shapes of laboratory test and numerical modeling. Both are nearly circular. The base diameter of the slumped slurry can be estimated to be around 350 mm in the photo, whereas the numerical modeling shows a base diameter of 355 mm. The numerical model of LS-DYNA has been proved to be able to simulate the rheological behavior of tailings slurry.

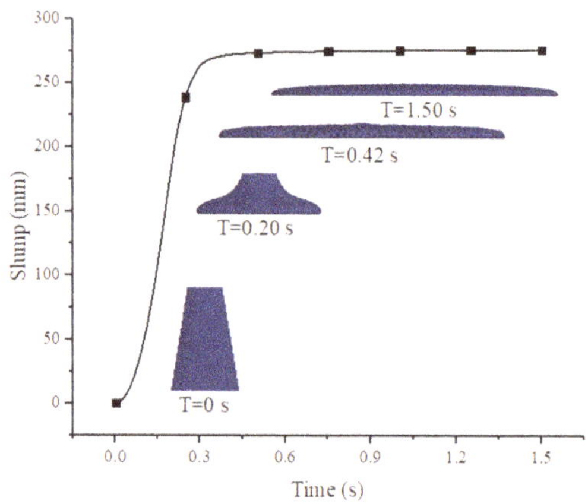

Figure A2. The numerical time-history of the slump.

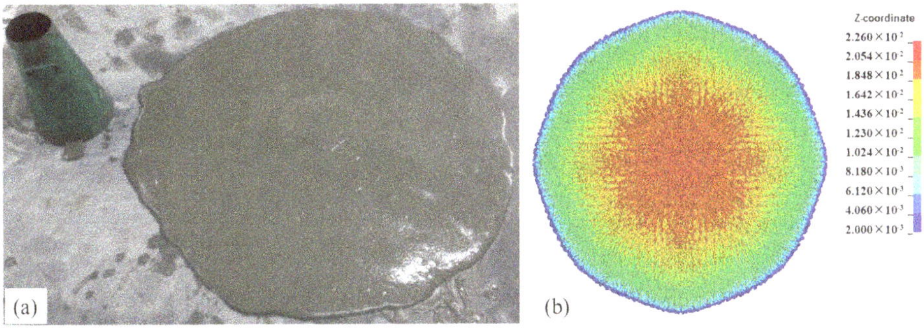

Figure A3. Final shape of the tailings slurry after the slump tests: (**a**) a photo of the laboratory test; (**b**) numerical result (Z-coordinate corresponds to the thickness of tailings slurry in meters).

Appendix B. Sensitivity Analyses of SPH Simulations

In the numerical modeling with SPH, the numerical results can be affected by several controlling model parameters. Particle approximation theory was set as FORM = 15 in this study, according to the suggestion for fluid simulation in LS-DYNA user's manual. Sensitivity analyses of other controlling parameters are necessary to obtain their optimal values and ensure stable numerical results. Figure A4 shows the sensitivity analyses of time step size, the number of particles, and the smoothing length. One sees that the optimal number of SPH particles is 343,000, while the optimal time step scale factor is 0.50. In case of the possible numerical instability caused by a strong splash of the tailings flow, as in the case of declination with 90 degrees, a value of 0.45 was taken for the time step scale factor. Regarding the smoothing length scale factor, the default value of 1.2 was used because the numerical results shown in the figure seemed to not be very sensitive to the variation of this value.

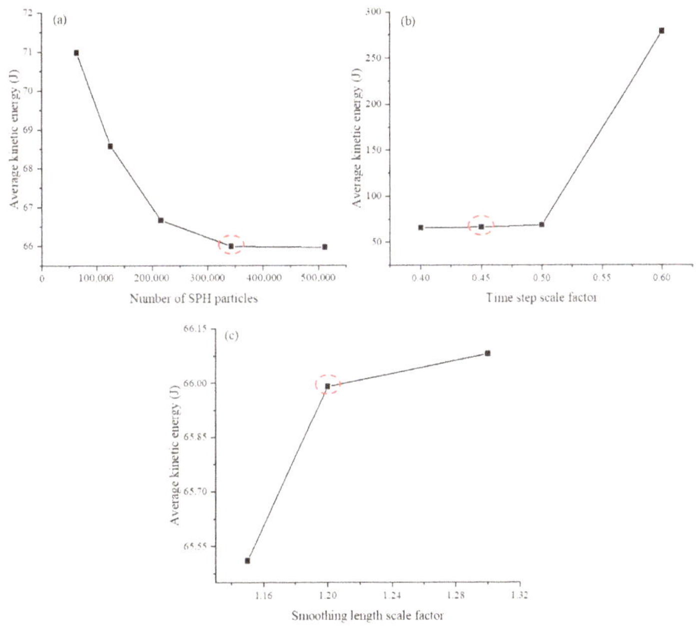

Figure A4. Convergence of average kinetic energy of tailings flow under different numerical parameters. (**a**) Mesh independence; (**b**) time step scale factor; (**c**) smoothing length scale factor.

References

1. Jones, H.; Boger, D.V. Sustainability and waste management in the resource industries. *Ind. Eng. Chem. Res.* **2012**, *51*, 10057–10065. [CrossRef]
2. Hasanpour, A.; Istrati, D.; Buckle, I. Coupled SPH–FEM Modeling of Tsunami-Borne Large Debris Flow and Impact on Coastal Structures. *J. Mar. Sci. Eng.* **2021**, *9*, 1068. [CrossRef]
3. Aubertin, M.; Li, L.; Arnoldi, S.; Belem, T.; Bussière, B.; Benzaazoua, M.; Simon, R. Interaction between backfill and rock mass in narrow stopes. *Soil Rock Am.* **2003**, *2*, 1157–1164.
4. Li, L.; Aubertin, M.; Simon, R.; Bussière, B.; Belem, T. Modeling arching effects in narrow backfilled stopes with FLAC. *FLAC Numer. Modeling Geomech.* **2003**, 211–219.
5. Li, L.; Aubertin, M.; Belem, T. Formulation of a three dimensional analytical solution to evaluate stresses in backfilled vertical narrow openings. *Can. Geotech. J.* **2005**, *42*, 1705–1717. [CrossRef]
6. Li, L.; Aubertin, M. A modified solution to assess the required strength of exposed backfill in mine stopes. *Can. Geotech. J.* **2012**, *49*, 994–1002. [CrossRef]
7. Li, L.; Aubertin, M. An improved method to assess the required strength of cemented backfill in underground stopes with an open face. *Int. J. Min. Sci. Technol.* **2014**, *24*, 549–558. [CrossRef]
8. Li, L. Analytical solution for determining the required strength of a side-exposed mine backfill containing a plug. *Can. Geotech. J.* **2014**, *51*, 508–519. [CrossRef]
9. Li, L. Generalized solution for mining backfill design. *Int. J. Geomech.* **2014**, *14*, 04014006. [CrossRef]
10. Liu, G.; Li, L.; Yang, X.; Guo, L. Stability analyses of vertically exposed cemented backfill: A revisit to Mitchell's physical model tests. *Int. J. Min. Sci. Technol.* **2016**, *26*, 1135–1144. [CrossRef]
11. Liu, G.; Li, L.; Yang, X.; Guo, L. Required strength estimation of a cemented backfill with the front wall exposed and back wall pressured. *Int. J. Min. Miner. Eng.* **2018**, *9*, 1–20. [CrossRef]
12. Yang, P.; Li, L.; Aubertin, M. A new solution to assess the required strength of mine backfill with a vertical exposure. *Int. J. Geomech.* **2017**, *17*, 04017084. [CrossRef]
13. Li, X.; Zhou, S.; Zhou, Y.; Min, C.; Cao, Z.; Du, J.; Luo, L.; Shi, Y. Durability evaluation of phosphogypsum-based cemented backfill through drying-wetting cycles. *Minerals* **2019**, *9*, 321. [CrossRef]
14. Pagé, P.; Li, L.; Yang, P.; Simon, R. Numerical investigation of the stability of a base-exposed sill mat made of cemented backfill. *Int. J. Rock Mech. Min. Sci.* **2019**, *114*, 195–207. [CrossRef]
15. Zhou, S.; Li, X.; Zhou, Y.; Min, C.; Shi, Y. Effect of phosphorus on the properties of phosphogypsum-based cemented backfill. *J. Hazard. Mater.* **2020**, *399*, 122993. [CrossRef]
16. Keita, A.M.T.; Jahanbakhshzadeh, A.; Li, L. Numerical analysis of the stability of arched sill mats made of cemented backfill. *Int. J. Rock Mech. Min. Sci.* **2021**, *140*, 104667. [CrossRef]
17. Wang, R.; Zeng, F.; Li, L. Stability analyses of side-exposed backfill considering mine depth and extraction of adjacent stope. *Int. J. Rock Mech. Min. Sci.* **2021**, *142*, 104735. [CrossRef]
18. Wang, R.; Li, L. Time-dependent stability analyses of side-exposed backfill considering creep of surrounding rock mass. *Rock Mech. Rock Eng.* **2022**, 1–25. [CrossRef]
19. Tardif-Drolet, M.; Li, L.; Pabst, T.; Zagury, G.J.; Mermillod-Blondin, R.; Genty, T. Revue de la réglementation sur la valorisation des résidus miniers hors site au Québec. *Environ. Rev.* **2020**, *28*, 32–44. [CrossRef]
20. Vick, S.G. *Planning, Design, and Analysis of Tailings Dams*; BiTech Publishers Ltd.: Vancouver, BC, Canada, 1990.
21. Fourie, A. Preventing catastrophic failures and mitigating environmental impacts of tailings storage facilities. *Procedia Earth Planet. Sci.* **2009**, *1*, 1067–1071. [CrossRef]
22. Villavicencio, G.; Espinace, R.; Palma, J.; Fourie, A.; Valenzuela, P. Failures of sand tailings dams in a highly seismic country. *Can. Geotech. J.* **2014**, *51*, 449–464. [CrossRef]
23. Stefaniak, K.; Wróżyńska, M. On possibilities of using global monitoring in effective prevention of tailings storage facilities failures. *Environ. Sci. Pollut. Res.* **2018**, *25*, 5280–5297. [CrossRef] [PubMed]
24. Edraki, M.; Baumgartl, T.; Manlapig, E.; Bradshaw, D.; Franks, D.M.; Moran, C.J. Designing mine tailings for better environmental, social and economic outcomes: A review of alternative approaches. *J. Clean. Prod.* **2014**, *84*, 411–420. [CrossRef]
25. Adiansyah, J.S.; Rosano, M.; Vink, S.; Keir, G. A framework for a sustainable approach to mine tailings management: Disposal strategies. *J. Clean. Prod.* **2015**, *108*, 1050–1062. [CrossRef]
26. Bussière, B. Colloquium 2004: Hydrogeotechnical properties of hard rock tailings from metal mines and emerging geoenvironmental disposal approaches. *Can. Geotech. J.* **2007**, *44*, 1019–1052. [CrossRef]
27. Lyu, Z.; Chai, J.; Xu, Z.; Qin, Y.; Cao, J. A comprehensive review on reasons for tailings dam failures based on case history. *Adv. Civ. Eng.* **2019**, *2019*, 4159306. [CrossRef]
28. WISE. Chronology of Major Tailings Dam Failures. WISE Uranium Project. Available online: https://www.wise-uranium.org/mdaf.html (accessed on 16 June 2019).
29. Azam, S.; Li, Q. Tailings Dam Failures: A Review of the Last One Hundred Years. *Geotech. News* **2010**, *28*, 50–54.
30. Oboni, F.; Oboni, C. Factual and Foreseeable Reliability of Tailings Dams and Nuclear Reactors: A Societal Acceptability Perspective. *Tailings Mine Waste* **2013**, 6–9.

31. Fell, R.; MacGregor, P.; Stapledon, D.; Bell, G.; Foster, M. *Geotechnical Engineering of Dams*, 2nd ed.; CRC Press: Boca Raton, FL, USA, 2014.
32. Kossoff, D.; Dubbin, W.E.; Alfredsson, M.; Edwards, S.J.; Macklin, M.G.; Hudson-Edwards, K.A. Mine tailings dams: Characteristics, failure, environmental impacts, and remediation. *Appl. Geochem.* **2014**, *51*, 229–245. [CrossRef]
33. Chovan, K.M.; Julien, M.R.; Ingabire, É.-P.; Masengo, E.; Lépine, T.; James, M.; Lavoie, P. Risk assessment for tailings management. *CIM J.* **2021**, *12*, 9–24. [CrossRef]
34. Van Niekerk, H.J.; Viljoen, M.J. Causes and consequences of the Merriespruit and other tailings-dam failures. *Land Degrad. Dev.* **2005**, *16*, 201–212. [CrossRef]
35. PBC. Independent Expert Engineering Investigation and Review Panel: Report on Mount Polley Tailings Storage Facility Breach. 2015. Available online: https://www.mountpolleyreviewpanel.ca/final-report (accessed on 30 January 2015).
36. Berchtold, A.E.; Price, M.H.H. *Responsible Mining in British Colombia: Guidelines towards Best Practice*; SkeenaWild Conservation Trust: Terrace, BC, Canada, 2018.
37. Eisenhammer, S.; Nogueira, M. A view of the Samarco mine, owned by Vale SA and BHP Billiton Ltd., in Mariana, Brazil. REUTERS/Washington Alves. 12 April 2016. Available online: https://www.reuters.com/world/americas/socio-environmental-damage-brazils-samarco-dam-disaster-high-67-bln-study-says-2021-11-03/ (accessed on 4 November 2021).
38. Aires, U.R.V.; Santos, B.S.M.; Coelho, C.D.; da Silva, D.D.; Calijuri, M.L. Changes in land use and land cover as a result of the failure of a mining tailings dam in Mariana, MG, Brazil. *Land Use Policy* **2018**, *70*, 63–70. [CrossRef]
39. Rotta, L.H.S.; Alcântara, E.; Park, E.; Negri, R.G.; Lin, Y.N.; Bernardo, N.; Mendes, T.S.G.; Filho, C.R.S. The 2019 Brumadinho tailings dam collapse: Possible cause and impacts of the worst human and environmental disaster in Brazil. *Int. J. Appl. Earth Obs. Geoinf.* **2020**, *90*, 102119.
40. Blight, G.E. Destructive mudflows as a consequence of tailings dyke failures. *Proc. Inst. Civ. Eng. Geotech. Eng.* **1997**, *125*, 9–18. [CrossRef]
41. Glotov, V.E.; Chlachula, J.; Glotova, L.P.; Little, E. Causes and environmental impact of the gold-tailings dam failure at Karamken, the Russian Far East. *Eng. Geol.* **2018**, *245*, 236–247. [CrossRef]
42. ACB. *Dam Safety*; Canadian Dam Association/Association Canadienne des Barrages: Markham, ON, Canada, 2007.
43. Jing, X.; Chen, Y.; Williams, D.J.; Serna, M.L.; Zheng, H. Overtopping failure of a reinforced tailings dam: Laboratory investigation and forecasting model of dam failure. *Water* **2019**, *11*, 315. [CrossRef]
44. Li, S.; Yuan, L.; Yang, H.; An, H.; Wang, G. Tailings dam safety monitoring and early warning based on spatial evolution process of mud-sand flow. *Saf. Sci.* **2020**, *124*, 104579. [CrossRef]
45. Blight, G.E. *Geotechnical Engineering for Mine Waste Storage Facilities*; CRC Press: Boca Raton, FL, USA, 2010.
46. Aubertin, M.; Bussière, B.; Bernier, L. *Environnement et Gestion des Rejets Miniers*; Presses Internationales Polytechnique: Montréal, QC, Canada, 2002.
47. Benzaazoua, M.; Bussière, B.; Demers, I.; Aubertin, M.; Fried, É.; Blier, A. Integrated mine tailings management by combining environmental desulphurization and cemented paste backfill: Application to mine Doyon, Quebec, Canada. *Min. Eng.* **2008**, *21*, 330–340. [CrossRef]
48. Bussière, B.; Guittonny, M. (Eds.) *Hard Rock Mine Reclamation From Prediction to Management of Acid Mine Drainage*; CRC Press: Boca Raton, FL, USA, 2021.
49. Zhang, B.; Muraleetharan, K.K.; Liu, C.Y. Liquefaction of unsaturated sands. *Int. J. Geomech.* **2016**, *16*, D4015002. [CrossRef]
50. Yuan, B.; Wang, F.Y.; Jin, Y.J. Study on the model for tailing dam breaking and its application. *China Safety Sci. J.* **2008**, *18*, 169–172. (In Chinese)
51. Chen, D.Q.; He, F.; Wang, L.G. Numerical calculation of the railings dam failure in Fengcheng City. *Met. Mine* **2009**, *10*, 74–80. (In Chinese)
52. Guo, C.Y.; Tang, Z.Y. Study on the tailings dam break model. *J. Saf. Sci. Technol.* **2010**, *6*, 63–67. (In Chinese)
53. Li, H.G.; Wang, H.G.; Liu, H. Discussions on maximum impact distance of noncoal mine tailings dam failure. *Nonferrous Met. Eng. Res.* **2011**, *32*, 12–15. (In Chinese)
54. Zhuo, Q.F.; Yuan, W.J.; Lin, J. Analysis for the tailings dam break in Fujian Province. *Min. Metall. Eng.* **2011**, *31*, 16–19. (In Chinese)
55. Wang, L.L.; Jiang, G.S.; Lu, X.J. Application of dam failure model to the risk evaluation of tailing dams. *Saf. Environ. Eng.* **2012**, *19*, 33–36. (In Chinese) [CrossRef]
56. Yuan, Z.Y. Numerical calculation of tailings dam-break and its influence on downstream areas. *Met. Mine* **2012**, *8*, 156–159. (In Chinese)
57. Xu, K.; Zhao, Y.S.; Zhang, Q. Numerical analysis for the tailings dam break. *Ind. Saf. Environ. Prot.* **2012**, *38*, 28–43. (In Chinese)
58. Jin, J.X.; Liang, L.; Wu, F.Y. Dam break simulation of tailings dam and forecast of impact range. *Met. Mine* **2013**, *3*, 141–144. (In Chinese)
59. Tang, Y.L.; Cao, X.Y.; Yin, T.T. On the dam breaking emergency saving measures of a mining-tailing stockpile at the upstream Dahuofang Reservoir. *J. Saf. Environ.* **2013**, *13*, 180–185. (In Chinese)
60. Nie, Z.D.; Liu, J.; Chen, P.; Guo, Y.; Liu, W.; Xiong, J.L.; Liu, Y. A case study on environmental risk assessment method of mine tailings collapse. *Environ. Sci. Technol.* **2014**, *37*, 550–554. (In Chinese)

61. Liu, H. An Experimental and Numerical Study of Runout from a Tailings Dam Failure. Ph.D. Thesis, The University of Western Australia, Crawley, Australia, 2018.
62. Souza, T.F.; Teixeira, S.H.C. Simulation of tailings release in dam break scenarios using physical models. *REM-Int. Eng. J.* **2019**, *72*, 385–393. [CrossRef]
63. Mahdi, A.; Shakibaeinia, A.; Dibike, Y.B. Numerical modelling of oil-sands tailings dam breach runout and overland flow. *Sci. Total Environ.* **2020**, *703*, 134568. [CrossRef] [PubMed]
64. Yu, D.; Tang, L.; Chen, C. Three-dimensional numerical simulation of mud flow from a tailing dam failure across complex terrain. *Nat. Hazards Earth Syst. Sci.* **2020**, *20*, 727–741. [CrossRef]
65. Chen, X.; Jing, X.; Chen, Y.; Pan, C.; Wang, W. Tailings dam break: The influence of slurry with different concentrations downstream. *Front. Earth Sci.* **2021**, *9*, 726336. [CrossRef]
66. Zhang, Y. Leakage of tailing sands in Heilongjiang continued: 10 interception dams have been set up to stop the leakage, and there is no safety risk to the tailings. Available online: https://baijiahao.baidu.com/s?id=1662654431312090219&wfr=spider&for=pc. (accessed on 31 March 2020). (In Chinese)
67. Soares-Frazão, S.; Zech, Y. Experimental study of dam-break flow against an isolated obstacle. *J. Hydraul. Res.* **2007**, *45* (Suppl. 1), 27–36. [CrossRef]
68. Soares-Frazão, S.; Zech, Y. Dam-break flow through an idealised city. *J. Hydraul. Res.* **2008**, *46*, 648–658. [CrossRef]
69. Ozmen-Cagatay, H.; Kocaman, S. Dam-break flow in the presence of obstacle: Experiment and CFD simulation. *Eng. Appl. Comput. Fluid Mech.* **2011**, *5*, 541–552. [CrossRef]
70. Xu, X. An improved SPH approach for simulating 3D dam-break flows with breaking waves. *Comput. Methods Appl. Mech. Eng.* **2016**, *311*, 723–742. [CrossRef]
71. Issakhov, A.; Zhandaulet, Y.; Nogaeva, A. Numerical simulation of dam break flow for various forms of the obstacle by VOF method. *Int. J. Multiph. Flow* **2018**, *109*, 191–206. [CrossRef]
72. Saghi, H.; Lakzian, E. Effects of using obstacles on the dam-break flow based on entropy generation analysis. *Eur. Phys. J. Plus* **2019**, *134*, 237. [CrossRef]
73. Chen, H.-X.; Li, J.; Feng, S.-J.; Gao, H.-Y.; Zhang, D.-M. Simulation of interactions between debris flow and check dams on threedimensional terrain. *Eng. Geol.* **2019**, *251*, 48–62. [CrossRef]
74. Jeong, S.; Lee, K. Analysis of the impact force of debris flows on a check dam by using a coupled Eulerian-Lagrangian (CEL) method. *Comput. Geotech.* **2019**, *116*, 103214. [CrossRef]
75. Huang, Y.; Jin, X.; Ji, J. Effects of Barrier Stiffness on Debris Flow Dynamic Impact—I: Laboratory Flume Test. *Water* **2022**, *14*, 177. [CrossRef]
76. Huang, Y.; Jin, X.; Ji, J. Effects of barrier stiffness on debris flow dynamic Impact—II: Numerical simulation. *Water* **2022**, *14*, 182. [CrossRef]
77. Abdelrazek, A.M.; Kimura, I.; Shimizu, Y. Simulation of three-dimensional rapid free-surface granular flow past different types of obstructions using the SPH method. *J. Glaciol.* **2016**, *62*, 335–347. [CrossRef]
78. Wang, G.; Tian, S.; Hu, B.; Xu, Z.; Chen, J.; Kong, X. Evolution pattern of tailings flow from dam failure and the buffering effect of Debris Blocking Dams. *Water* **2019**, *11*, 2388. [CrossRef]
79. Zeng, Q.-Y.; Pan, J.-P.; Sun, H.-Z. SPH simulation of structures impacted by tailings debris flow and its application to the buffering effect analysis of debris checking dams. *Math. Probl. Eng.* **2020**, *2020*, 9304921. [CrossRef]
80. Wang, S.; Mei, G.; Xie, X.; Guo, L. The influence of the instantaneous collapse of tailings pond on downstream facilities. *Adv. Civ. Eng.* **2021**, *2021*, 4253315. [CrossRef]
81. Hallquist, J.O. *LS-DYNA Keyword User's Manual*; Version 970; Livermore Software Technology Corporation: Livermore, CA, USA, 2007.
82. Lucy, L.B. A numerical approach to the testing of the fission hypothesis. *Astron. J.* **1977**, *82*, 1013–1024. [CrossRef]
83. Gingold, R.A.; Monaghan, J.J. Smoothed particle hydrodynamics: Theory and application to non-spherical stars. *Mon. Not. R. Astron. Soc.* **1977**, *181*, 375–389. [CrossRef]
84. Shao, S.; Lo, E.Y.M. Incompressible SPH method for simulating Newtonian and non-Newtonian flows with a free surface. *Adv. Water Resour.* **2003**, *26*, 787–800. [CrossRef]
85. Antoci, C.; Gallati, M.; Sibilla, S. Numerical simulation of fluid–structure interaction by SPH. *Comput. Struct.* **2007**, *85*, 879–890. [CrossRef]
86. Atif, M.M.; Chi, S.-W.; Grossi, E.; Shabana, A.A. Evaluation of breaking wave effects in liquid sloshing problems: ANCF/SPH comparative study. *Nonlinear Dyn.* **2019**, *97*, 45–62. [CrossRef]
87. Liu, G.R.; Liu, M.B. *Smoothed Particle Hydrodynamics: A Meshfree Particle Method*; World Scientific Publishing Co Pte: Singapore, 2003.
88. Yang, Y.; Li, J. SPH-FE-Based Numerical simulation on dynamic characteristics of structure under water waves. *J. Mar. Sci. Eng.* **2020**, *8*, 630. [CrossRef]
89. Jiang, H.; Zhao, H.; Gao, K.; Wang, O.; Wang, Y.; Meng, D. Numerical investigation of hard rock breakage by high-pressure water jet assisted indenter impact using the coupled SPH/FEM method. *Powder Technol.* **2020**, *376*, 176–186. [CrossRef]
90. Koneshwaran, S.; Thambiratnam, D.P.; Gallage, C. Blast response of segmented bored tunnel using coupled SPH–FE method. *Structures* **2015**, *2*, 58–71. [CrossRef]
91. Pashkov, A. Simulation of the forming of large double curvature parts on contact-type shot peening installations. *Russ. Metall.* **2021**, *13*, 1821–1828. [CrossRef]

92. Gao, F. *Study on Affections of Pipeline Diameter to the Transportation Characteristics of Backfilling Slurry of Unclassified Tailings Inner Pipe*; Hebei United University: Tangshan, China, 2015.
93. Dinçer, A.E. Investigation of the sloshing behavior due to seismic excitations considering two-way coupling of the fluid and the structure. *Water* **2019**, *11*, 2664. [CrossRef]
94. Sanapala, V.M.R.; Velusamy, K.; Patnaik, B.S.V. Numerical simulation of parametric liquid sloshing in a horizontally baffled rectangular container. *J. Fluids Struct.* **2018**, *76*, 229–250. [CrossRef]
95. Calvetti, F.; di Prisco, C.G.; Vairaktaris, E. DEM assessment of impact forces of dry granular masses on rigid barriers. *Acta Geotech.* **2017**, *12*, 129–144. [CrossRef]
96. Aubertin, M. *Waste Rock Disposal to Improve the Geotechnical and Geochemical Stability of Piles*; 23rd World Mining Congress and Expo 2013; Canadian Institute of Mining, Metallurgy and Petroleum (CIM): Westmount, QC, Canada, 2013.
97. Pashias, N.; Boger, D.V.; Summers, J.; Glenister, D.J. A fifty cent rheometer for yield stress measurement. *J. Rheol.* **1996**, *40*, 1179–1189. [CrossRef]
98. Zheng, J.; Li, L. Experimental study of the "short-term" pressures of uncemented paste backfill with different solid contents for barricade design. *J. Clean. Prod.* **2020**, *275*, 123068. [CrossRef]
99. Shen, W.; Zhao, T.; Zhao, J.; Dai, F.; Zhou, G.G.D. Quantifying the impact of dry debris flow against a rigid barrier by DEM analyses. *Eng. Geol.* **2018**, *241*, 86–96. [CrossRef]
100. Giannaros, E.; Kotzakolios, A.; Kostopoulos, V.; Campoli, G. Hypervelocity impact response of CFRP laminates using smoothed particle hydrodynamics method: Implementation and validation. *Int. J. Impact Eng.* **2019**, *123*, 56–69. [CrossRef]
101. Lee, S.; Hong, J.W. Parametric studies on smoothed particle hydrodynamic simulations for accurate estimation of open surface flow force. *Int. J. Nav. Archit. Ocean Eng.* **2020**, *12*, 85–101. [CrossRef]

processes

MDPI

Article

Applicability of Constitutive Models to Describing the Compressibility of Mining Backfill: A Comparative Study

Ruofan Wang *, Feitao Zeng and Li Li

Research Institute on Mines and Environment (RIME UQAT-Polytechnique), Department of Civil, Geological and Mining Engineering, École Polytechnique de Montréal, C.P. 6079 Succursale Centre-Ville, Montréal, QC H3C 3A7, Canada; feitao.zeng@polymtl.ca (F.Z.); li.li@polymtl.ca (L.L.)
* Correspondence: ruofan.wang@polymtl.ca

Abstract: The compressibility of mining backfill governs its resistance to the closure of surrounding rock mass, which should be well reflected in numerical modeling. In most numerical simulations of backfill, the Mohr–Coulomb elasto-plastic model is used, but is constantly criticized for its poor representativeness to the mechanical response of geomaterials. Finding an appropriate constitutive model to better represent the compressibility of mining backfill is critical and necessary. In this paper, Mohr–Coulomb elasto-plastic model, double-yield model, and Soft Soil model are briefly recalled. Their applicability to describing the backfill compressibility is then assessed by comparing numerical and experimental results of one-dimensional consolidation and consolidated drained triaxial compression tests made on lowly cemented backfills available in the literature. The comparisons show that the Soft Soil model can be used to properly describe the experimental results while the application of the Mohr–Coulomb model and double-yield model shows poor description on the compressibility of the backfill submitted to large and cycle loading. A further application of the Soft Soil model to the case of a backfilled stope overlying a sill mat shows stress distributions close to those obtained by applying the Mohr–Coulomb model when rock wall closure is absent. After excavating the underlying stope, rock wall closure is generated and exercises compression on the overlying backfill. Compared to the results obtained by applying the Soft Soil model, an application of the Mohr–Coulomb model tends to overestimate the stresses in the backfill when the mine depth is small and underestimate the stresses when the mine depth is large due to the poor description of fill compressibility. The Soft Soil model is recommended to describe the compressibility of uncemented or lightly cemented backfill with small cohesions under external compressions associated with rock wall closure.

Keywords: mining backfill; compressibility; constitutive models; numerical modeling

Citation: Wang, R.; Zeng, F.; Li, L. Applicability of Constitutive Models to Describing the Compressibility of Mining Backfill: A Comparative Study. *Processes* **2021**, *9*, 2139. https://doi.org/10.3390/pr9122139

Academic Editor: Haiping Zhu

Received: 25 October 2021
Accepted: 22 November 2021
Published: 26 November 2021

1. Introduction

Backfill is being considered as an integral part of several underground mining methods. It is used as working platform in overhand cut-and-fill mining method or for creating safer working space in underhand cut-and-fill mining method. Using mine waste as underground mining backfill helps to minimize the surface disposal of mine waste [1–4]. However, the main objective of backfilling the mined-out spaces is to effectively control the rock wall closure and maintain the regional ground stability [5–9]. The compressibility of backfill plays an important role in resisting the closure of surrounding rock walls associated with adjacent extraction or/and creep behavior. Previous studies showed that a backfill of low compressibility can carry significant stresses generated by walls convergence and provide considerable support to surrounding rocks [6,10–13]. In underground mines, especially with overhand cut-and-fill or open stoping methods, however, the commonly used backfill is uncemented or has a low-cement content. The compressibility is large under low compression state and decreases as the compression increases. Understanding and properly describing the compressibility of mining backfill is thus of specific interest for mining industry to evaluate fill performance and stability of underground structures.

Numerical modeling provides an efficient and cost-effective method to study the complex mechanical behavior of backfill. Nonetheless, the reliability and applicability of a numerical model largely depends on the capability of applied constitutive model. There are many constitutive models of geomaterial proposed in literatures with various levels of complexity [14–17]. Until now, the Mohr–Coulomb elasto-plastic model is the most used one to simulate the mining backfill due to the simplicity and clear physical meaning of model parameters [18–25]. The justification of the Mohr–Coulomb model is usually attributed to the good fit with the shear strength of backfill [26–30]. However, it is well-known that the Mohr–Coulomb elasto-plastic model suffers from its linearly elasticity and the neglect of volumetric yielding. In reality, geomaterial can have a nonlinear behavior before yielding while a mining backfill can become denser upon a large compression generated by wall closure. Given the restrictions of the Mohr–Coulomb elasto-plastic model, it remains unknown which constitutive model should be applied to better represent the compressibility of mining backfill.

There are few research studies devoted toward identifying a constitutive model applicable to describing the compressibility of mining backfill. Among these studies, Oliver and Landriault [31] investigated the convergence resistance of backfill by simulating an oedometer test of dense sand with the Mohr–Coulomb model and the strain-softening model. Different values of Young's modulus and Poisson's ratio were applied. Results show that numerical model remains elastic over the full strain range except when a very small Poisson's ratio is applied. The predicted compressive stresses obtained with the two constitutive models are almost identical and significantly smaller than the experimental results. Clark [32] reproduced the non-linear stress–strain response of the dewatered tailings backfill in uniaxial compression tests with a cap model in FLAC. The input parameters for the cap model were obtained by applying the curve fitting on all the experimental results. The good agreements between the numerical model and experimental results do not mean that the calibrated numerical model can be used to correctly predict the mechanical behavior of the backfill under an untested stress condition. Fourie et al. [33] performed a finite element analysis by making use of the linearly elastic, Mohr–Coulomb, Drucker-Prager, and modified Cam-Clay models to simulate a backfilled stope at depth of 2000 m below the ground surface. The hanging wall convergence for the modified Cam-Clay model was found to be 11% larger than the results obtained with the other three models due to the plastic volumetric strain of backfill. The numerical results were not further compared with physical test data. Lagger et al. [34] used the double-yield model in FLAC^{3D} to simulate the oedometer test of pea gravel as a filling material. The cap pressure of the double-yield model was calibrated based on the experimental results in a stress range of 0 to 6 MPa. Within this range, numerical results reasonably correlate with the test data, but the comparisons for the higher load stage was not shown. Therefore, more works are necessary to identify a suitable constitutive model to describe the compressibility of mining backfill, particularly by analyzing its predictive capability and comparing with physical results.

In this study, the Mohr–Coulomb elasto-plastic, double-yield, and Soft Soil models were recalled. Their abilities to describe the compressibility of backfill were assessed by comparing the numerical results obtained using FLAC^{3D} with the experimental results of one-dimensional consolidation and consolidated drained triaxial compression tests of lowly cemented backfill available in the literature. Some unknown model parameters were calibrated based on part of the experimental results, and the calibrated models were applied to predict the other part of the experimental results. The identified model was then benchmarked with the Mohr–Coulomb model in simulating a backfilled stope, overlying a sill mat at different mine depths. The applicability of the identified model and the significance of modeling fill compressibility will be shown and discussed. In addition, validations of FLAC^{3D} against analytical solutions of a cylinder hole in the linearly elastic and Mohr–Coulomb material and the sensitivity analyses of the numerical models are provided in the Appendices A–C.

2. Commonly Used Constitutive Models in Geotechnical Engineering

For the sake of completeness, a few constitutive models commonly used in geotechnical engineering, including the Mohr–Coulomb elasto-plastic, double-yield, and Soft Soil models, are briefly recalled. Compression stresses are positive and tension is negative. All strength parameters are in terms of effective stresses.

2.1. Mohr–Coulomb Elasto-Plastic Model

The Mohr–Coulomb elasto-plastic model considers a material as linearly elastic and perfectly plastic once the stress state reaches a state of yield [35]. It is the most commonly used constitutive model in modeling the mechanical behavior of mining backfill.

Figure 1 shows the envelope of the Mohr–Coulomb model in p–q space (Figure 1a) and the typical stress–strain relation (Figure 1b). In the figure, ϕ and c are the friction angle and cohesion, respectively; ε_q is the deviatoric strain; p and q are the mean and deviatoric stresses, respectively, expressed as:

$$p = \frac{(\sigma_1 + \sigma_2 + \sigma_3)}{3} \quad (1)$$

$$q = \sqrt{\frac{(\sigma_1 - \sigma_2)^2 + (\sigma_2 - \sigma_3)^2 + (\sigma_3 - \sigma_1)^2}{2}} \quad (2)$$

where σ_1, σ_2, and σ_3 are the major, intermediate and minor principal stresses, respectively.

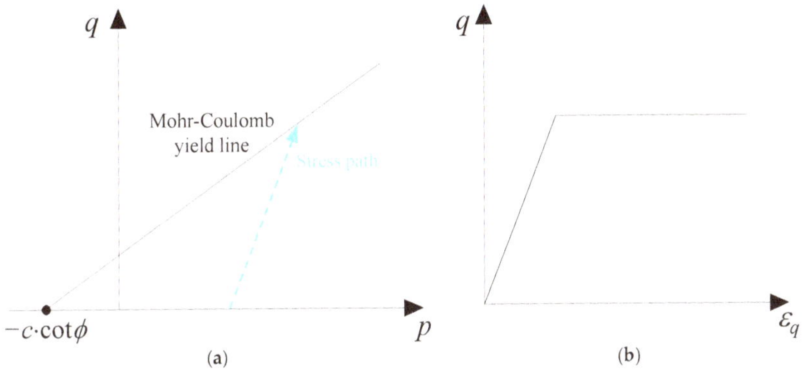

Figure 1. Schematic presentation of the Mohr–Coulomb elasto-plastic model: (**a**) yield surface in p–q space; (**b**) stress–strain relationship.

The linear stress–strain relation in the elastic regime (below the envelope) is described using a constant Young's modulus E and Poisson's ratio ν and assumed to follow Hooke's law. Once the stress state reaches the yield envelop defined by the Mohr–Coulomb criterion, infinite plastic shear strain occurs under a constant load.

The stress–strain relation in the elastic regime is expressed as:

$$\varepsilon_x = \frac{1}{E}\left[\sigma_x - \nu\left(\sigma_y + \sigma_z\right)\right] \quad (3)$$

$$\varepsilon_y = \frac{1}{E}\left[\sigma_y - \nu\left(\sigma_x + \sigma_z\right)\right] \quad (4)$$

$$\varepsilon_z = \frac{1}{E}\left[\sigma_z - \nu\left(\sigma_x + \sigma_y\right)\right] \quad (5)$$

$$\gamma_{xy} = \frac{2(1+\nu)}{E}\tau_{xy},\ \gamma_{yz} = \frac{2(1+\nu)}{E}\tau_{yz},\ \gamma_{xz} = \frac{2(1+\nu)}{E}\tau_{xz} \tag{6}$$

where $\sigma_x, \sigma_y, \sigma_z, \tau_{xy}, \tau_{yz}, \tau_{xz}$ are the components of stress tensor; $\varepsilon_x, \varepsilon_y, \varepsilon_z, \gamma_{xy}, \gamma_{yz}, \gamma_{xz}$ are the components of the strain tensor.

The well-known Mohr–Coulomb criterion is expressed as follows, to relate shear strength τ and the corresponding normal stress σ [36,37]:

$$\tau = \sigma \cdot \tan\phi + c \tag{7}$$

or as follows, in terms of principal stresses:

$$\frac{(\sigma_1 - \sigma_3)}{2} - \frac{(\sigma_1 + \sigma_3)}{2} \cdot \sin\phi - c \cdot \cos\phi = 0 \tag{8}$$

A 3D generalization of Equation (8) in terms of stress invariants is given as [38]:

$$q + \left(\frac{3\sin\phi}{\sin\theta_l \cdot \sin\phi - \sqrt{3}\cos\theta_l}\right)p + \left(\frac{3\cos\phi}{\sin\theta_l \cdot \sin\phi - \sqrt{3}\cos\theta_l}\right)c = 0 \tag{9}$$

where θ_l is the Lode angle.

The Mohr–Coulomb criterion correlates well with the shear strength of backfill, but tends to overestimate the tensile strength [39]. A tensile strength T smaller than that calculated by applying the equation of Mohr–Coulomb, called tension cut-off and, thus, usually applies for mining backfill. An associated or non-associated flow rule can be applied by varying the value of dilation angle ψ to model the plastic volume change due to shearing. The Mohr–Coulomb model does not capture the plastic volumetric strain under isotropic compression.

2.2. Double-Yield Model

The double-yield model was built in FLAC [40], which involves shear and tensile yield criteria of the Mohr–Coulomb model, and a volumetric yield surface. Its stress–strain relation in the elastic region is described by Hooke's law. Figure 2 illustrates the schematic yield surface and stress–strain behavior of the double-yield model.

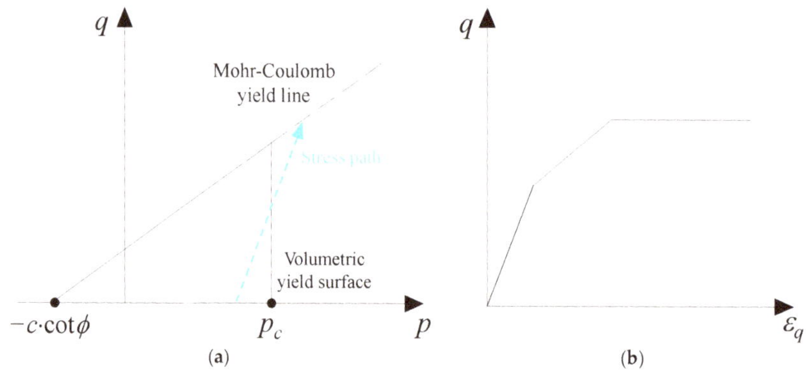

Figure 2. Schematic presentation of the double-yield model: (**a**) yield surface in p–q space; (**b**) stress–strain relationship.

The volumetric yield surface (or cap) of the double-yield model shown in Figure 2a is independent on the deviatoric stress, and is defined as:

$$p - p_c = 0 \tag{10}$$

where p_c is the current cap pressure (or preconsolidation pressure).

The double-yield model has an associated volumetric flow rule and its hardening rule relates to the plastic volumetric strain ε_v^p through a defined piecewise-linear function. The prescribed piecewise-linear function is flexible, but needs to be calibrated based on the results of physical tests. The bulk modulus K in the double-yield model is proportional to the derivative of cap hardening function as:

$$K = R\frac{\mathrm{d}p_c}{\mathrm{d}\varepsilon_v^p} \tag{11}$$

where R is a constant.

Equation (11) indicates that the elastic modulus of the double-yield model is dependent on a piecewise-linear function of the cap pressure, which explains the varied slope of the stress–strain curve in the elastic region, as shown in Figure 2b. Compared with the Mohr–Coulomb model, the double-yield model involves a volumetric yield surface, which enables accounting the plastic volumetric strain due to the mean stress. The double-yield model has been adopted in some studies to simulate the mechanical performance of considerably compressible backfill material [34,41].

2.3. Soft Soil Model

The Soft Soil model is an advanced Cam-Clay type model [14,42] based on the critical state concept [43] and captures the irreversible void change accompanying the soil deformation. Figure 3a shows the relation between the volumetric strain ε_v and mean stress p in the Soft Soil model. It is postulated that ε_v linearly reduces with the increase of p along a normal consolidation line (NCL) in the semi-logarithmic space. The NCL has a slope of λ^*. For unloading and reloading, ε_v varies following an elastic swelling line (SL) with a slope of κ^*. In the figure, p_0 is a reference value of mean stress. ε_v^n is the reference volumetric strain corresponding to (p_0 + c·cotϕ) on the NCL and ε_v^s is the reference volumetric strain corresponding to (p_0 + c·cotϕ) on the SL. The equations for NCL and SL in the Soft Soil model are given as:

$$\varepsilon_v = \varepsilon_v^n - \lambda^* \cdot \ln\left(\frac{p + c \cdot \cot\phi}{p_0 + c \cdot \cot\phi}\right) \tag{12}$$

$$\varepsilon_v = \varepsilon_v^s - \kappa^* \cdot \ln\left(\frac{p + c \cdot \cot\phi}{p_0 + c \cdot \cot\phi}\right) \tag{13}$$

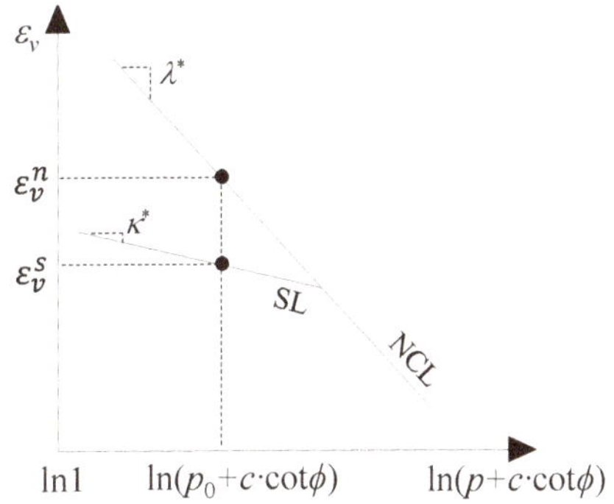

(a)

Figure 3. *Cont.*

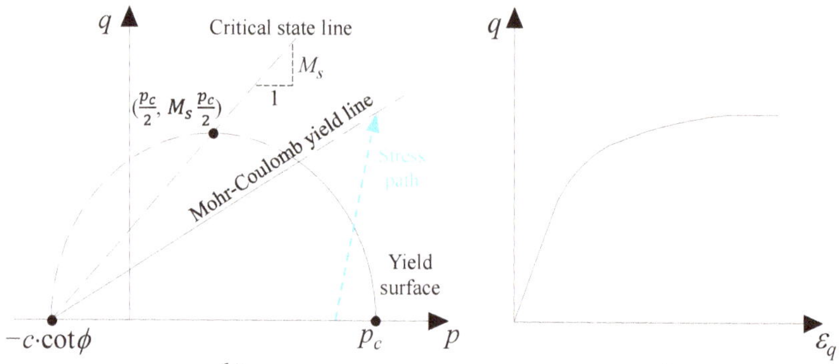

(b) (c)

Figure 3. Schematic presentation of the Soft Soil model: (**a**) logarithmic relation between the volumetric strain and mean stress; (**b**) yield surface in p–q space; (**c**) stress–strain relationship.

Figure 3b,c show the yield surface and stress–strain relation of the Soft Soil model. The yield surface consists of the envelope of the Mohr–Coulomb model and an elliptical cap. The elliptical cap has an apex on a critical state line as shown in Figure 3b and its formulation is expressed as:

$$\frac{q^2}{M_s{}^2(p + c\cdot\cot\phi)} + (p - p_c) = 0 \tag{14}$$

where M_s is the slope of the critical state line in p–q space, which can be calculated based on the flow rule of the modified Cam-Clay model [42,44] as:

$$M_s = 3\sqrt{\frac{(1-K_0)^2}{(1+2K_0)^2} + \frac{(1-K_0)(1-2v)\left(\frac{\lambda^*}{\kappa^*}-1\right)}{(1+2K_0)(1-2v)\frac{\lambda^*}{\kappa^*} - (1-K_0)(1+v)}} \tag{15}$$

where K_0 is the coefficient of earth pressure at-rest in normally consolidated condition.

The Soft Soil model employs an associated flow rule for the volumetric yield surface. The hardening of the volumetric yield surface is attributed to the plastic volumetric strain ε_v^p and is defined as:

$$\frac{dp_c}{d\varepsilon_v^p} = \frac{p_c}{\lambda^* - \kappa^*} \tag{16}$$

The elastic modulus in the Soft Soil model is mean stress-dependent expressed as:

$$K = \frac{p + c\cdot\cot\phi}{\kappa^*} \tag{17}$$

The Soft Soil model can be considered as a combination of the Mohr–Coulomb criterion and the modified Cam-Clay model. The critical state line in the Soft Soil model controls the shape of yield surface while the shear strength is defined by the Mohr–Coulomb envelope. The Soft Soil model has been used in a few studies to analyze the compressibility of soft clay [45,46], but has rarely been applied for mining backfill.

3. Comparisons between Numerical Models and Laboratory Tests

The applicability of constitutive models presented above to describing the compressibility of mining backfill, is identified by modeling one-dimensional consolidation and consolidated drained triaxial compression tests made on lowly cemented backfill available in the literature using FLAC^{3D} [40].

3.1. Comparison with One-Dimensional Consolidation Tests

Pierce [26] conducted one-dimensional consolidation tests on Golden Giant paste backfill. The backfill samples were made by mixing the tailings with 3% binder by weight and cured for 28 days. Samples were casted in a rigid metal cylinder, which also acted as a confining ring in the consolidation tests. The backfill samples have a diameter of 75 mm and a height of around 37.5 mm. The measured properties include a density ρ of 2013 kg/m^3, a porosity n of 49%, a cohesion c of 40 kPa and a friction angle ϕ of 41°. During the consolidation tests of Pierce [26], a porous stone was put under the cylinder to allow drainage and a platen was placed on the top of the cylinder for the incremental load. Figure 4 shows the physical model and the laboratory results of the applied stress–compressive strain curve of one-dimensional consolidation tests conducted by Pierce [26]. The loading path has four stages, involving an increase from 0 to 4 kN in an increment of 0.5 kN, a decrease from 4 to 1 kN in an increment of 1 kN, an increase from 1 to 6 kN in an increment of 1 kN, and an increase from 6 to 12 kN in an increment of 2 kN.

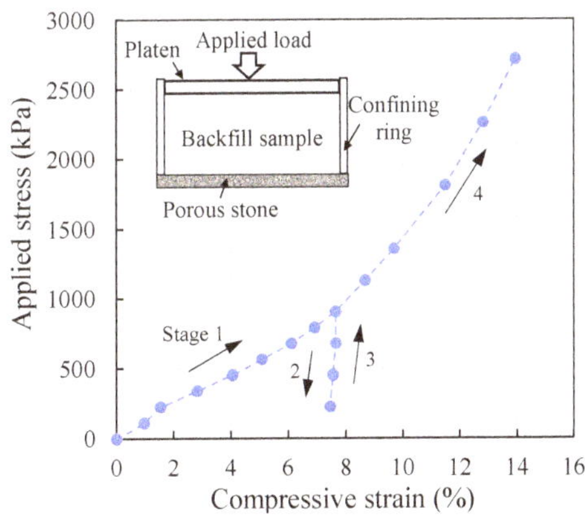

Figure 4. Physical model and stress–strain curve of one-dimensional consolidation tests on Golden Giant paste backfill with a binder content of 3% and a curing time of 28 days conducted by Pierce [26].

Figure 5 shows a numerical model of a one-dimensional consolidation test constructed with FLAC^{3D}, which involves a backfill sample, a cylinder (cell) and a loading platen. The platen is built to allow loading on a displacement boundary of the top surface. The numerical model shown in Figure 5 has the identical dimensions as the physical model of Pierce [26]. The mesh size of the numerical model is 3 mm based on the sensitivity analyses. Both the cylinder and platen are modeled as linear elastic material with a Young's modulus of 200 GPa and a Poisson's ratio of 0.3. The Mohr–Coulomb, double-yield, and Soft Soil models are applied for the backfill sample to make comparisons. Interface elements are applied between the cylinder and backfill. The normal and shear stiffness of the interface are determined based on an equation recommended in the FLAC^{3D} manual [40]. The interface friction angle ϕ_i is taken as 2/3 of ϕ, while its cohesion c_i is assumed equal to c of backfill [47,48]. The displacements on the bottom of the model are restricted and other boundaries are set free. The same loading path in Pierce [26] tests is applied in the numerical simulations while the average normal displacement on the top of platen is recorded to calculate the compressive strain for different applied stresses.

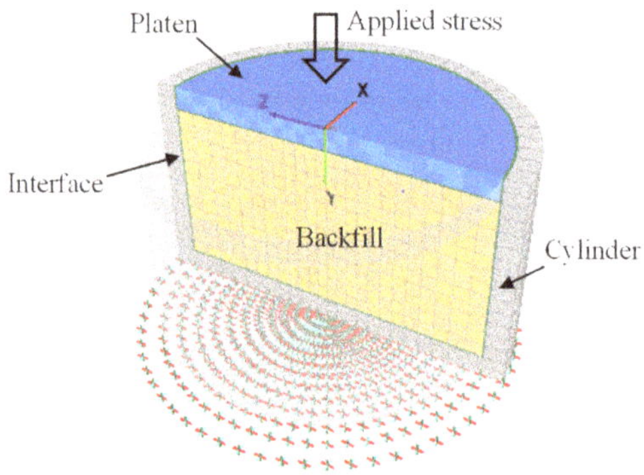

Figure 5. A numerical model of one-dimensional consolidation tests built with $FLAC^{3D}$.

In the numerical simulations, the Poisson's ratio of backfill is related to its friction angle through $\nu = (1 - \sin\phi)/(2 - \sin\phi)$ by considering a unique value of at-rest earth pressure coefficient K_0 [49,50].The tensile strength T of backfill is taken as 1/10 of its uniaxial compressive strength (UCS) [39]. The initial value of void ratio e_{ini} is calculated based on the measured porosity. However, some parameters remain unknown including K and shear modulus G for the Mohr–Coulomb model, R and the piecewise-linear function of p_c for the double-yield model, λ^*, κ^*, and p_c for the Soft Soil model. These parameters are determined by calibrating the numerical results based on part of laboratory results of Pierce [26] associated with the loading paths 1 and 2 as shown in Figure 4. Table 1 summarizes all parameters applied for different constitutive models. Numerical models with the calibrated parameters are then called the calibrated models, which are further applied to predict the other part of laboratory results associated with loading paths 3 and 4.

Table 1. Parameters of different constitutive models applied for backfill in numerical simulations of one-dimensional consolidation tests with ρ = 2013 kg/m^3, ϕ_i = 27°, c_i = 40 kPa.

Constitutive Models	Parameters									
Mohr-Coulomb	K (kPa)	G (kPa)	ϕ (°)	c (kPa)	ψ (°)	T (kPa)				
	5388	3141	41	40	0	17.6				
Double-yield	K_{max} (GPa)	G_{max} (GPa)	ϕ (°)	c (kPa)	ψ (°)	T (kPa)	R			
	50	29.2	41	40	0	17.6	24.1			
	Prescribed piecewise-linear function for cap (kPa) hardening in terms of (ε_v^p, P_c) (0, 0); (0.008, 12); (0.0094, 14.2); (0.0103, 15.5); (0.0119, 30.5); (0.0178, 69.59); (0.0181, 72); (0.0246, 87.5); (0.0273, 92.46); (0.0301, 97.5); (0.0336, 120); (0.0393, 144.71); (0.0449, 169.2); (0.0479, 187); (0.0492, 193.73); (0.0573, 237); (0.0592, 246.89); (0.0611, 257); (0.0631, 285.97); (0.0689, 294.84); (0.0714, 310); (0.0741, 325); (0.0889, 455.85)									
Soft Soil	ν	ϕ (°)	c (kPa)	ψ (°)	T (kPa)	κ^*	λ^*	K_0	p_c (kPa)	e_{ini}
	0.26	41	40	0	17.6	0.0052	0.051	0.34	127	0.961

Note: K_{max} and G_{max} denote the upper limits of the bulk and shear modulus. K and G for the Mohr–Coulomb model, R and piecewise-linear function of p_c for the double-yield model, λ^*, κ^*, and p_c for the Soft Soil model are calibrated based on the experimental results.

Figure 6 shows the comparisons between the laboratory results of one-dimensional consolidation tests reported by Pierce [26] and the numerical results by applying the Mohr–Coulomb (Figure 6a), double-yield (Figure 6b), and Soft Soil (Figure 6c) models for backfill. In Figure 6a, one sees the compressive strain of the Mohr–Coulomb model linearly increases as the applied stress increases. For the unloading stage, the stress–strain curve of the Mohr–Coulomb model is almost parallel to that in the loading stage. The minor scatter between the curves of loading and unloading is attributed to the yield of fill-wall interface. However, the experimental strain shows nonlinear relation with the applied stress in the test while only a small component of compressive strain is reversible at the unloading stage. The poor agreement between numerical and laboratory results is explained as that the Mohr–Coulomb model simulates a constant elastic modulus and does not capture the volumetric yield of backfill.

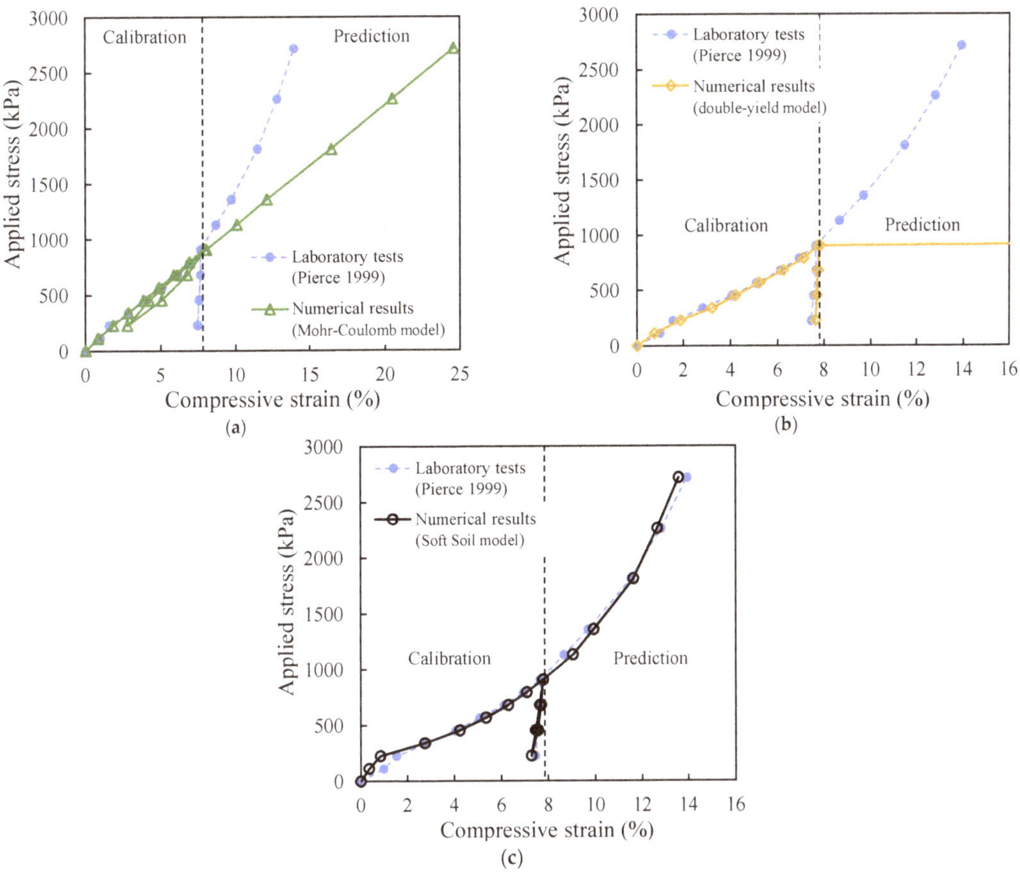

Figure 6. Comparisons between the experimental results of one-dimensional consolidation tests reported by Pierce [26] and the numerical results by applying the (**a**) Mohr–Coulomb; (**b**) double-yield; and (**c**) Soft Soil models for backfill with parameters in Table 1.

Figure 6b shows that the calibrated results of the double-yield model correlate well with the laboratory results, but the predicted compressive strain steeply increases as the applied stress further increases. It is because that the prescribed piecewise-linear function of the cap pressure in the double-yield mode is calibrated, based on part of the laboratory results. The prescribed function is flexible and can result in a good fit between the numerical and test results. However, when the applied stress exceeds the range of prescribed function, the double-yield model demonstrates infinite plastic volumetric strain as shown in Figure 6b. The predictive capability of the double-yield model is thus limited when the test data for calibration are insufficient. For the Soft Soil model, Figure 6c illustrates that both the described and predicted numerical results agree well with the laboratory results. Minor difference between the described results of the Soft Soil model and test results is seen when the applied stress is smaller than 230 kPa. This is because that the cementation in the backfill increases its primary stiffness at a small stress level. As the applied stress increases, the cement bond yields as shown by a drop of fill stiffness in Figure 6c. The mechanical behavior of cemented backfill then approaches an uncemented condition. In the Soft Soil models, the large primary stiffness caused by the cement bond at the small stress level can be pseudo-simulated by using an overconsolidation state, though their mechanisms are different. The Soft Soil model is thus deemed capable of describing the compressibility of lightly cemented or uncemented backfill in a confined compression condition.

3.2. Comparison with Consolidated Drained Triaxial Compression Tests

Rankine [28] conducted consolidated drained triaxial compression tests on Cannington paste backfill. The backfill samples have a diameter of 38 mm, a height of 76 mm, a cement content of 2%, a solid content of 74% by weight, and were cured for 28 days. The density of backfill is 2091 kg/m^3 and the porosity is 51.2%. Figure 7 shows the physical model and deviatoric stress–strain curve of consolidated drained triaxial compression tests under confining pressures of 100, 200, 500 kPa performed by Rankine [28].

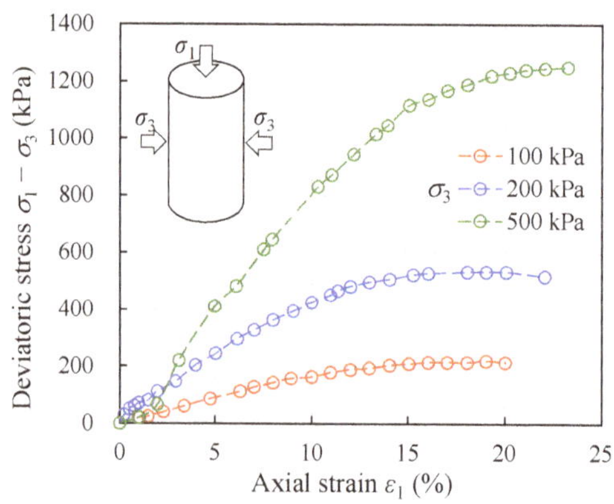

Figure 7. Physical model and deviatoric stress–strain curve under different confining pressures of consolidated drained compression triaxial tests on Cannington paste backfill, with a cement content of 2% and a curing time of 28 days performed by Rankine [28].

Figure 8 illustrates a numerical model of backfill sample built with FLAC^{3D} in simulations of consolidated drained triaxial compression tests. The numerical model has same sizes as the samples of Rankine [28], while the mesh size of the numerical model is 2 mm based on the sensitivity analyses. The Mohr–Coulomb, double-yield, and Soft Soil models are applied for backfill. The normal displacements on the bottom of the numerical model are restricted. The initial state is modeled by applying the confining stress normal to the surface of the sample after which the displacement is reset to zero. A normal velocity of 1×10^{-7} m/step is then applied on the top surface to simulate the compression. The normal stress and the axial displacement on the top surface of the numerical model are recorded during the calculation.

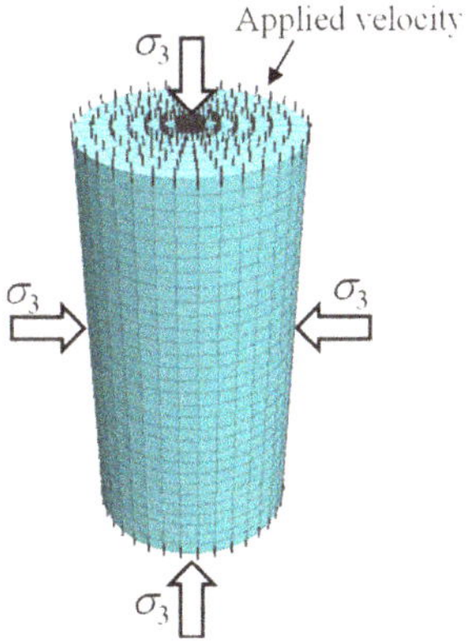

Figure 8. A numerical model of consolidated drained triaxial compression tests built with $FLAC^{3D}$.

In the simulations, the Poisson's ratio is related to the friction angle of backfill considering a unique value of the at-rest earth pressure coefficient K_0. The ratio of tensile strength T to UCS of backfill is taken as 0.41 according to the experimental results of Rankine [28]. e_{ini} is calculated as 1.05 based on the measured porosity. Calibrations based on the laboratory results under confining pressures of 100 and 200 kPa are performed to obtain some unknown parameters involving c and ϕ, K and G for the Mohr–Coulomb model, R and the piecewise-linear function of p_c for the double-yield model, λ^*, κ^*, and p_c for the Soft Soil model. Table 2 shows material parameters used for numerical simulations. The calibrated numerical models are then applied to predict the laboratory results of Rankine [28] under a confining pressure of 500 kPa.

Table 2. Parameters of different constitutive models applied for backfill in numerical simulations of the consolidated drained triaxial compression tests with ρ = 2091 kg/m^3.

Constitutive Models	Parameters									
Mohr-Coulomb	K (kPa)	G (kPa)	ϕ (°)	c (kPa)	ψ (°)	T (kPa)				
	2935	1203	32	14.73	0	21.8				
Double-yield	K_{max} (GPa)	G_{max} (GPa)	ϕ (°)	c (kPa)	ψ (°)	T (kPa)	R			
	50	20.5	32	14.73	0	21.8	2			
	Prescribed piecewise-linear function for cap (kPa) hardening (ε_v^p, P_c) (0, 0); (0.052, 50); (0.103, 100); (0.155, 150); (0.196, 190); (0.2, 200); (0.218, 250); (0.237, 300); (0.256, 350); (0.274, 400); (0.312, 500); (0.376, 670); (0.383, 690); (0.387, 700)									
Soft Soil	ν	ϕ (°)	c (kPa)	ψ (°)	T (kPa)	κ^*	λ^*	K_0	p_c (kPa)	e_{ini}
	0.32	32	14.73	0	21.8	0.0078	0.135	0.47	50	1.05

Note: c and ϕ, K and G for the Mohr–Coulomb model, R and the piecewise-linear function of p_c for the double-yield model, λ^*, κ^*, and p_c for the Soft Soil model are calibrated based on the experimental results.

Figure 9 shows the comparisons between the laboratory results of consolidated drained triaxial compression tests conducted by Rankine [28] and numerical results under different confining pressures by applying the Mohr–Coulomb (Figure 9a), double-yield (Figure 9b), and Soft Soil (Figure 9c) models for backfill. Figure 9a illustrates that the calibrated and predicted strength of the Mohr–Coulomb model are close to the laboratory results. However, the elastic modulus of the Mohr–Coulomb model is constant while the stiffness of mining backfill increases as the confining pressure increases. The Mohr–Coulomb model thus largely underestimates the stress magnitude at a given strain under a confining pressure of 500 kPa. Meanwhile, it overestimates the strain at failure under a confining pressure of 500 kPa by predicting a value of 38% while the experimental result shows a value of 20%.

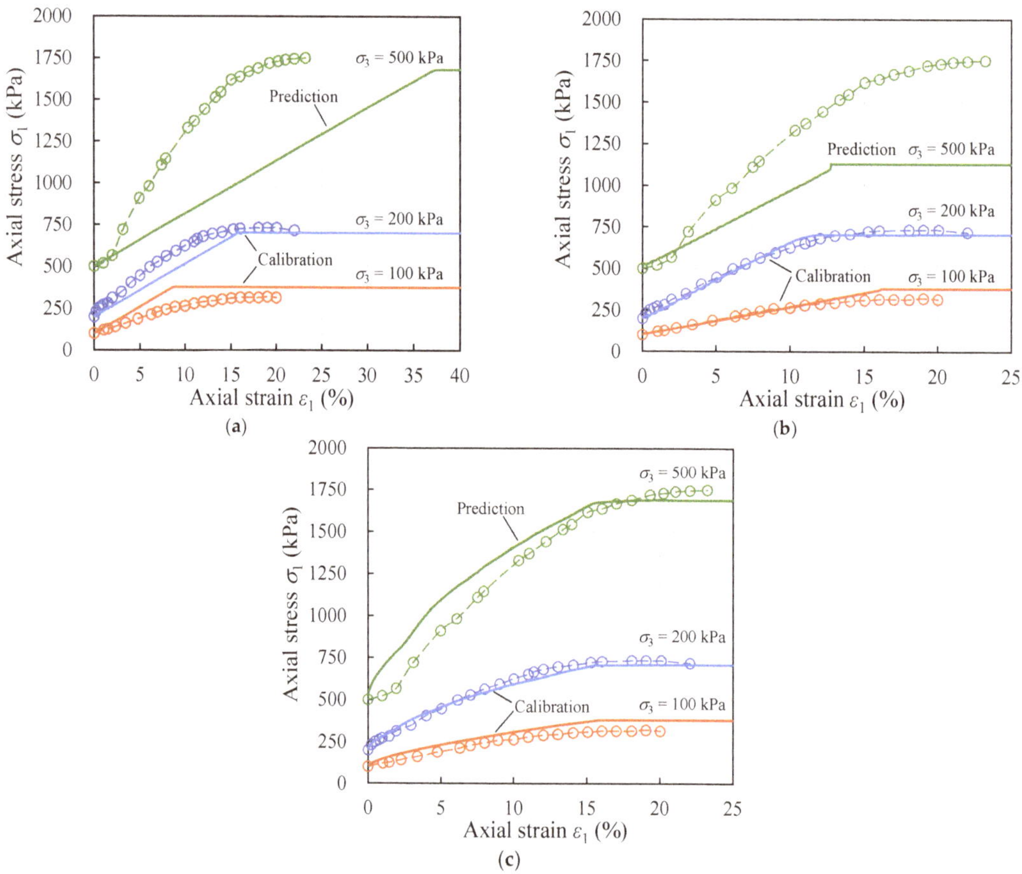

Figure 9. Comparisons between the experimental results (dash lines with points) of consolidated drained triaxial compression tests reported by Rankine [28] and the numerical results (solid lines) under different confining pressures by applying the (**a**) Mohr–Coulomb; (**b**) double-yield and (**c**) Soft Soil models for backfill with parameters in Table 2.

Figure 9b shows that the double-yield model can reasonably describe the laboratory results for the confining pressures of 100 and 200 kPa. However, the predicted strength and stiffness of the double-yield model under a confining pressure of 500 kPa are very different from the experimental results. It is explained as that the infinite volumetric plastic strain occurs once the applied stress exceeds the upper bond of prescribed piecewise-linear function of the cap in the double-yield model. The predictive capability of the double-yield model is thus limited. Figure 9c shows that the described and predicted results of the Soft Soil model reasonably agree with the laboratory results. Based on the comparisons between numerical results and laboratory tests, the Soft Soil model is identified superior to the Mohr–Coulomb and the double-yield model in describing the compressibility of mining backfill with slight cementation (or uncemented backfill). In order to further exhibit the applicability of the Soft Soil model, it will be benchmarked with respect to the Mohr–

Coulomb model in modeling a typical backfilled stope overlying a sill mat at different mine depths.

4. Simulations of Backfilled Stope Overlying a Sill Mat

In underhand cut-and-fill mining, uncemented or lightly cemented backfill is used to fill the mined-out upper stope overlying a sill mat. During the extraction of underlying stope, the sill mat will act as an artificial roof, which makes the stress distribution within the overlying backfilled stope significant for its stability. Sobhi and Li [11] analyzed this problem with PLAXIS^{2D} only using the Mohr–Coulomb constitutive model to simulate the backfilled stope. The compressibility of backfill under the rock wall closure associated underlying extraction was thus not properly considered by Sobhi and Li [11]. In this section, the problem of a backfilled stope overlying a sill mat at different mine depths D of 200 and 1000 m are numerically investigated with FLAC^{3D}. Emphasis is placed on the comparisons between numerical results predicted by applying the Mohr–Coulomb and Soft Soil models. Figure 10 shows a physical model and a plane strain numerical model (D = 200 m) of the problem. The symmetry plane (x = 0) is taken into account by considering half of the model. The excavations have a width B of 6 m. The overlying stope has a height H of 10 m and is filled with uncemented backfill. A gap of 0.5 m is left on the top of the backfill to represent the poor contact between fill and stope roof. The sill mat has a height H_s of 3 m, while the underlying stope is 13.5 m in height. The domain size of the numerical model is a distance from the origin to the boundaries of the model. Based on the sensitivity analyses, the numerical model is constructed with the optimal domain and mesh sizes of 300 and 0.25 m.

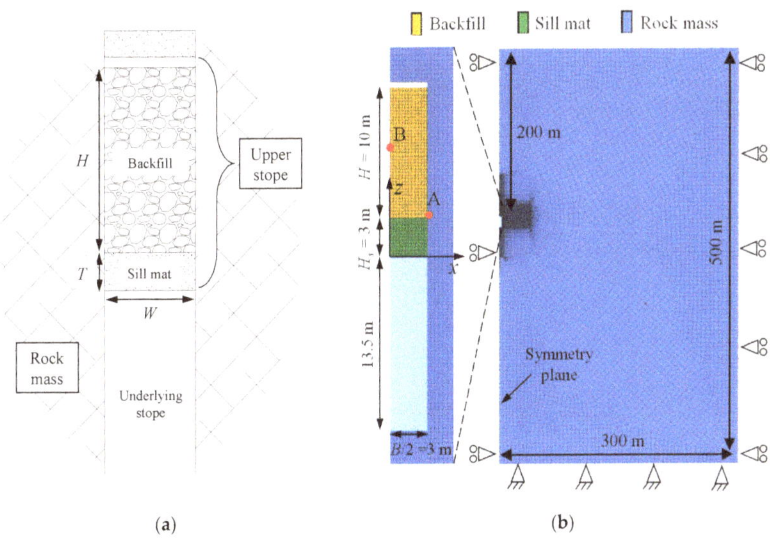

Figure 10. (**a**) Physical model and (**b**) numerical model built with FLAC^{3D}of an undercut below a sill mat with overlying backfill.

The rock mass and sill mat obey the Mohr–Coulomb model while the overlying backfill is modeled with different constitutive models. The rock mass is characterized by unit weight γ_R = 27 kN/m^3, Young's modulus E_R = 42 GPa, Poisson's ratio ν_R = 0.25, cohesion c_R = 9.4 MPa, friction angle ϕ_R = 38°, and dilation angle ψ_R = 0°. The sill mat is characterized by unit weight γ_s = 20 kN/m^3, Young's modulus E_s = 1.5 GPa, Poisson's ratio ν_s = 0.3, cohesion c_s = 5 MPa, friction angle ϕ_s = 35°, and dilation angle ψ_s = 0°. Table 3 provides the material parameters for the overlying uncemented backfill in which the same parameters are applied in the Mohr–Coulomb and Soft Soil models, where possible. The shear strength parameters (i.e., c_i and ϕ_i) of fill–rock interfaces are considered equal to those of backfill by assuming rough rock walls.

Table 3. Parameters of the Mohr–Coulomb and Soft Soil models applied for overlying uncemented backfill with unit weight γ = 18 kN/m^3.

Constitutive Models	Parameters									
Mohr-Coulomb	K (MPa)	G (MPa)	ϕ (°)	c (kPa)	ψ (°)	T (kPa)				
	250	115	35	0	0	0				
Soft Soil	ν	ϕ (°)	c (kPa)	ψ (°)	κ^* $\times 10^{-3}$	λ^* $\times 10^{-3}$	T (kPa)	K_0	p_c (kPa)	e_{ini}
	0.3	35	0	0	0.2	1	0	0.43	1	0.9

In the numerical model, the displacement along the third direction (y-axis) is constrained to simulate a two-dimensional plane strain condition. The top boundary of the numerical model is set free to simulate the ground surface while normal displacement is restricted on the lateral boundaries. For the bottom boundary, the displacements are constrained in all directions. Numerical simulations are conducted at mine depths D of 200 and 1000 m respectively. The lateral earth pressure coefficient K_r = 2 is employed by considering the typical stress regime of the Canadian Shield [51]. In numerical simulations, the overlying stope is excavated after obtaining the initial equilibrium state. The displacement is then reset to zero and overlying stope is sequentially backfilled with 1 m per layer. This is followed by excavating the underlying stope in one step to expose the sill mat. Figure 11 shows the displacement distributions in the numerical model at each step as references with the Soft Soil model at a mine depth of 200 m.

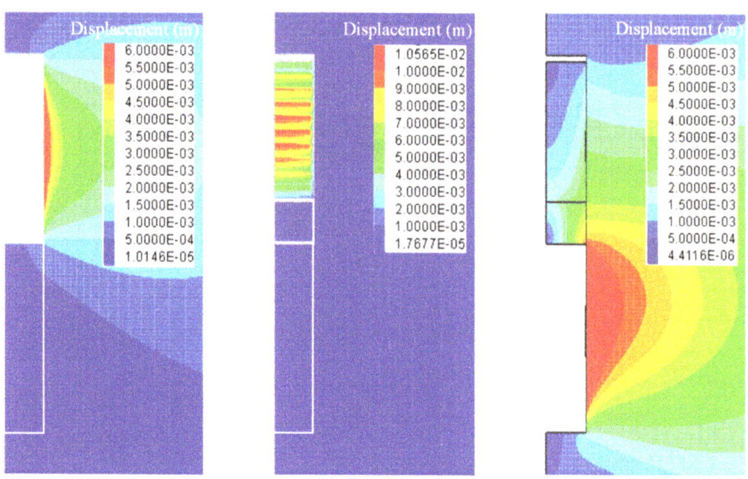

(a) (b) (c)

Figure 11. Distributions of the displacement with the Soft Soil model at a mine depth of 200 m for different simulation steps of (**a**) excavating the overlying stope; (**b**) backfilling the mined-out overlying stope; and (**c**) extracting the underlying stope.

Figure 12 shows the iso-contours of bulk modulus in the overlying backfill after placement by applying different constitutive models. In the figure, one sees the bulk modulus of the Mohr–Coulomb model is 250 MPa and is a constant independent on the stope height. The bulk modulus of the Soft Soil model around the middle height of the stope is around 250 MPa, but its value moderately increases along the height of the backfilled stope, because the elastic modulus of the Soft Soil model is mean stress dependent, as given by Equation (17).

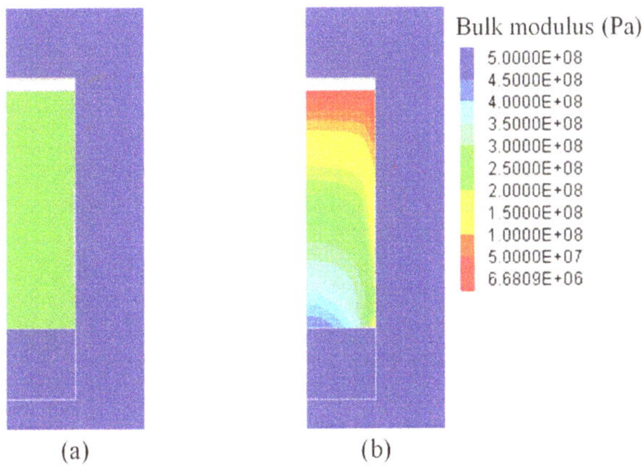

Figure 12. Distributions of the bulk modulus in overlying backfill after placement simulated with the (**a**) Mohr–Coulomb and (**b**) Soft Soil models.

Figure 13 illustrates the variation of the vertical and horizontal stresses along the vertical central line (VCL) of the overlying backfill before the excavation of underlying stope. Results shown in Figure 13 are independent on different mine depths because the backfill is placed after the rock wall displacement (i.e., delayed placement). In the figure, one sees that both the vertical and horizontal stresses increase smoothly along the stope height while the arching effect is evident by comparing with the overburden stresses. The stress distributions in the backfilled stope prior to the underlying excavation by applying the Mohr–Coulomb and Soft Soil models are almost identical. At the lower part of the stope, the stress state of the Soft Soil model is slightly larger than that of the Mohr–Coulomb model. The results shown in Figure 13 agree well with the results reported by Sobhi and Li [11].

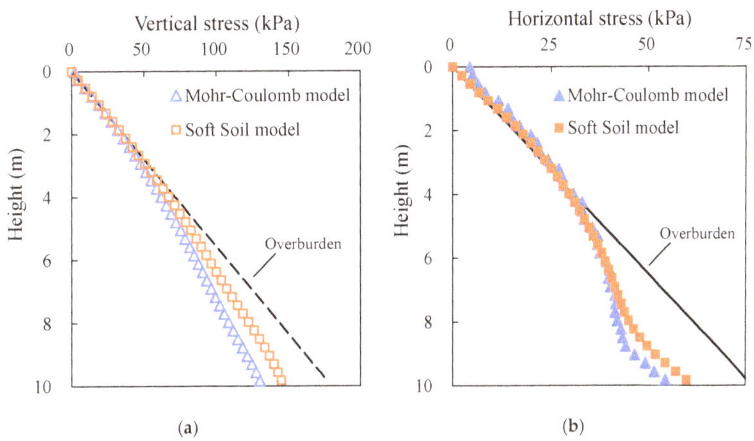

Figure 13. Variation of the (**a**) vertical and (**b**) horizontal stresses along the VCL of overlying backfilled stope before the underlying extraction.

Figure 14 shows the variation of vertical and horizontal stresses along the VCL of overlying backfill after excavating the underlying stope at a mine depth $D = 200$ m. Under the influence of rock wall closure associated with the underlying extraction, the vertical and horizontal stresses in the overlying backfill increase compared to the results shown in Figure 13. However, the vertical and horizontal stresses of the Soft Soil model are smaller than those of the Mohr–Coulomb model below the stope height of 2 m. The value of vertical stress at the bottom of overlying backfill is 256 kPa for the Mohr–Coulomb model and is 169 kPa for the Soft Soil model. The different results of two constitutive models are explained as that the Soft Soil model simulates plastic volumetric strain of backfill under the compression from rock walls. The backfill needs to be compacted with certain amount of compressive strain before it can sustain large compressive stress. This feature is not captured by the Mohr–Coulomb model, which can thus overestimate the stress state in the overlying stope when the rock wall closure is small at a shallow mine depth. Since the stability of sill mat largely depends on the stresses within the overlying backfill, using the Mohr–Coulomb model may further cause an inaccurate estimation on the required strength of sill mat.

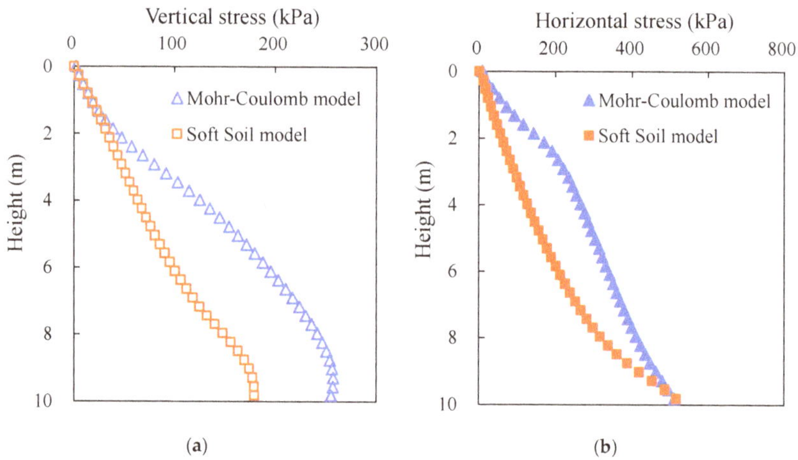

Figure 14. Variations of the (**a**) vertical and (**b**) horizontal stresses along the VCL of overlying backfill after excavating the underlying stope at a mine depth of 200 m.

Figure 15 shows the variations of vertical and horizontal stresses along the VCL of overlying backfill after the excavation of underlying stope at a mine depth $D = 1000$ m. As the mine depth increases from 200 to 1000 m, the rock wall closure associated with underlying excavation becomes larger which increases the vertical and horizontal stresses within the overlying backfill. The stress distribution predicted by the Soft Soil model is similar to that of the Mohr–Coulomb model above the stope height of 6 m as shown in Figure 15. However, the stresses of the Soft Soil model rapidly increase as the stope height increases. At the bottom of overlying backfill, the vertical and horizontal stresses of the Soft Soil model reach 2.2 and 7.5 MPa, which are significantly larger than the values of 0.9 and 2.5 MPa as predicted by the Mohr–Coulomb model. The different results of two constitutive models shown in Figure 15 are attributed to that the Soft Soil model accounts the volumetric hardening and pressure-dependent behavior of backfill. With a significant rock wall closure at a large mine depth, the mining backfill demonstrates large volumetric plastic strain, during which it becomes harder with a large elastic modulus, resulting in an increase in stresses generated by rock wall closure. The Mohr–Coulomb model does not simulate the volumetric hardening of backfill, which, thus, underestimates the stresses in backfilled stopes when the walls convergence is significant at a large mine depth.

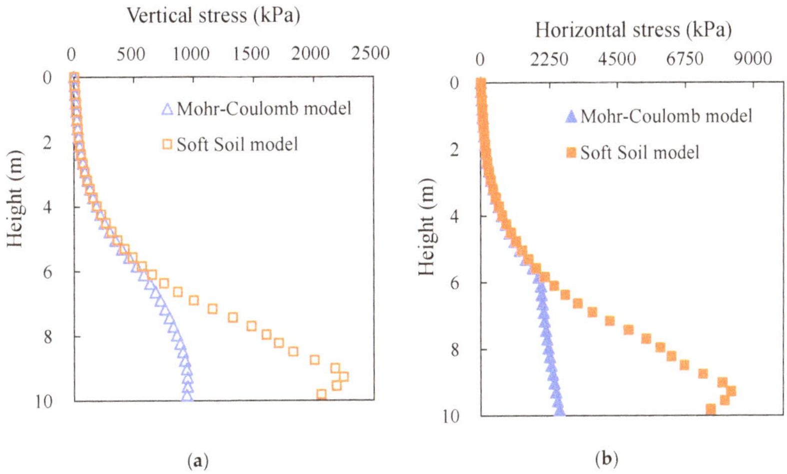

Figure 15. Variation of the (**a**) vertical and (**b**) horizontal stresses along the VCL of overlying backfill after excavating the underlying stope at a mine depth of 1000 m.

Numerical simulations have shown that the Mohr–Coulomb and the Soft Soil models predict similar stress distributions in an isolated backfilled stope when the rock wall closure is absent. However, the stresses within backfill simulated with two constitutive models can be different when closure of surround rock walls is applied. In this condition, the commonly used Mohr–Coulomb model can under- or overestimate the stresses due to the neglect of volumetric yield and pressure-dependent behavior of backfill. The Soft Soil model is deemed more applicable to simulating uncemented or lightly cemented backfill when its compressibility or closure resistance is a dominant factor.

5. Discussion

By comparing the numerical results against the experimental results of one-dimensional consolidation and consolidated drained triaxial compression tests, the Soft Soil model is identified as capable of describing the compressibility of mining backfill. Nonetheless, the Soft Soil model should not be applied for backfill with a very large cohesion. This is partially because that its elliptical yield surface crosses the p-axis at a value of $c{\cdot}\cot\phi$ on the left side and has an apex on the critical state line as shown in Figure 3b. It implies that the lower bond of preconsolidation pressure in the Soft Soil model is determined by the value of cohesion. When a large cohesion is applied, the backfill modeled with the Soft Soil model is over-consolidated with a significant preconsolidation pressure, which can be unrealistic in some conditions. Another reason is that the slope of the critical state line (Equation (15)) in the Soft Soil model is calculated based on the flow rule of the modified Cam-Clay model, which considers a nil cohesion [42,44].Therefore, the Soft Soil model is deemed applicable to modeling the uncemented or lightly cemented backfill with a small (or nil) cohesion value.

For cemented backfill, cementation increases its primary stiffness at low stress condition by binding together fill particles. Experimental results show that the cement bond can yield as the applied stress increases after which the mechanical behavior of cemented backfill approaches an uncemented condition [26,52]. The results in Figure 6c show that the effect of cementation on fill stiffness at low stress levels can be pseudo-simulated using an over-consolidated state in the Soft Soil model. However, one should note that the mechanisms of cementation and overconsolidation are different. More effort is needed to investigate the effect of cement content and curing time on the compressibility of mining backfill and incorporate it in a constitutive model [52–54].

In numerical simulations, the Poisson's ratio of backfill relates to the friction angle as $\nu = (1 - \sin\phi)/(2 - \sin\phi)$, which is based on a unique value of at-rest earth pressure coefficient K_0 [49,50]. Such equation is practical in numerical modeling with an elastoplastic model. Previous studies have proposed several forms of empirical equation to define the relationship between ν and ϕ [55,56]. More works are needed to investigate this aspect based on experimental results. Meanwhile, since the stress–strain curve of soil-like material is highly nonlinear, how to determine the Poisson's ratio of backfill based on the experimental results for numerical modeling is a problem, and needs to be studied in future works.

The Soft Soil model postulates a perfectly plastic behavior when the stress state reaches the Mohr–Coulomb yield line. The post-peak behavior of backfill is affected by the cement content and confining pressure level [27,29]. Experimental results show that mining backfill with large cement content demonstrates strain softening under small confining pressures at the post-peak stage. The large confining pressure can also result in a strain hardening behavior of mining backfill. The post-peak behavior of backfill is not analyzed in this study.

This study focusses on modeling the compressibility of backfill in the long-term condition while the pore water pressure is not considered. More effort is thus necessary to evaluate the hydraulic conductivity and the effects of pore water pressure and drainage condition on the compressibility of backfill in the short-term condition [57–59].

The simulations of a backfilled stope overlying a sill mat indicate the applicability of Soft Soil model and the significance of modeling fill compressibility. The commonly used Mohr–Coulomb model tends to under- or overestimate the stress states in a backfilled stope when the walls closure is applied due to the poor description of the fill compressibility. Although the capability of the Soft Soil model has been tested against some consolidation and triaxial tests, field measurements are still needed when available to make further verifications.

6. Conclusions

The Mohr–Coulomb elasto-plastic, double-yield, and Soft Soil constitutive models are recalled and evaluated for the applicability to describing the compressibility of mining backfill. Numerical results with different constitutive models in FLAC^{3D} are compared with one-dimensional consolidation and consolidated drained triaxial compression tests made on lowly cemented backfill available in the literature. Part of the experimental results is used to calibrate some model parameters and the calibrated models are applied to predict the other part of the test results. Based on the comparisons, the Soft Soil model shows the satisfactory description of the experimental results while its prediction is also quite good. The prevalently used Mohr–Coulomb model demonstrates poor correlations with the experimental results due to the neglect of volumetric yield and pressure-dependent behavior of backfill. The double-yield model accurately describes the experimental results based on the calibration, but its predictive capability is limited when the test results are insufficient.

The comparisons between Soft Soil and Mohr–Coulomb models in simulating a backfilled stope overlying a sill mat at different mine depths show similar stress distributions when rock wall closure is absent. However, when the rock wall closure associated with the underlying extraction is applied, application of the Soft Soil model shows that the Mohr–Coulomb model tends to overestimate the stresses in backfill at a shallow mine depth and underestimate the stresses at a large mine depth due to the poor description of the fill compressibility. The Soft Soil model is recommended to describe the compressibility of uncemented or lightly cemented backfill with small cohesions under external compressions associated with rock wall closure.

Author Contributions: R.W.: conceptualization, numerical modeling, analysis, literature, writing, and editing of the original draft. F.Z.: editing of the original draft. L.L.: project administration, methodology, supervision, editing of the original draft. All authors have read and agreed to the published version of the manuscript.

Funding: This work was financially supported by the Natural Sciences, and Engineering Research Council of Canada (RGPIN-2018-06902), Mitacs Elevate Postdoctoral Fellowship (IT12569 and IT12570), China Scholarship Council (201706420059), and industrial partners of the Research Institute on Mines and the Environment (RIME UQAT-Polytechnique; http://rime-irme.ca/ accessed on 24 November 2021). The authors are grateful for their support.

Institutional Review Board Statement: Not applicable.

Informed Consent Statement: Not applicable.

Data Availability Statement: Data are contained within the article.

Acknowledgments: The authors acknowledge financial support from the Natural Sciences and Engineering Research Council of Canada, Mitacs Elevate Postdoctoral Fellowship, China Scholarship Council, and industrial partners of RIME UQAT-Polytechnique.

Conflicts of Interest: The authors declare no conflict of interest.

List of Symbols

a	radius of the hole
B	width
c	cohesion
c_i	interface cohesion
c_R	cohesion of rock mass
c_s	cohesion of sill mat
D	mine depth
E	Young's modulus
e_{ini}	initial value of void ratio
E_R	Young's modulus of rock mass
E_s	Young's modulus of sill mat
G	shear modulus
G_{max}	upper limit of shear modulus
H	height
H_s	height of sill mat
K	bulk modulus
K_0	coefficient of earth pressure at-rest
K_{max}	upper limit of bulk modulus
K_r	lateral earth pressure coefficient
M_s	slope of critical state line
n	porosity
p	mean stress
p_0	reference mean stress
p_c	cap pressure
P_0	isotropic in-situ stress
P_{in}	internal pressure
q	deviatoric stress
R	constant
R_0	radius of yield zone around hole
T	tensile strength
ε_q	deviatoric strain
ε_v	volumetric strain
ε_v^p	plastic volumetric strain
ε_v^n	reference volumetric strain on normal consolidation line
ε_v^s	reference volumetric swelling line
$\varepsilon_x, \varepsilon_y, \varepsilon_z$	components of normal strain
$\gamma_{xy}, \gamma_{yz}, \gamma_{xz}$	components of shear strain
σ	normal stress
σ_1	major principal stress
σ_2	intermediate principal stress
σ_3	minor principal stress

σ_r	radial stress
σ_θ	tangential stress
σ_z	normal stress along third direction
σ_{re}	radial stress at the elastic-plastic interface
σ_x, σ_y, σ_z	components of normal stress
σ_x^∞, σ_y^∞, σ_z^∞	normal stress components of in-situ stress field
τ	shear strength
$\tau_{r\theta}$, $\tau_{\theta z}$, τ_{zr}	shear stresses around cylinder hole
τ_{xy}, τ_{yz}, τ_{xz}	components of shear stress
τ_{xy}^∞, τ_{xz}^∞, τ_{yz}^∞,	shear stress components of in-situ stress field
r, θ	cylindrical coordinates
U, V, W	components of displacement
ν	Poisson's ratio
ν_R	Poisson's ratio of rock mass
ν_s	Poisson's ratio of sill mat
ϕ	friction angle
ϕ_i	interface friction angle
ϕ_R	friction angle of rock mass
ϕ_s	friction angle of sill mat
ρ	density
ψ	dilation angle
ψ_R	dilation angle of rock mass
ψ_s	dilation angle of sill mat
λ^*	slope of normal consolidation line
κ^*	slope of swelling line
θ_l	lode angle
γ	unit weight
γ_R	unit weight of rock mass
γ_s	unit weight of sill mat

Appendix A Validation of FLAC^{3D} against Analytical Solutions of Stresses and Displacements around a Cylinder Hole in the Linearly Elastic Material

FLAC^{3D} can be validated against analytical solutions of stresses and displacements around an infinite cylinder hole in the infinite linearly elastic material. The problem can be analyzed in a plane strain condition. The physical model of this problem is shown in Figure A1. The origin locates at the central point of the model. In the figure, a is the radius of the hole. Domain size is the distance from the hole to the model boundary. r and θ are the cylindrical coordinates. The model is characterized by Young's modulus E and Poisson's ratio ν.

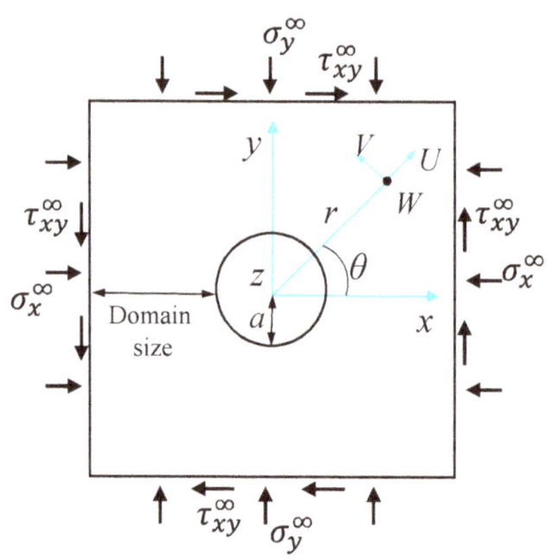

Figure A1. Physical plane strain model of a cylinder hole in an infinite linearly elastic material.

When the infinite hole is subject to a stress field composed of σ_x^∞, σ_y^∞, σ_z^∞, τ_{xy}^∞, τ_{xz}^∞, τ_{yz}^∞, the analytical solutions for stresses around the hole are given as [60,61]:

$$\sigma_r = \frac{\sigma_x^\infty + \sigma_y^\infty}{2}\left(1 - \frac{a^2}{r^2}\right) + \frac{\sigma_x^\infty - \sigma_y^\infty}{2}\left(1 + 3\frac{a^4}{r^4} - 4\frac{a^2}{r^2}\right)\cos 2\theta + \tau_{xy}^\infty\left(1 + 3\frac{a^4}{r^4} - 4\frac{a^2}{r^2}\right)\sin 2\theta \quad \text{(A1)}$$

$$\sigma_\theta = \frac{\sigma_x^\infty + \sigma_y^\infty}{2}\left(1 + \frac{a^2}{r^2}\right) - \frac{\sigma_x^\infty - \sigma_y^\infty}{2}\left(1 + 3\frac{a^4}{r^4}\right)\cos 2\theta - \tau_{xy}^\infty\left(1 + 3\frac{a^4}{r^4}\right)\sin 2\theta \quad \text{(A2)}$$

$$\sigma_z = -2\nu\left(\sigma_x^\infty - \sigma_y^\infty\right)\frac{a^2}{r^2}\cos 2\theta - 4\nu\tau_{xy}^\infty\frac{a^2}{r^2}\sin 2\theta + \sigma_z^\infty \quad \text{(A3)}$$

$$\tau_{r\theta} = -\frac{\sigma_x^\infty - \sigma_y^\infty}{2}\left(1 - 3\frac{a^4}{r^4} + 2\frac{a^2}{r^2}\right)\sin 2\theta + \tau_{xy}^\infty\left(1 - 3\frac{a^4}{r^4} + 2\frac{a^2}{r^2}\right)\cos 2\theta \quad \text{(A4)}$$

$$\tau_{\theta z} = \left(-\tau_{xz}^\infty \sin\theta + \tau_{yz}^\infty \cos\theta\right)\left(1 + \frac{a^2}{r^2}\right) \quad \text{(A5)}$$

$$\tau_{zr} = \left(\tau_{xz}^\infty \cos\theta + \tau_{yz}^\infty \sin\theta\right)\left(1 - \frac{a^2}{r^2}\right) \quad \text{(A6)}$$

where σ_r is the radial stress; σ_θ is the tangential stress; σ_z is the normal stress along third direction (z-axis); $\tau_{r\theta}$, $\tau_{\theta z}$, and τ_{zr} are the shear stresses around infinite cylinder hole based on cylindrical coordinates.

The analytical solutions for displacements around the infinite cylinder hole were given by Li [62], as follows:

$$U = \frac{1}{E}\left\{\frac{\sigma_x^\infty + \sigma_y^\infty}{2}(1+\nu) + \left[-(1+\nu)\frac{a^2}{r^2} + 4\left(1 - \nu^2\right)\right]\left(\frac{\sigma_x^\infty - \sigma_y^\infty}{2}\cos 2\theta + \tau_{xy}^\infty \sin 2\theta\right)\right\}\frac{a^2}{r} \quad \text{(A7)}$$

$$V = -\frac{(1+\nu)}{E}\left[\frac{a^2}{r^2} + 2(1-2\nu)\right]\left(\frac{\sigma_x^\infty - \sigma_y^\infty}{2}\sin 2\theta - \tau_{xy}^\infty \cos 2\theta\right)\frac{a^2}{r} \quad \text{(A8)}$$

$$W = \frac{2(1+\nu)}{E}\left(\tau_{xz}^\infty \cos\theta + \tau_{yz}^\infty \sin\theta\right)\frac{a^2}{r} \quad \text{(A9)}$$

where U, V, and W are the components of displacement in the directions of r, θ, and z (third direction), respectively.

Figure A2 shows the corresponding plane strain numerical model built with FLAC^{3D}. In the numerical model, hole radius a = 1 m, E = 10 GPa, ν = 0.25. The applied stress components include σ_x^∞ = 15 MPa, σ_y^∞ = 10 MPa, and τ_{xy}^∞ = 3 MPa. The displacement along the third direction (z-axis) is restricted. In order to ensure stable numerical results, the domain size and mesh size of the numerical model need to be determined based on the sensitivity analyses.

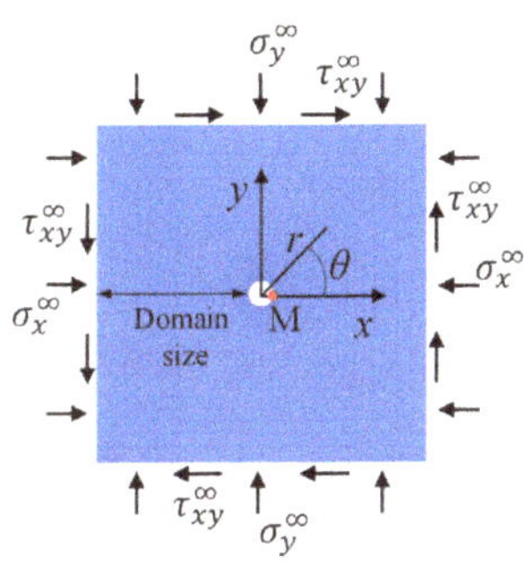

Figure A2. Plane strain numerical model of a cylinder hole in an infinite linearly elastic material built with FLAC^{3D}.

The numerical results of radial displacement and tangential stress are obtained at point M in Figure A2. Figure A3 shows the variations of radial displacement and tangential stress at point M as functions of the domain size. The variation of numerical results reduces as the domain size increases from 1 to 20 m, and become stable when the domain size is larger than 10 m. Therefore, the optimal domain size is determined as 12 m that is 6 times the size of the hole.

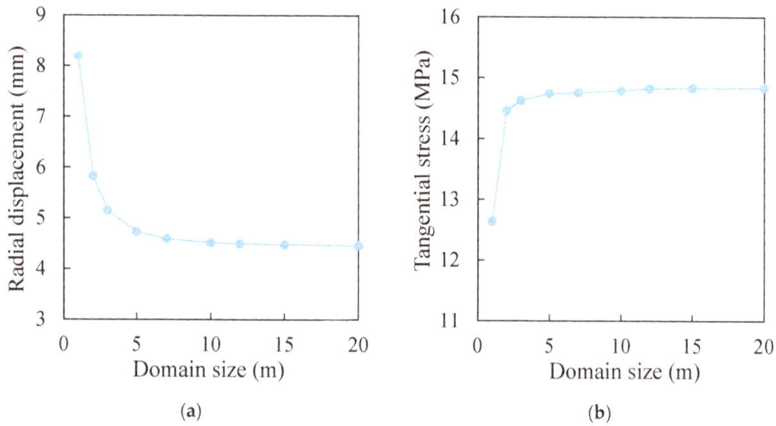

Figure A3. Variations of (**a**) radial displacement and (**b**) tangential stress at point M as functions of domain size.

Figure A4 shows the variations of radial displacement and tangential stress at point M as functions of mesh size. The mesh size ranges from 1 to 0.01 m. The numerical results become stable when the mesh size is smaller than 0.1 m. Further reduction of the mesh size will not greatly change the results. Therefore, the optimal mesh size is determined as 0.05 m to ensure stable results.

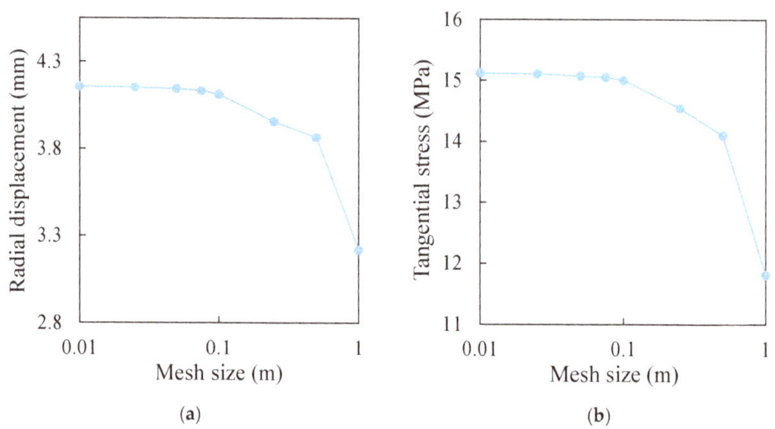

Figure A4. Variations of (**a**) radial displacement and (**b**) tangential stress at point M as functions of mesh size.

Numerical simulations are then conducted by using the optimal domain and mesh sizes. Figure A5 shows the comparisons between the numerical results of σ_r, σ_θ, $\tau_{r\theta}$, U, V long x-axis with the analytical solutions. The good correlations between the numerical and analytical results validate the FLAC^{3D}.

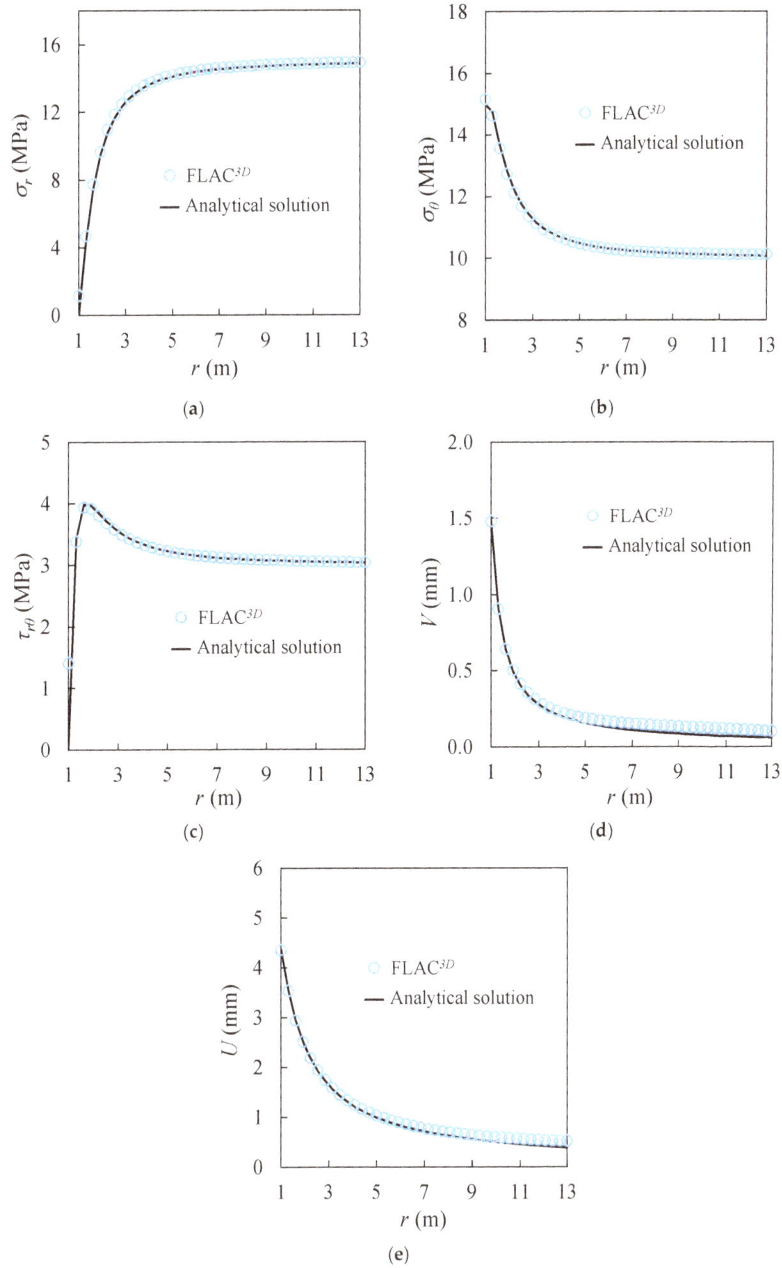

Figure A5. Comparisons between the numerical results and the analytical results of (**a**) σ_r, (**b**) σ_θ, (**c**) $\tau_{r\theta}$, (**d**) V, (**e**) U long x-axis around a cylinder hole in the linearly elastic material.

Appendix B Validation of FLAC^{3D} against Analytical Solutions of Stresses and Displacements around a Cylinder Hole in the Mohr–Coulomb Material

FLAC^{3D} can be further validated against analytical solutions of stresses and displacements around an infinite cylinder hole in the infinite Mohr–Coulomb (MC) material.

Figure A6 shows the plane strain physical model of the problem subject to the isotropic in-situ stress P_0. The origin is on the central point of the hole. Due to axial symmetry, only one-quarter of the model is considered. In Figure A6, a is the radius of the hole. The model is characterized by Young's modulus E, Poisson's ratio ν, cohesion c, friction angle ϕ, and dilation angle ψ.

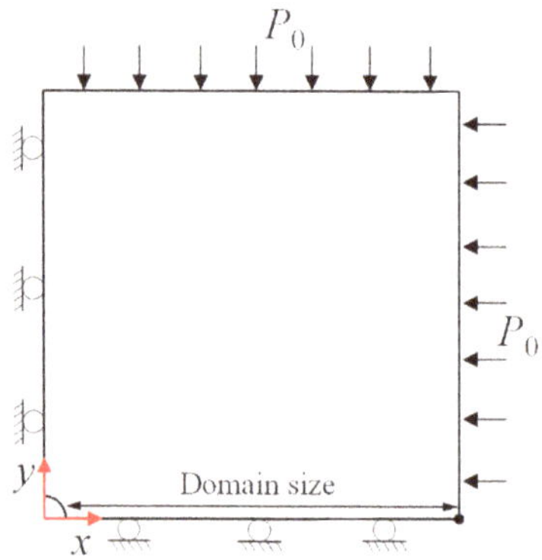

Figure A6. Physical plane strain model of a cylinder hole in an infinite MC material.

According to Salencon's analytical solutions [63], the radius of yield zone R_0 around the cylinder hole can be calculated as:

$$R_0 = a\left(\frac{2}{K_p+1}\frac{P_0+\frac{q_{MC}}{K_p-1}}{P_{in}+\frac{q_{MC}}{K_p-1}}\right)^{1/(K_p-1)} \tag{A10}$$

where P_{in} is the internal pressure; $q_{MC} = 2c\cdot\tan(45^\circ + \phi/2)$; $K_p = (1 + \sin\phi)/(1 - \sin\phi)$.

The radial stress σ_r, tangential stress σ_θ, and radial displacement U in the elastic zone are given as:

$$\sigma_r = P_0 - (P_0 - \sigma_{re})\left(\frac{R_0}{r}\right)^2 \tag{A11}$$

$$\sigma_\theta = P_0 + (P_0 - \sigma_{re})\left(\frac{R_0}{r}\right)^2 \tag{A12}$$

$$U = \frac{{R_0}^2}{2G}\left(P_0 - \frac{2P_0 - q_{MC}}{K_p+1}\right)\frac{1}{r} \tag{A13}$$

where σ_{re} is the radial stress at the elastic-plastic interface and is given as:

$$\sigma_{re} = \frac{1}{K_p+1}(2P_0 - q_{MC}) \tag{A14}$$

The stresses and radial displacement in the plastic zone are given as:

$$\sigma_r = -\frac{q_{MC}}{K_p-1} + \left(P_{iner} + \frac{q_{MC}}{K_p-1}\right)\left(\frac{r}{a}\right)^{(K_p-1)} \tag{A15}$$

$$\sigma_\theta = -\frac{q_{MC}}{K_p-1} + K_p\left(P_{iner} + \frac{q_{MC}}{K_p-1}\right)\left(\frac{r}{a}\right)^{(K_p-1)} \tag{A16}$$

$$U = \frac{r}{2G}\left[(2v-1)\left(P_0+\frac{q_{MC}}{K_p-1}\right) + \frac{(1-v)({K_p}^2-1)}{K_p+K_\psi}\left(P_{in}+\frac{q_{MC}}{K_p-1}\right)\left(\frac{R_0}{a}\right)^{(K_p-1)}\left(\frac{R_0}{a}\right)^{(K_\psi+1)} + \left(\frac{(1-v)(K_pK_\psi+1)}{K_p+K_\psi} - v\right)\left(P_{in}+\frac{q_{MC}}{K_p-1}\right)\left(\frac{r}{a}\right)^{(K_p-1)}\right] \tag{A17}$$

where $K_\psi = (1 + \sin\psi)/(1 - \sin\psi)$.

Figure A7 shows the plane strain numerical model built with FLAC^{3D}. In the numerical model, hole radius a = 1 m, P_0 = 10 MPa. The MC material is characterized by E = 10 GPa, ν = 0.25, c = 1 MPa, ϕ = 35°, ψ = 0°. The displacement along the third direction (z-axis) is restricted. Normal displacement on the left boundary and the bottom are prohibited to simulate symmetric planes. Sensitivity analyses are then performed to determine optimal domain and mesh sizes to ensure stable numerical results.

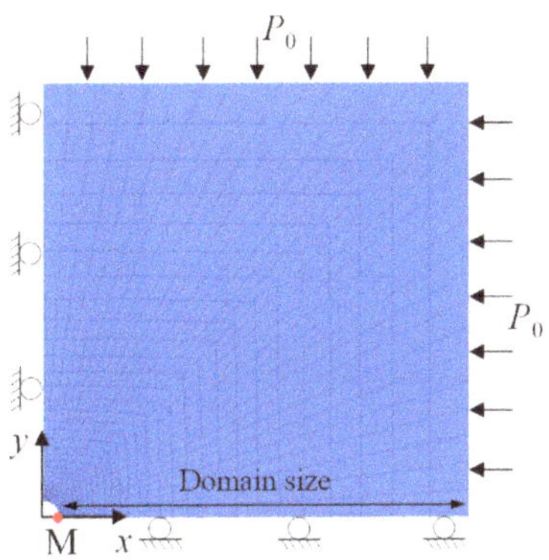

Figure A7. Plane strain numerical model of a cylinder hole in an infinite MC material built with FLAC^{3D}.

The numerical results of radial displacement and tangential stress are obtained at point M in Figure A7. Figure A8 shows the variations of radial displacement and tangential stress at point M as functions of domain size ranging from 1 to 30 m. The numerical results become stable when the domain size is larger than 12 m. To be conservative, the optimal domain size is determined as 20 m.

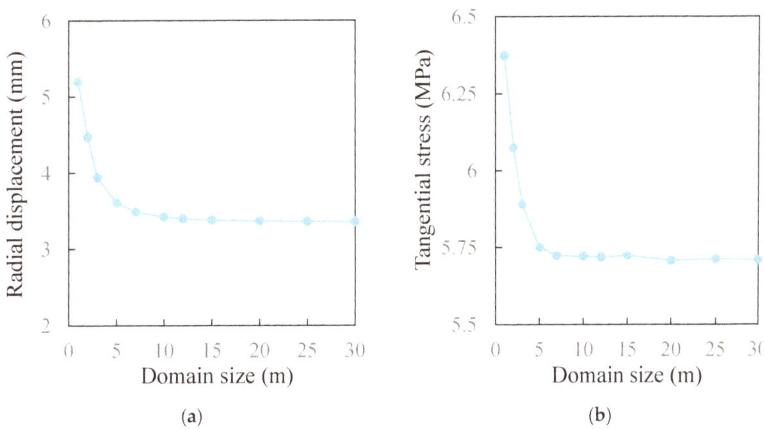

Figure A8. Variations of (**a**) radial displacement and (**b**) tangential stress at point M as functions of domain size.

Figure A9 shows the variations of radial displacement and tangential stress at point M as functions of mesh size. As the mesh size reduces from 1 to 0.005 m, the variation of numerical results decreases. The numerical results become stable when the mesh size reduces to 0.03 mm. Further reduction of the mesh size will not greatly affect the results. Therefore, the optimal mesh size is determined as 0.02 m to ensure stable results.

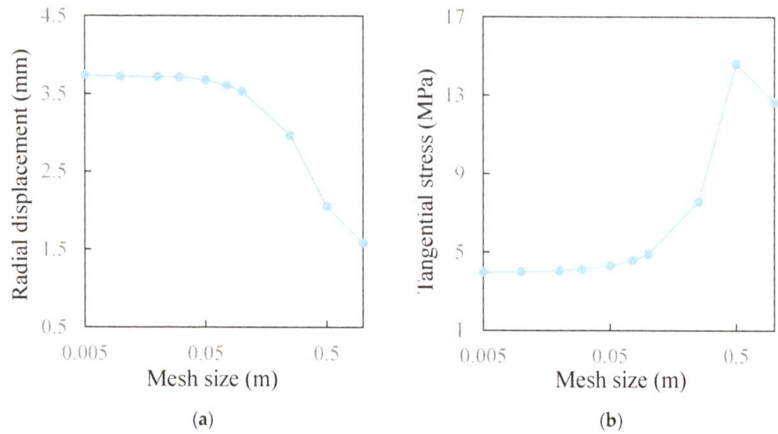

Figure A9. Variations of (**a**) radial displacement and (**b**) tangential stress at point M as functions of mesh size.

The optimal domain and mesh sizes are then used to conduct numerical simulations of the problem. Figure A10 shows the comparisons between the numerical results of stresses and radial displacement long x-axis with the analytical solutions. The good agreements between the numerical and analytical results indicate that the FLAC^{3D} is validated.

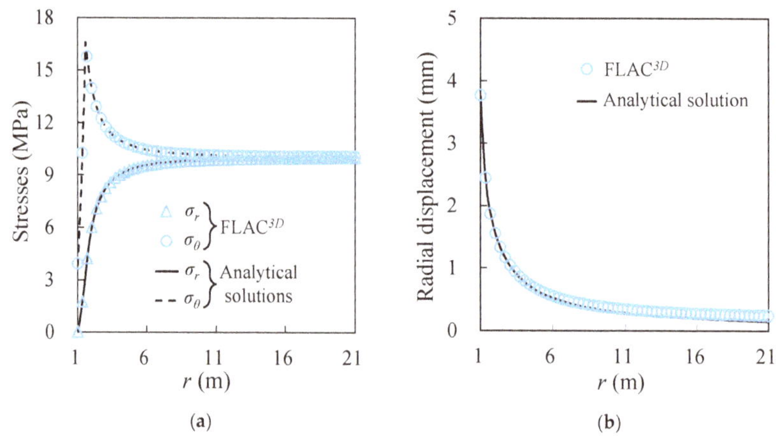

Figure A10. Comparisons between the numerical results and the analytical results of (a) σ_r and σ_θ, (b) U along x-axis around a cylinder hole in the MC material.

Appendix C Sensitivity Analyses of Domain and Mesh Sizes in the Numerical Simulations

Sensitivity analyses are conducted to determine the optimal mesh size for the numerical models of consolidation and triaxial tests in this study. For the numerical model of a backfilled stope overlying a sill mat, both optimal domain and mesh sizes are determined based on the sensitivity analyses.

For the one-dimensional consolidation simulation, the physical and numerical models are illustrated in Figures 4 and 5. The material parameters are provided in Table 1. Figure A11 shows the variation of compressive strain under an applied stress of 500 kPa with different constitutive models as a function of the mesh size. The values of mesh sizes vary between 20 and 2 mm. The numerical results are considered stable when the mesh size is smaller than 5 mm.

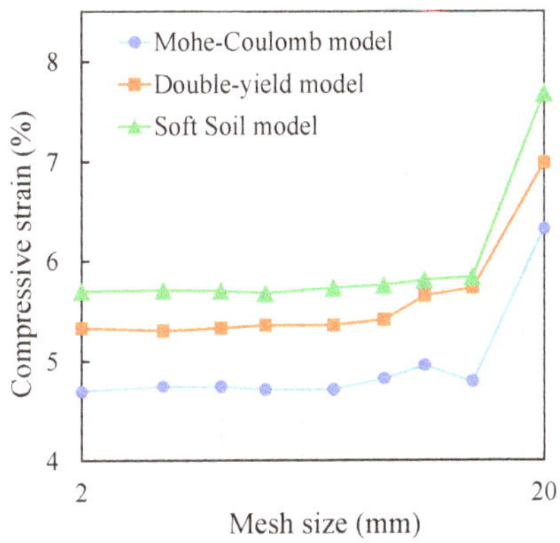

Figure A11. Variation of compressive strain under an applied stress of 500 kPa in one-dimensional consolidation simulations as functions of the mesh size.

For the numerical simulations of consolidated drained triaxial compression tests, the physical and numerical models are shown in Figures 7 and 8, while the material parameters are provided in Table 2. Figure A12 shows the variation of axial stress under an axial strain of 5% and a confining pressure of 200 kPa with different constitutive models as functions of the mesh size. The values of the mesh sizes range from 10 to 1 mm. The numerical results are considered stable when the mesh size reduces to 3 mm.

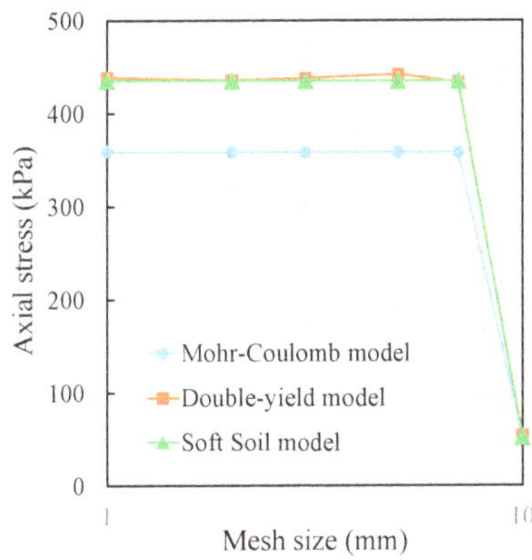

Figure A12. Variation of axial stress under an axial strain of 5% and a confining pressure of 200 kPa in triaxial compression simulations as functions of the mesh size.

For the numerical model of a backfilled stope overlying a sill mat shown in Figure 10b, Mohr–Coulomb model is applied for backfill to conduct the sensitivity analyses at a mine depth of 1000 m. The material parameters are provided in Table 3. In the sensitivity analyses, two indicators are analyzed for different domains and mesh sizes. One is the total displacement of surrounding rock mass at Point A in Figure 10b (one corner of sill mat) after extracting the overlying stope. The other indicator is the horizontal stress after excavating the underlying stope at Point B in Figure 10b, which is at the middle height on the VCL of backfill. Figure A13 shows the variation of total displacement at Point A and horizontal stress at Point B as functions of domain sizes ranging from 35 to 550 m. The numerical results become stable when the domain size is larger than 200 m.

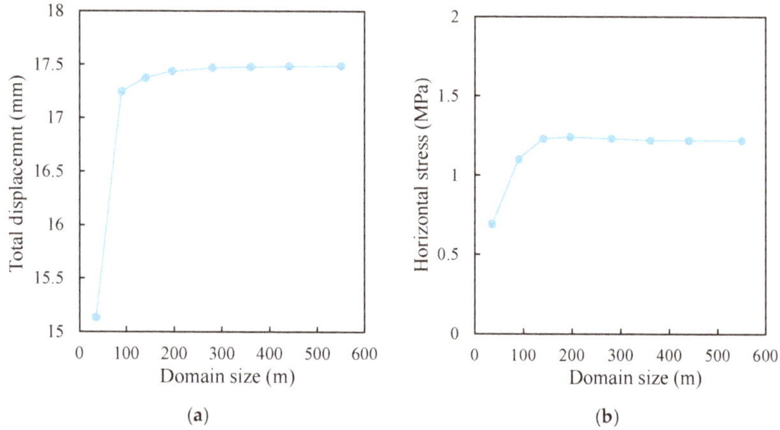

Figure A13. Variation of (**a**) total displacement at Point A after extracting the overlying stope and (**b**) vertical stress at Point B after excavating the underlying stope as functions of domain size.

Figure A14 shows the variation of total displacement at Point A and horizontal stress at Point B as functions of the mesh size, which ranges from 5 to 0.1 m. The numerical results become stable when the mesh size reduces to 0.5 m.

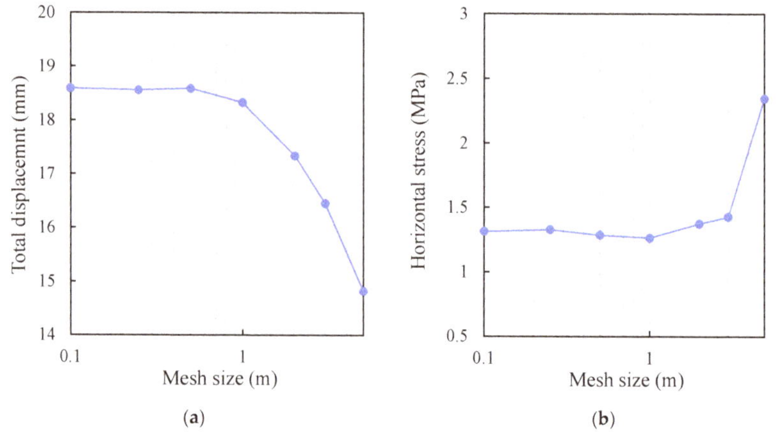

Figure A14. Variation of (**a**) total displacement at Point A after extracting the overlying stope and (**b**) horizontal stress at Point B after excavating the underlying stope as functions of the mesh size.

References

1. Hassani, F.; Archibald, J. *Mine Backfill*; Canadian Institute of Mine, Metallurgy and Petroleum: Montreal, QC, Canada, 1998.
2. Benzaazoua, M.; Bussière, B.; Demers, I.; Aubertin, M.; Fried, É.; Blier, A. Integrated mine tailings management by combining environmental desulphurization and cemented paste backfill: Application to mine Doyon, Quebec, Canada. *Min. Eng.* **2008**, *21*, 330–340. [CrossRef]
3. Stephan, G. Cut-and-fill mining. In *SME Mining Engineering Handbook*, 3rd ed.; Society for Mining, Metallurgy and Exploration: Littleton, CO, USA, 2011; pp. 1365–1373.
4. Kortnik, J. Green mining–use of hydraulic backfill in the Velenje Coal Mine. In Proceedings of the 13th International Symposium on Mining with Backfill, Katowice, Poland, 25–28 May 2021; pp. 422–432.
5. Potvin, Y.; Thomas, E.; Fourie, A. *Handbook on Mine Fill*; Australian Centre for Geomechanics: Perth, Australia, 2005.
6. Newman, C.R.; Agioutantis, Z.G. Stress redistribution around single and multiple stope-and-fill operations. In Proceedings of the 52nd US Rock Mechanics/Geomechanics Symposium, Seattle, WA, USA, 17–20 June 2018.

7. Zhao, T.; Ma, S.; Zhang, Z. Ground control monitoring in backfilled strip mining under the metropolitan district: Case study. *Int. J. Geomech.* **2018**, *18*, 05018003. [CrossRef]
8. Vasichev, S. Ground control with backfill and caving in deep-level mining of gently dipping ore bodies. In Proceedings of the IOP Conference Series: Earth and Environmental Science, Subsurface Management, Exploration and Mining Technologies, Economics, Geoecology, Novosibirsk, Russia, 24–26 April 2019; p. 012022.
9. Wang, R.; Zeng, F.; Li, L. Stability analyses of side-exposed backfill considering mine depth and extraction of adjacent stope. *Int. J. Rock Mech. Min. Sci.* **2021**, *142*, 104735. [CrossRef]
10. Falaknaz, N.; Aubertin, M.; Li, L. Numerical investigation of the geomechanical response of adjacent backfilled stopes. *Can. Geotech. J.* **2015**, *52*, 1507–1525. [CrossRef]
11. Sobhi, M.A.; Li, L. Numerical investigation of the stresses in backfilled stopes overlying a sill mat. *J. Rock Mech. Geotech. Eng.* **2017**, *9*, 490–501. [CrossRef]
12. Raffaldi, M.J.; Seymour, J.B.; Richardson, J.; Zahl, E.; Board, M. Cemented paste backfill geomechanics at a narrow-vein underhand cut-and-fill mine. *Rock Mech. Rock Eng.* **2019**, *52*, 4925–4940. [CrossRef] [PubMed]
13. Qi, C.; Fourie, A. Numerical investigation of the stress distribution in backfilled stopes considering creep behaviour of rock mass. *Rock Mech. Rock Eng.* **2019**, *52*, 3353–3371. [CrossRef]
14. Brinkgreve, R.B.J. Selection of soil models and parameters for geotechnical engineering application. In Proceedings of the Geo-Frontiers Congress, Austin, TX, USA, 24–26 January 2005; pp. 69–98.
15. Helinski, M.; Fahey, M.; Fourie, A. Numerical modeling of cemented mine backfill deposition. *J. Geotech. Geoenviron. Eng.* **2007**, *133*, 1308–1319. [CrossRef]
16. Li, L.; Aubertin, M.; Shirazi, A. Implementation and application of a new elastoplastic model based on a multiaxial criterion to assess the stress state near underground openings. *Int. J. Geomech.* **2010**, *10*, 13–21. [CrossRef]
17. Cui, L.; Fall, M. A coupled thermo–hydro-mechanical–chemical model for underground cemented tailings backfill. *Tunn. Undergr. Space Technol.* **2015**, *50*, 396–414. [CrossRef]
18. Aubertin, M.; Li, L.; Arnoldi, S.; Belem, T.; Bussière, B.; Benzaazoua, M.; Simon, R. Interaction between backfill and rock mass in narrow stopes. In Proceedings of the 12th Panamerican Conference on Soil Mechanics and Geotechnical Engineering and 39th U.S. Rock Mechanics Symposium, Cambridge, MA, USA, 22–26 June 2003; pp. 1157–1164.
19. Li, L.; Aubertin, M.; Simon, R.; Bussière, B.; Belem, T. Modeling arching effects in narrow backfilled stopes with FLAC. In Proceedings of the 3rd International Symposium on FLAC and FLAC3D Numerical Modelling in Geomechanics, Sudbury, ON, Canada, 22–24 October 2003; pp. 211–219.
20. Pirapakaran, K.; Sivakugan, N. Arching within hydraulic fill stopes. *Geotech. Geol. Eng.* **2007**, *25*, 25–35. [CrossRef]
21. Emad, M.Z.; Mitri, H.S.; Henning, J.G. Effect of blast vibrations on the stability of cemented rockfill. *Int. J. Min. Reclam. Environ.* **2012**, *26*, 233–243. [CrossRef]
22. Li, L.; Aubertin, M. An improved method to assess the required strength of cemented backfill in underground stopes with an open face. *Int. J. Min. Sci. Technol.* **2014**, *24*, 549–558. [CrossRef]
23. Liu, G.; Li, L.; Yang, X.; Guo, L. Numerical analysis of stress distribution in backfilled stopes considering interfaces between the backfill and rock walls. *Int. J. Geomech.* **2017**, *17*, 06016014. [CrossRef]
24. Pagé, P.; Li, L.; Yang, P.; Simon, R. Numerical investigation of the stability of a base-exposed sill mat made of cemented backfill. *Int. J. Rock Mech. Min. Sci.* **2019**, *114*, 195–207. [CrossRef]
25. Keita, A.M.T.; Jahanbakhshzadeh, A.; Li, L. Numerical analysis of the stability of arched sill mats made of cemented backfill. *Int. J. Rock Mech Min. Sci.* **2021**, *140*, 104667. [CrossRef]
26. Pierce, M.E. Laboratory and Numerical Analysis of the Strength and Deformation Behaviour of Paste Backfill. Master's Thesis, Queen's University, Kingston, ON, Canada, 1999.
27. Belem, T.; Benzaazoua, M.; Bussière, B. Mechanical behaviour of cemented paste backfill. In Proceedings of the 53rd Canadian Geotechnical Conference, Montreal, QC, Canada, 15–18 October 2000; pp. 373–380.
28. Rankine, R.M. The Geotechnical Characterisation and Stability Analysis of BHP Billiton's Cannington Mine Paste fill. Ph.D. Thesis, James Cook University, Douglas, Australia, 2004.
29. Fall, M.; Belem, T.; Samb, S.; Benzaazoua, M. Experimental characterization of the stress–strain behaviour of cemented paste backfill in compression. *J. Mater. Sci.* **2007**, *42*, 3914–3922. [CrossRef]
30. Jafari, M.; Shahsavari, M.; Grabinsky, M. Drained triaxial compressive shear response of cemented paste backfill (CPB). *Rock Mech. Rock Eng.* **2021**, *54*, 3309–3325. [CrossRef]
31. Oliver, P.H.; Landriault, D. The convergence resistance of mine backfills. In Proceedings of the 4th International Symposium on mining with backfill, Montreal, QC, Canada, 2–5 October 1989; pp. 433–436.
32. Clark, I.H. The cap model for stress path analysis of mine backfill compaction processes. In Proceedings of the 7th International Conference on Computer Methods and Advances in Geomechanics, Cairns, Australia, 6–10 May 1991; pp. 1293–1298.
33. Fourie, A.B.; Gürtunca, R.G.; De Swardt, G.; Wendland, E. An evaluation of four constitutive models for the simulation of backfill behaviour. In Proceedings of the 5th International Symposium on Mining with Backfill, Johannesburg, South Africa, 7–9 June 1993; pp. 33–38.
34. Lagger, M.; Henzinger, M.R.; Schubert, W. Numerical investigations on pea gravel using a nonlinear constitutive model. In Proceedings of the ISRM International Symposium-EUROCK, Cappadocia, Turkey, 29–31 August 2016; pp. 521–526.

35. Labuz, J.F.; Zang, A. Mohr–Coulomb failure criterion. *Rock Mech. Rock Eng.* **2012**, *45*, 975–979. [CrossRef]
36. Mohr, O. Welche Umstände bedingen die Elastizitätsgrenze und den Bruch eines Materials. *Zeit. des Ver. Deut. Ing.* **1900**, *44*, 1524–1530.
37. Coulomb, C.A. Sur une application des règles de mximis et mnimis à qelques poblèmes de satique relatits à l'architecture. *Académie R. Des Sci. Mémoires De Mathématique Et De Phys.* **1773**, *7*, 343–382.
38. Pietruszczak, S. *Fundamentals of Plasticity in Geomechanics*; CRC Press: Boca Raton, FL, USA, 2010.
39. Mitchell, R.J.; Wong, B.C. Behaviour of cemented tailings sands. *Can. Geotech. J.* **1982**, *19*, 289–295. [CrossRef]
40. Itasca. *FLAC3D—Fast Lagrangian Analysis of Continua in 3 Dimensions*; User's Guide; Version 7.0; Itasca Consulting Group: Minneapolis, MN, USA, 2019.
41. Antonov, D. Optimization of the Use of Cement in Backfilling Operations. Ph.D. Thesis, École Polytechnique de Montréal, Montréal, QC, Canada, 2005.
42. Brinkgreve, R.B.J. Geomaterial Models and Numerical Analysis of Softening. Ph.D. Thesis, Delft University of Technology, Delft, The Netherlands, 1996.
43. Roscoe, K.H.; Schofield, A.N.; Wroth, C.P. On the yielding of soils. *Géotechnique* **1958**, *8*, 22–53. [CrossRef]
44. Burland, J.B. Correspondence on the yielding and dilation of clay. *Géotechnique* **1965**, *15*, 211–214. [CrossRef]
45. Neher, H.; Wehnert, M.; Bonnier, P. An evaluation of soft soil models based on trial embankments. In Proceedings of the 10th International Conference on Computer Methods and Advances in Geomechanics, Tucson, AZ, USA, 7–12 January 2001; pp. 373–379.
46. Kahlström, M. Plaxis 2D Comparison of Mohr-Coulomb and Soft Soil Material Models. Master's Thesis, Luleå University of Technology, Luleå, Sweden, 2013.
47. Canadian Geotechnical Society. *Canadian Foundation Engineering Manual*; BiTech Publishers Ltd.: Richmond, BC, Canada, 1978.
48. Bowles, L. *Foundation Analysis and Design*, 5th ed.; McGraw-hill: New York, NY, USA, 1996.
49. Duncan, J.M.; Bursey, A. Soil modulus correlations. In *Foundation Engineering in the Face of Uncertainty*; James, L.W., Kok-Kwang, P., Mohamad, H., Eds.; American Society of Civil Engineers: Reston, VA, USA, 2013; pp. 321–336.
50. Yang, P.; Li, L.; Aubertin, M. Theoretical and numerical analyses of earth pressure coefficient along the centerline of vertical openings with granular fills. *Appl. Sci.* **2018**, *8*, 1721. [CrossRef]
51. Herget, G. *Stresses in Rock*; AA Balkema: Rotterdam, The Netherlands, 1988.
52. Stewart, J.M.; Clark, I.H.; Morris, A.N. Assessment of fill quality as a basis for selecting and developing optimal backfill systems for South African gold mines. In Proceedings of the International Conference on Gold, Johannesburg, South Africa, 15–17 September 1986; pp. 255–270.
53. Liu, M.D.; Carter, J.P. Virgin compression of structured soils. *Géotechnique* **1999**, *49*, 43–57. [CrossRef]
54. Jafari, M.; Shahsavari, M.; Grabinsky, M. Experimental study of the behavior of cemented paste backfill under high isotropic compression. *J. Geotech. Geoenviron. Eng.* **2020**, *146*, 06020019. [CrossRef]
55. Duncan, J.M.; Williams, G.W.; Sehn, A.L.; Seed, R.B. Estimation earth pressures due to compaction. *J. Geotech. Eng.* **1991**, *117*, 1833–1847. [CrossRef]
56. Yang, P. Investigation of the Geomechanical Behavior of Mine Backfill and Its Interaction with Rock Walls and Barricades. Ph.D. Thesis, École Polytechnique de Montréal, Montréal, QC, Canada, 2016.
57. Pariseau, W.G. Coupled three-dimensional finite element modeling of mining in wet ground. In Proceedings of the 3rd Canadian Conference on Computer Applications in the Mineral Industry, Montreal, QC, Canada, 22–25 October 1995; pp. 283–292.
58. Godbout, J.; Bussière, B.; Belem, T. Evolution of cemented paste backfill saturated hydraulic conductivity at early curing time. In Proceedings of the Diamond Jubilee Canadian Geotechnical Conference and the 8th Joint CGS/IAH-CNC Groundwater Conference, Ottawa, ON, Canada, 21–25 October 2007.
59. Fall, M.; Adrien, D.; Célestin, J.; Pokharel, M.; Touré, M. Saturated hydraulic conductivity of cemented paste backfill. *Miner Eng.* **2009**, *22*, 1307–1317. [CrossRef]
60. Hiramatsu, Y.; Oka, Y. Analysis of stress around a circular shaft or drift excavated in ground in a three dimensional stress state. *J. Min. Metall. Inst. Jpn.* **1962**, *78*, 93–98.
61. Hiramatsu, Y.; Oka, Y. Determination of the stress in rock unaffected by boreholes or drifts, from measured strains or deformations. *Int. J. Rock Mech. Min. Sci. Geomech. Abstr.* **1968**, *5*, 337–353. [CrossRef]
62. Li, L. Étude Expérimentale Du Comportement Hydromécanique D'une Fracture. Ph.D. Thesis, Institut de Physique du Globe de Paris, Université Paris 7, Paris, France, 1997.
63. Salencon, J. Contraction Quasi-statique d'une cavité à symétrie sphérique ou cylindrique dans un milieu élastoplastique. *Ann. Ponts Chaussés* **1969**, *4*, 231–236.

processes MDPI

Article

Analysis on Water Inrush Prevention Mechanism of Paste-Filled Floor above Confined Water

Qingliang Chang, Xingjie Yao *, Jianzhuang Qin, Mengda Li, Yizhe Wang, Huaqiang Zhou, Ying Xu * and Yuantian Sun

Key Laboratory of Deep Coal Resource Mining of the Ministry of Education, School of Mines, China University of Mining and Technology, Xuzhou 221116, China; zkdcql@cumt.edu.cn (Q.C.); TS21020037A31@cumt.edu.cn (J.Q.); TS21020124P21@cumt.edu.cn (M.L.); TS21020178P21@cumt.edu.cn (Y.W.); 2242@cumt.edu.cn (H.Z.); yuantiansun@cumt.edu.cn (Y.S.)

* Correspondence: TS21020065A31TM@cumt.edu.cn (X.Y.); 4983@cumt.edu.cn (Y.X.)

Abstract: To study the influence of filling step and advancing distance on the deformation and failure of a working face floor, a mechanical model based on elastic foundation beam theory is established. The deflection and bending moment curves of the floor under different filling steps and advancing distance are obtained by Maple. Then, a fluid–solid coupling model of paste-filling mining on confined water is established by FLAC^{3D}. The effects of different filling steps and advancing distance on the floor displacement, stress, and plastic zone of the floor are analyzed. The results show that there is a "concave" quadratic relationship between the filling step and the maximum displacement of the floor, and there is a "convex" quadratic relationship between the advancing distance and the maximum displacement of the floor. The maximum stress of the floor increases linearly with the increase in filling distance and tends to be stable with the increase in advancing distance. Moreover, the increase in filling steps will lead to the continuous increase in longitudinal failure. This study could guide paste-filling mining above confined water.

Keywords: confined water; paste filling mining; filling step; advancing distance; floor failure

Citation: Chang, Q.; Yao, X.; Qin, J.; Li, M.; Wang, Y.; Zhou, H.; Xu, Y.; Sun, Y. Analysis on Water Inrush Prevention Mechanism of Paste-Filled Floor above Confined Water. *Processes* **2022**, *10*, 274. https://doi.org/10.3390/pr10020274

Academic Editor: Haiping Zhu

Received: 15 October 2021
Accepted: 25 January 2022
Published: 29 January 2022

1. Introduction

With the extension of the mining depth of coal resources in China, water inrush from the coal mine floor is becoming more serious [1]. In this case, the problem of safe mining above confined water needs to be solved urgently [2]. Therefore, relevant researchers have carried out a series of studies on the failure characteristics and laws of the floor on confined water [3,4]. The "Key layer theory" is proposed, and then, the water inrush criterion of stope floor is established [5]. The theory of "original rock tension fracture" and "zero failure" further analyzes the maximum failure depth of the mining floor [6,7]. The theory of "strong seepage channel" shows that whether water inrush occurs in the floor is based on the water inrush channel [8,9]. The "rock water stress relationship" means that the water inrush was the result of the joint action of rock, water, and stress [10,11]. The "dominant surface" theory of water inrush shows that the coal-bearing stratum is the key to control floor stability [12–16]. In another aspect, with the accelerated development of urban underground resources and the massive construction of traffic engineering over the world, more traffic tunnels and underground engineering have been built with the characteristic of high water pressure alongside complex geological structures such as fault fracture zones, water inrush areas, weak rock mass status, etc. Facing these challenges, many scholars have conducted a considerable amount of research on reducing economic losses, injuries, and deaths and have obtained abundant and beneficial research results [17–21].

The parameters affecting the floor stability of paste-filling face mainly include filling strength, filling rate, filling step, etc. Relevant researchers have conducted a series of studies on the floor failure of paste filling on confined water [22–26]. Some researchers have analyzed the influencing factors of floor failure in paste-filling face from the perspective of filling strength and filling rate [27,28]. Many researchers analyzed the relationship between floor failure depth and mining height [29–31]. Previous studies have established the evaluation formula of paste-filling floor and the relationship between safety factor and filling parameters [32,33]. However, the effects of filling step, advancing distance, and floor failure on floor stability have not been studied thoroughly.

Therefore, taking the paste-filling mining project on confined water in the 11,607 areas of the working face of the Daizhuang coal mine as the background, this paper focuses on analyzing the deformation characteristics and failure of the floor under different filling steps and advancing distance. The research results will help to understand the mine floor water prevention. It is also the development and supplement of filling mining theory.

2. Project Overview

The Daizhuang coal mine is located in the north of the Jining coalfield. According to the characteristics of the water inrush properties in different areas of 1160 mining areas, the mine puts forward a regional paste-filling mining scheme to solve the problem. The paste-filling mining method is adopted. Generally, solid wastes such as gangue and construction waste are processed and crushed, mixed with cement and water, stirred into paste slurry without critical flow rate and bleeding, and then transported to the filling site by self-weight or pump pressure. In this project, the water and solid content of fill material are about 20% and 80%, respectively. The particle size distribution curve of fill materials is given in Figure 1.

The average buried depth of the 11,607 areas of the working face is 543 m. No. 16 coal is mainly mined. The average thickness of the coal seam is 2 m, and the average inclination angle is 7°. The aquifers have a great impact on its mining, which includes 13th ash and Ordovician ash. The upper boundary of the 13th ash aquifer is 24 m away from the coal seam floor, and the upper boundary of the Ordovician ash aquifer is 53 m away from the coal seam floor. These two aquifers constitute the direct and indirect water-filled aquifers of No. 16 coal floors. Table 1 shows the lithologic characteristics of the roof and floor and hydrological characteristics of the floor aquifer around No. 16 coal.

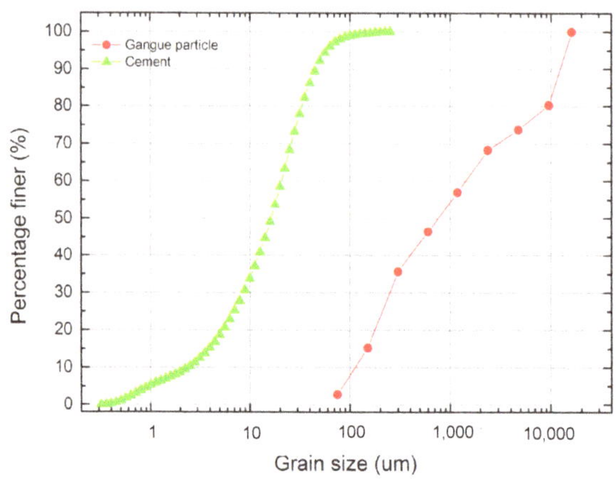

Figure 1. The distribution of the particles of filling materials.

Table 1. Characteristics of rock strata around No. 16 coal seam.

Roof and Floor Name	Rock Properties	Platts Hardness Coefficient	Thickness/m	Lithological Characteristics
Basic roof	Fine-grained sandstone–siltstone	4.4–8.2	10	Gray, thin-medium-thick layered, mainly sandstone, containing mica fragments, small cross bedding, and parallel bedding.
Direct roof	Limestone	14.0–15.9	5.3	Gray, gray–brown, partially muddy, with gentle bedding, well-developed fissures, broken core in the upper part, and more crinoid stems and fossils in the lower part.
Direct floor	Mudstone	3.8–4.0	0.4	Gray, clumpy, jagged fracture, with plant root fossils.
Basic floor	Limestone	6.0–7.8	0.5	Gray–brown, containing fossils of the family sidaidae, with cracks developed, filled with calcite, and argillaceous.
Thirteen ash	Distance from coal seam floor 24 m	0.32–0.48	4.03 m	There is a mudstone interlayer locally, unit water flow q = 0.0001–0.180 L/s·m, Permeability coefficient K = 0.29–0.35 m/d, water quality type is SO_4-Ca(Mg) type, salinity 0.760–2.496 g/L.
Ordovician limestone	Distance from coal seam floor 53 m	0.13	46.1 m	The unit water flow q = 0.0161–3.098 L/s·m, the permeability coefficient K = 0.289–2.476 m/d, the water quality type is $SO_4{\cdot}HCO_3$–Ca(Mg) type, and the salinity is 1.18–2.97 g/L.

3. Theoretical Analysis

In this part, the elastic foundation beam theory is applied to calculate the bending deformation of the floor. Due to the length of the floor along the strike direction of the filling working face being much greater than the span in the inclined direction, it is approximately considered that the deformation of the floor is close and continuous. The floor and waterproof layer can be simplified as a mechanical analysis model of the elastic foundation beam. According to the actual situation, the corresponding pressure is applied to the floor, and the stress state of the floor under different filling steps and advance distance is simulated. The deflection and bending moment curve equation of the floor on the filling surface is established, and then, the relationship of the filling step, advance distance, the deformation, and failure of the floor is analyzed.

According to the coordinate system established in Figure 2, the deflection equation of the floor is analyzed step by step, in which q_1 is the superimposed pressure of confined water, q_2 is the deadweight pressure of a waterproof layer, q_3 is the self-weight pressure of the filling body, q_4 is the deadweight pressure of the coal body and, $q_6, q_7, q_8, q_9, q_{10}, q_{11}$, and q_{12} represent the overlying pressure of the corresponding section floor, respectively.

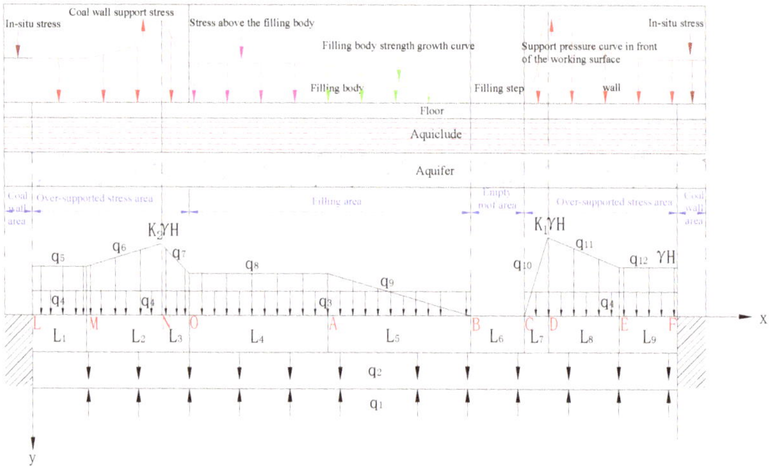

Figure 2. Analysis model of overburden state and floor stress of paste-filling face.

(1) For LM segment analysis:

$$\mathrm{EI}\frac{d^4 w_1(x)}{dx^4} + K_1 w_1(x) = q_{LM}(x) \tag{1}$$

where q_{LM} is the superposition of the overlying pressure of the floor of the LM section, the self-weight pressure of the coal body, the self-weight pressure of the aquifer, and the pressure-bearing water pressure. That is:

$$q_{LM} = q_{lm1} + q_{lm2} + q_{lm4} + q_5 \tag{2}$$

where K_1—foundation coefficient of floor, $\mathrm{N{\cdot}m^{-3}}$, and q_{lm}—pressure on LM section, $\mathrm{N{\cdot}m^{-2}}$.

The characteristic coefficient α is taken $\sqrt[4]{\frac{K_1}{4EI}}$. The deflection of the floor of the LM section is as follows:

$$w_1(x) = e^{-\alpha x}(A_1 \cos(\alpha x) + B_1 \sin(\alpha x)) + e^{\alpha x}(C_1 \cos(\alpha x) + D_1 \sin(\alpha x)) + \frac{q_{LM}(\mathrm{x})}{K_1} \tag{3}$$

When x tends to infinity, the deformation of the floor is a certain value. Due to the long floor at the left end, the floor is used as a semi-infinite beam. At this point, C1 = 0, D1 = 0, and the upper formula becomes:

$$w_1(x) = e^{\alpha x}(A_1 \cos(\alpha x) + B_1 \sin(\alpha x)) + \frac{q_{LM}(\mathrm{x})}{K_1} \tag{4}$$

(2) For MN segment analysis:

$$\mathrm{EI}\frac{d^4 w_1(x)}{dx^4} + K_1 w_2(x) = q_{MN}(x) + q_6(x) \tag{5}$$

where q_{MN} is the superposition of the dead weight pressure of the coal body, the deadweight pressure of the waterproof layer, and the confined water pressure in the MN section. That is:

$$q_{MN} = q_{mn1} + q_{mn2} + q_{mn4} \tag{6}$$

$$q_6(x) = a_1 x + b_1 \tag{7}$$

$$w_2(x) = e^{-\alpha x}(A_2\cos(\alpha x) + B_2\sin(\alpha x)) + e^{\alpha x}(C_2\cos(\alpha x) + D_2\sin(\alpha x)) + \frac{q_{MN}(\mathrm{x})}{K_1} + \frac{q_6(\mathrm{x})}{K_1}. \tag{8}$$

(3) For NO section analysis, establish the floor deflection curve equation:

$$\mathrm{EI}\frac{d^4w_1(x)}{dx^4} + K_1w_3(x) = q_{NO}(x) + q_7(x) \tag{9}$$

where q_{NO} is the superposition of the self-weight pressure of the filling body of the NO section, the self-weight pressure of the water barrier, and the pressure-bearing water pressure, which is:

$$q_{NO} = q_{no1} + q_{no2} + q_{no4} + q_7 \tag{10}$$

$$q_7(x) = a_2x + b_2 \tag{11}$$

$$w_3(x) = e^{-\alpha x}(A_3\cos(\alpha x) + B_3\sin(\alpha x)) + e^{\alpha x}(C_3\cos(\alpha x) + D_3\sin(\alpha x)) + \frac{q_{NO}(\mathrm{x})}{K_1} + \frac{q_7(\mathrm{x})}{K_1} \tag{12}$$

(4) For OA section analysis, establish the floor deflection curve equation:

$$\mathrm{EI}\frac{d^4w_1(x)}{dx^4} + K_1w_4(x) = q_{OA}(x) \tag{13}$$

Among them, it is the superposition of the overburden pressure of the OA section, the self-weight pressure of the filling body, the self-weight pressure of the aquifer, and the pressure-bearing water pressure, which is:

$$q_{OA} = q_{oa1} + q_{oa2} + q_{oa3} + q_8 \tag{14}$$

$$w_4(x) = e^{-\alpha x}(A_4\cos(\alpha x) + B_4\sin(\alpha x)) + e^{\alpha x}(C_4\cos(\alpha x) + D_4\sin(\alpha x)) + \frac{q_{OA}(\mathrm{x})}{K_1} \tag{15}$$

(5) For the AB section analysis, superimpose according to the superposition principle and establish the floor deflection curve equation:

$$\mathrm{EI}\frac{d^4w_2(x)}{dx^4} + K_1w_5(x) = q_{AB}(x) + q_9(x) \tag{16}$$

where q_{AB} is the superposition of the self-weight pressure of the filling body of the AB section, the self-weight pressure of the water barrier, and the pressure-bearing water pressure, which is:

$$q_{AB} = q_{ab1} + q_{ab2} + q_{ab3} \tag{17}$$

$$q_{9(x)} = a_3x + b_3 \tag{18}$$

$$w_5(x) = e^{-\alpha x}(A_5\cos(\alpha x) + B_5\sin(\alpha x)) + e^{\alpha x}(C_5\cos(\alpha x) + D_5\sin(\alpha x)) + \frac{q_{AB}(\mathrm{x})}{K_1} + \frac{q_9(\mathrm{x})}{K_1}. \tag{19}$$

(6) Analyze the BC section and establish the floor deflection curve equation:

$$\mathrm{EI}\frac{d^4w_3(x)}{dx^4} + K_1w_6(x) = q_{BC}(x) \tag{20}$$

where q_{BC} is the superposition of the self-weight pressure of the water barrier of the BC section and the pressure-bearing water pressure, which is:

$$q_{BC} = q_{bc1} + q_{bc2} \tag{21}$$

$$w_6(x) = e^{-\alpha x}(A_6\cos(\alpha x) + B_6\sin(\alpha x)) + e^{\alpha x}(C_6\cos(\alpha x) + D_6\sin(\alpha x)) + \frac{q_{BC}(x)}{K_1} \tag{22}$$

(7) Similar to the MN section, the CD section is analyzed, and the floor deflection can be obtained as:

$$w_7(x) = e^{-\alpha x}(A_7\cos(\alpha x) + B_7\sin(\alpha x)) + e^{\alpha x}(C_7\cos(\alpha x) + D_7\sin(\alpha x)) + \frac{q_{CD}(x)}{K_1} + \frac{q_{10}(x)}{K_1} \tag{23}$$

where q_{CD} is the superposition of the self-weight pressure of the coal body in the CD section, the self-weight load of the aquifer, and the pressure of pressurized water, which is:

$$q_{CD} = q_{cd1} + q_{cd2} + q_{cd4} \tag{24}$$

$$q_{10}(x) = a_4 x + b_4 \tag{25}$$

(8) Similar to the NO section, the DE section is analyzed, and the deflection of the floor can be obtained as:

$$w_8(x) = e^{-\alpha x}(A_8\cos(\alpha x) + B_8\sin(\alpha x)) + e^{\alpha x}(C_8\cos(\alpha x) + D_8\sin(\alpha x)) + \frac{q_{DE}(x)}{K_1} + \frac{q_{11}(x)}{K_1} \tag{26}$$

where q_{DE} is the superposition of the coal self-weight pressure of the DE section, the self-weight pressure of the aquifer, and the pressure-bearing water pressure, which is:

$$q_{DE} = q_{de1} + q_{de2} + q_{de4} \tag{27}$$

$$q_{11}(x) = a_5 x + b_5 \tag{28}$$

(9) Similar to the LM section, the EF section is analyzed, and the deflection of the floor can be obtained as:

$$w_9(x) = e^{-\alpha x}(C_9\cos(\alpha x) + D_9\sin(\alpha x)) + \frac{q_{EF}(x)}{K_1} \tag{29}$$

Since the rotation angle $\theta(x)$, bending moment $M(x)$, shear force $Q(x)$, and deflection $w(x)$ of the arbitrary cross-section of the floor are related to:

$$\begin{cases} \theta(x) = \frac{dw(x)}{dx} \\ M(x) = -EI\frac{dw^2(x)}{dx^2} \\ Q(x) = -EI\frac{dw^3(x)}{dx^3} \end{cases}. \tag{30}$$

From the continuity conditions between the various sections of the floor, one can obtain:

$$\begin{cases} w_1(A) = w_2(A) \\ \theta_1(A) = \theta_2(A) \\ M_1(A) = M_2(A) \\ Q_1(A) = Q_2(A) \\ w_2(B) = w_3(B) \\ \theta_2(B) = \theta_3(B) \\ M_2(B) = M_3(B) \\ Q_2(B) = Q_3(B) \end{cases} \begin{cases} w_3(C) = w_4(C) \\ \theta_3(C) = \theta_4(C) \\ M_3(C) = M_4(C) \\ Q_3(C) = Q_4(C) \\ w_4(D) = w_5(D) \\ \theta_4(D) = \theta_5(D) \\ M_4(D) = M_5(D) \\ Q_4(D) = Q_5(D) \end{cases} \begin{cases} w_5(E) = w_6(E) \\ \theta_5(E) = \theta_6(E) \\ M_5(E) = M_6(E) \\ Q_5(E) = Q_6(E) \\ w_6(E) = w_7(E) \\ w_6(E) = w_7(E) \\ M_6(E) = M_7(E) \\ Q_6(E) = Q_7(E) \end{cases} \begin{cases} w_7(E) = w_8(E) \\ \theta_7(E) = \theta_8(E) \\ M_7(E) = M_8(E) \\ Q_7(E) = Q_8(E) \\ w_8(E) = w_9(E) \\ w_8(E) = w_9(E) \\ M_8(E) = M_9(E) \\ Q_8(E) = Q_9(E) \end{cases} \tag{31}$$

Maple is currently one of the most general mathematics and engineering calculation software, and it is widely used in the fields of science and engineering. Using Maple and bringing the relevant working face parameters into Equations (1)–(31), the corresponding parameters can be obtained: A_1, B_1, A_2, B_2, C_2, D_2, A_3, B_3, C_3, D_3, A_4, B_4, C_4, D_4, A_5, B_5, C_5, D_5, A_6, B_6, C_6, D_6, A_7, B_7, C_7, D_7, A_8, B_8, C_8, D_8, C_9, and D_9.

According to the actual situation of the mine, taking the filling step of 4 m and the working face advancing to 48 m as an example, the stress concentration factor at the coal wall of the open cut (K_2) is taken as 1.8, and the stress concentration factor in front of the working face (K_1) is taken as 2.1 based on empirical experiences. The foundation coefficient of the coal seam floor waterproof layer is set as 0.7×10^8 N/m^3. The elastic modulus of the coal seam floor is based on a field test. The relevant parameters are as follows: E 3.5 GPa, $I = bh^3/12$, $b = 1$, $h = 4$, $L_1 = 20$ m, $L_2 = 32$ m, $L_3 = 8$ m, $L_4 = 34$ m, $L_5 = 10$ m, $L_6 = 4$ m, $L_7 = 8$ m, $L_8 = 32$ m, $L_9 = 20$ m, $q_1 = -6.7$ MPa, $q_2 = 1.25$ MPa, $q_3 = 0.02$ MPa, $q_4 = 0.03$ MPa, $q_5 = 14$ MPa, $q_6 = 0.125x + 19$ MPa, $q_7 = -0.75x + 12$ MPa, $q_8 = 12$ MPa, $q_9 = -1.2x + 52.8$ MPa, $q_{10} = 0.2625x - 126$ MPa, $q_{11} = -7/32x + 1064/32$ MPa. Substituting Formulas (1)–(11), the floor deflection curve and bending moment curve under this condition can be obtained.

3.1. Analysis of Deformation and Failure of the Filling Floor under Different Filling Steps

To compare the effect of the filling step on the damage of the floor, the filling steps of the working face are set as 1 m, 2 m, 3 m, and 4 m. The pressure gradually increases, and the supporting pressure coefficients are set as 1.8, 1.9, 2.0, and 2.1. When the working face is advanced to 48 m, the deflection curve of the floor is shown in Figure 3a, and the bending moment curve of the floor is shown in Figure 3b.

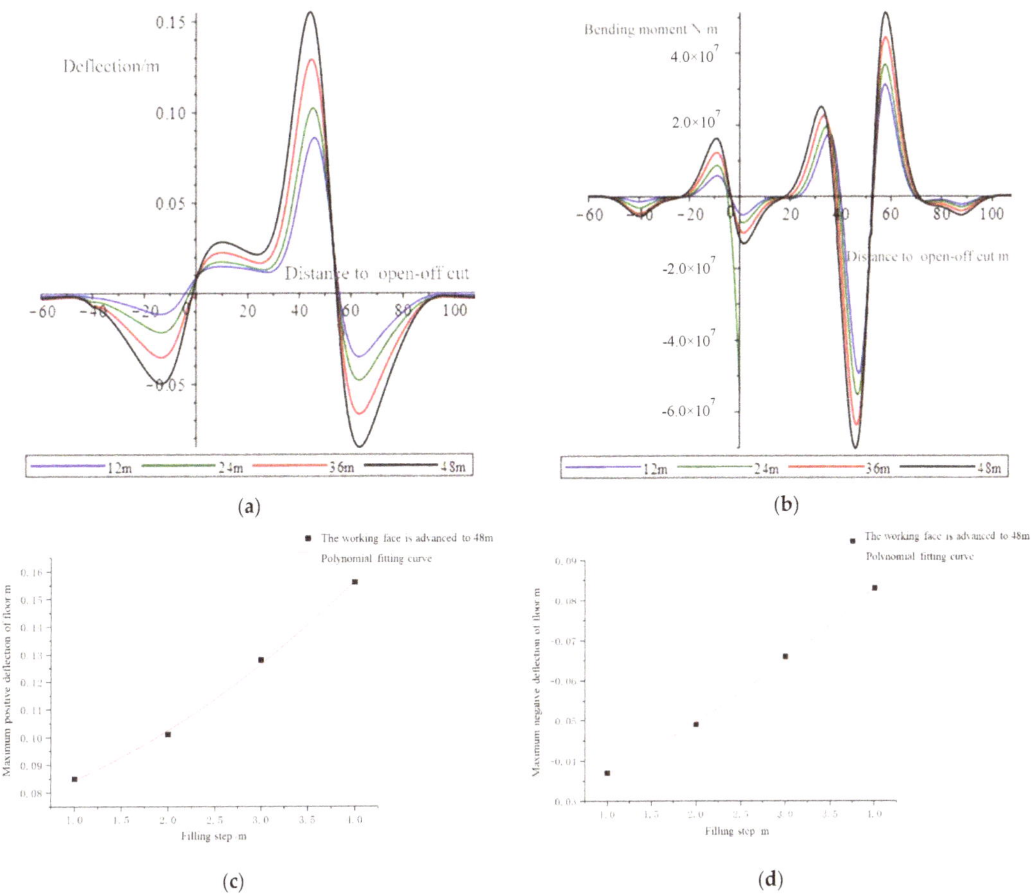

Figure 3. (**a**) Floor deflection under different filling steps; (**b**) Floor bending moment under different filling steps; (**c**) The relationship between the filling step and the maximum positive deflection of the floor; (**d**) The relationship between the filling step and the maximum negative deflection of the floor.

In Figure 3a, it can be seen that the impact of the change of the filling step on the floor is mainly reflected in the opening of the cut and the downward displacement of the floor at the coal wall. This can be explained by the bending moment diagram in Figure 3b. The bending moments at the open cut and the working face are positive, indicating that the two positions are subjected to compressive stress and therefore will produce downward displacement. In the empty roof area, the bending moment of the floor is negative, indicating that the floor in the empty top area is under tensile stress. With the increase in the filling step, the corresponding bending moments at each position of the floor continue to increase.

By fitting the maximum displacement of the floor, it is obtained that the maximum positive displacement of the floor and the filling step show a "concave" quadratic function relationship shown in Figure 3c. The function relationship is $y = 0.009x^2 + 0.003x + 0.0725$. The maximum negative displacement of the floor and filling step presents a "concave" quadratic function relationship shown in Figure 3d. The function relationship is $y = -0.00925x^2 - 0.00125x - 0.02625$. From the aforementioned two formulas, it can be concluded that with the increase in the filling step, the displacement and deformation of the floor will continue to increase. Therefore, the size of the filling step is closely related to the damage degree of the floor.

3.2. Analysis of Deformation and Failure of the Filling Floor under Different Advancing Distances

To compare the influence of advancing distance on the damage of the floor, the advancing distances of the working face are taken as 12 m, 24 m, 36 m, and 48 m. When the filling step of the working face is 4 m, the deflection curve of the floor is shown in Figure 4a. The bending moment curve is shown in Figure 4b.

The floor flection during the advancement of the filling surface is shown in Figure 4a, which can be regarded as the dynamic change process. Under the action of high stress on the coal walls, floor heave occurs in the goaf. With the advancement of the working face, the deformation and failure of the floor show the following tendency. First of all, the maximum heave of the goaf floor remains unchanged. The main reason is that the system of "filling body–empty roof–coal wall" is formed during the advancement of the working face. Secondly, as the working face advances, the floor heave of the filled area gradually decreases. The main reason is that the filling body plays a role in transmitting the pressure of the overburden. Under the action of the overburdened pressure, the deformation of the floor is suppressed and gradually compressed. Hence, the floor volume is reduced. Figure 4b can be seen as the dynamic change process of the floor bending moment during the advancing process of the filling surface. The side bending moment of the coal wall is positive, indicating that the coal body is compressed, and as the working face advances, the bending moment in the filled area is also positive. Therefore, the coal body in this area is also under pressure, which is the main reason for the gradual decrease in the floor heave in the filled area. As the advancing distance of the working face increases, the bending moment in front of the coal wall of the working face gradually increases and then stabilizes.

It is obtained that the maximum negative displacement of the floor and the advancing distance of the filling surface show a "convex" quadratic function relationship shown in Figure 4c. The function relationship is $y = -2.9514 \cdot 10^{-5}x^2 - 0.00276x - 0.02125$. There is a downward displacement of the floor before the work. The amount gradually increases and then stabilizes, indicating that the increase in the advancing distance will increase the destruction depth of the floor, but when it is advanced to a certain distance, the damage depth of the floor will no longer continue to increase and maintain a stable value.

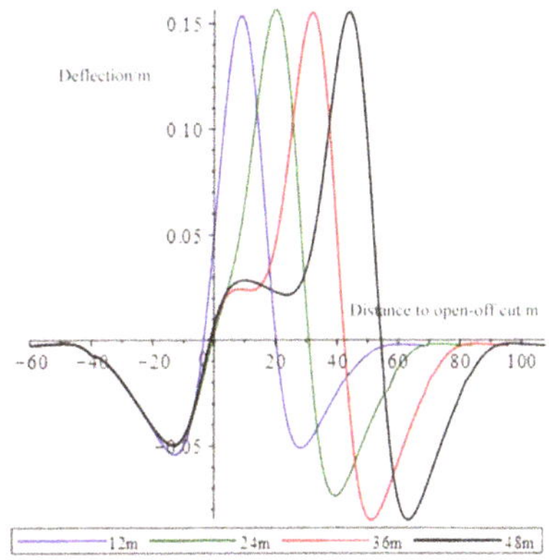

(a)

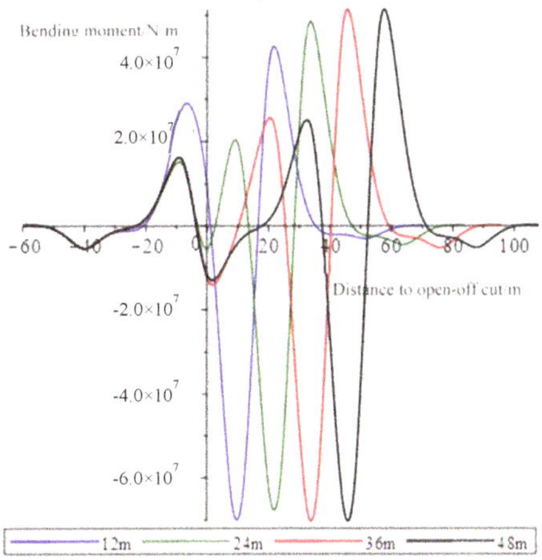

(b)

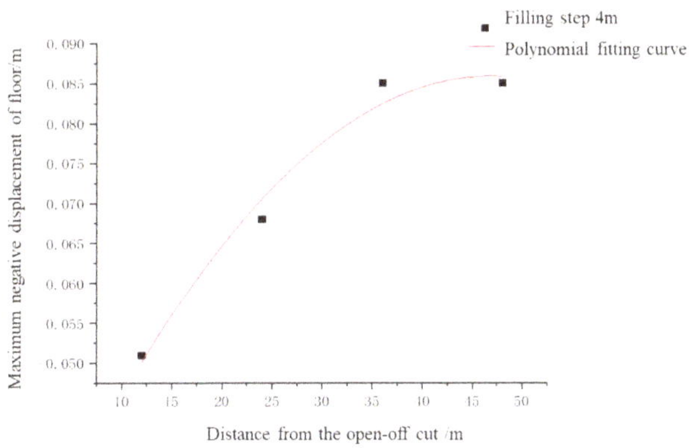

(c)

Figure 4. (**a**) Floor deflection at different advancing distances of paste-filling working face; (**b**) Bending moment of the floor under different advancing distances of paste-filling working face; (**c**) The relationship between the different advancing distances of the paste-filling working face and the maximum negative deflection of the floor.

4. Numerical Simulation

To verify the accuracy of the theoretical analysis in this paper, FLAC^{3D}5.0 is used to construct a numerical simulation for calculation. FLAC^{3D} 5.0 is a three-dimensional numerical simulation software developed based on the finite difference method, which has been widely used in the field of rock and soil mechanics. In this paper, combined with the actual situation of the 11,607 working face, a simplified model is established, as shown in Figure 5. Based on the field measurement, laboratory test, and GSI method, the rock mass properties are given in Table 2 and a numerical model is shown in Figure 6. Horizontal displacement constraints are imposed on the left and right boundaries of the model, and the bottom boundary of the model is fixed in the vertical direction. The top rock layer of the model is about 500 m away from the ground. Therefore, a vertical downward 12.5 MPa uniform pressure is applied to the top to simulate the overburden pressure. To consider the impact of confined water, the software built-in fluid–solid coupling model is used to give a confined water pressure of 3.2 MPa in the 13th ash aquifer and 3.5 MPa in the Ordovician aquifer according to the test results. To eliminate the influence of the boundary of the model, 76 m boundary coal pillars are left on both sides of the working face along the advancing direction, and the size of the model is 200 m × 100 m × 115 m. The rectangular grid elements are introduced in analysis models, and the accuracy and efficiency of such elements have been verified by many studies. The time step is reached when the maximal unbalanced force is lower than 1×10^{-5}. We set key rock mass layers (e.g., coal layer) with the mesh of 1 m, and far away that, we used the mesh of 2 m and 4 m. The rock mass parameters are set in Table 2. The Mohr–Coulomb strength criterion is used for rock mass. Although the Mohr–Coulomb criterion is not very appropriate for rock mass, it is a widely used model in which the parameters are easy to obtain and use considering the previous studies. Thus, in this study, we used the MC criterion to investigate the basic laws.

The setting of the strength parameters of the filling body is shown in Table 2. The filling rate is taken as 98%, and the mined-out area is filled continuously as the working face advances. The floor stress, displacement, and plastic zone under filling steps of 1, 2, 3, and 4 m and advancing distances of 12, 24, 36, and 48 m are simulated correspondingly.

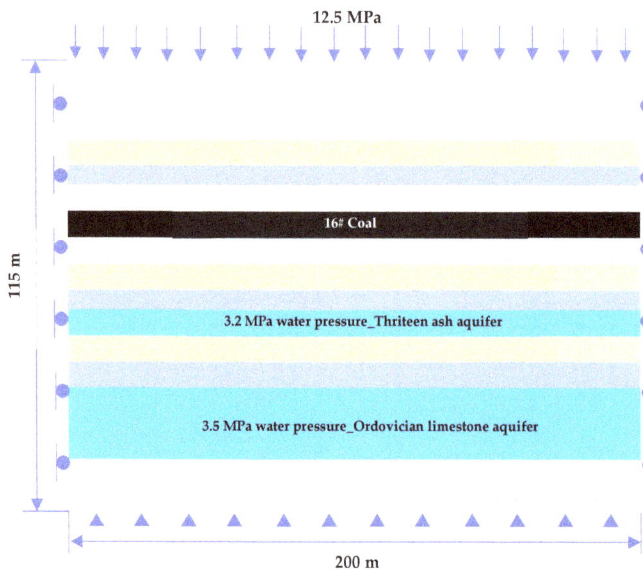

Figure 5. Geometric model diagram.

Table 2. Mechanical parameters of rock mass of the model.

Lithology	Density kg/m^3	Tensile Strength/MPa	Internal Friction Angle (°)	Cohesion/MPa	Bulk Modulus/GPa	Shear Modulus/GPa	GSI
Fine-grained sandstone	2740	0.64	38	7.41	1.2	0.87	82
Mudstone	2600	0.20	35	4.53	0.35	0.21	74
Siltstone	2740	0.79	38	7.41	1.31	0.94	82
Coal seam	1600	0.09	37	2.34	0.07	0.03	62
Thirteen ash	2780	0.57	40	8.69	1.68	1.31	89
Austrian gray	2780	0.57	40	8.69	1.68	1.31	89
Filling body	1990	0.10	24.3	0.79	0.39	0.23	/

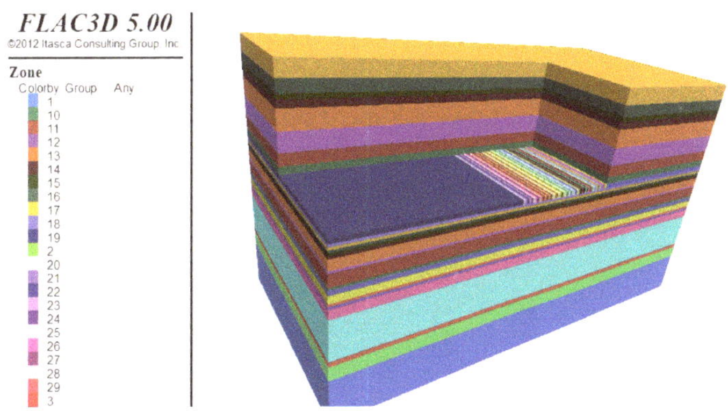

Figure 6. Numerical simulation model.

4.1. Floor Stress Influenced by Filling Step and Advancing Distance

The filling steps of the paste-filling working face are 1, 2, 3, and 4 m, and the advancing distances are 12, 24, 36, and 48 m. Figure 7 shows the vertical stress of the floor under different filling steps and advancing distances.

The stress of the floor of the paste-filling working face is shown in Figure 7. It can be seen that the stress on the floor above the work surface presents a general law that first increases suddenly and then gradually decreases to the original rock stress. The stress concentration of the floor in front of the coal body is often caused by the excavation of the coal mass, which is also the location where the floor failure begins to occur. The peak stress of the floor shows a regular change with the change of the filling step and the advancing distance. Figures 8 and 9 are obtained by fitting the maximum stress of the floor under different advancing distances and different filling steps.

As shown in Figure 8, when the filling step is fixed, the working surface will advance from 12 to 48 m as the paste is filled. The maximum stress of the floor keeps increasing, and the increasing amplitude keeps decreasing, showing a quadratic function relationship. When it is advanced to 36 m, it tends to a constant value. It shows that at 36 m, the "filling body–empty roof area–coal wall" forms a stable support system. As the advancing distance continues to increase, the maximum stress of the floor will not increase, and the floor will be damaged.

When the advancing distance is constant, the maximum stress of the floor changes significantly with the change of the filling step. As shown in Figure 9, the filling step and the maximum stress of the floor show a linear relationship. As the step length increases, the maximum stress of the floor will continue to increase, and the increase will gradually increase, indicating that the change of the filling step length has a greater impact on the damage degree of the floor. The main reason is that the increase in the filling step will increase the range of the empty roof area of each round of filling. For the "filling body–empty roof area–coal wall" support system, it is necessary to achieve stability. The concentrated stress on the front will inevitably increase and be transmitted to the floor, resulting in deformation and damage of the floor. Therefore, the size of the filling step will be more important for the destruction of the floor during the continuous advancement of the working face.

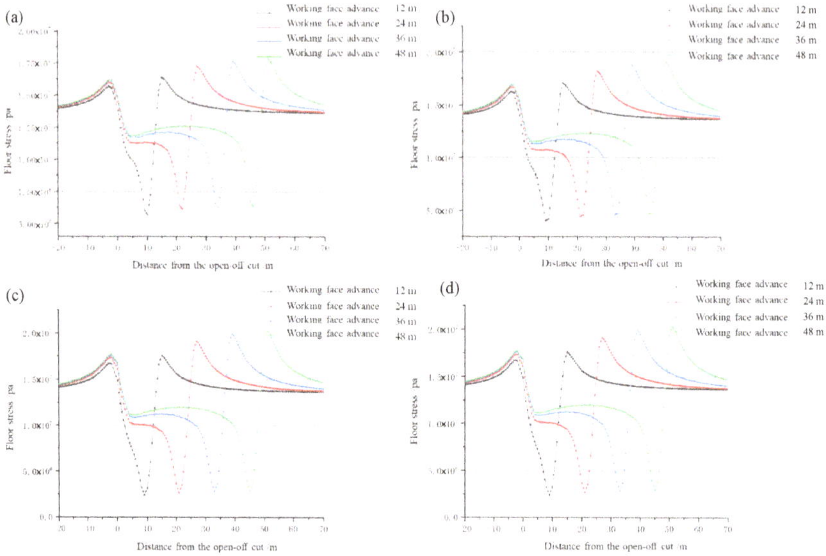

Figure 7. Abutment pressure in front of coal wall of the paste-filling face under different filling steps. (**a**) Filling step distance 1 m, (**b**) Filling step distance 2 m, (**c**) Filling step distance 3 m, (**d**) Filling step distance 4 m.

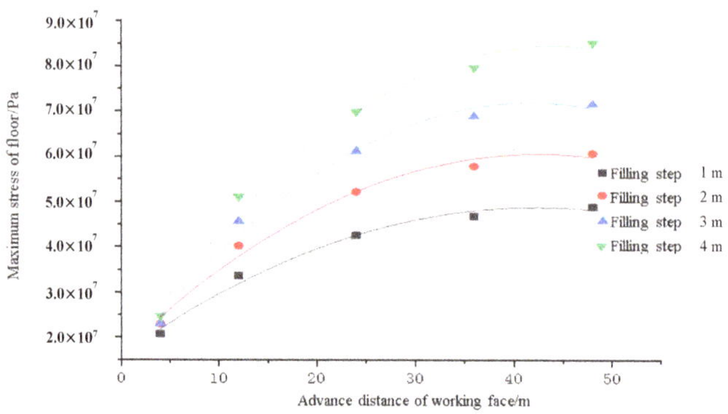

Figure 8. The relationship between the maximum stress of paste-filling face floor and the advancing distance of working face under different filling steps.

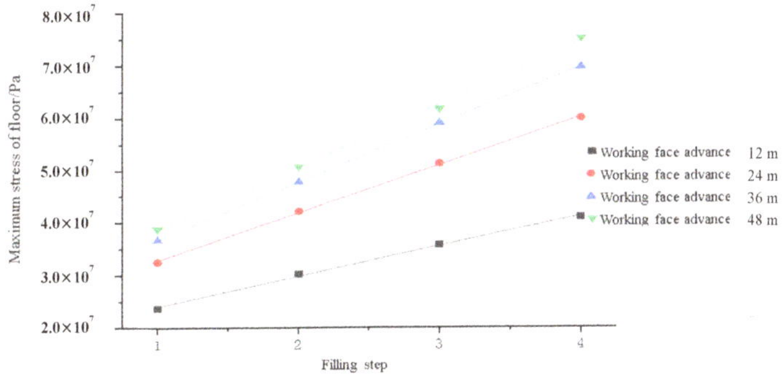

Figure 9. The relationship between the maximum stress of the paste-filling working face floor and the filling step under different advancing distances.

4.2. Floor Displacement Influenced by Filling Step and Advancing Distance

The filling steps of the paste filling working face are 1, 2, 3, and 4 m, and the advancing distances are 12, 24, 36, and 48 m. Under different filling steps, the displacement diagram is shown in Figure 10. Under different advancing distances of the working face, the relationship between the filling step and the maximum displacement of the floor is shown in Figure 11.

It can be seen from Figures 10 and 11 that the maximum positive displacement of the floor of the working face filled with paste often occurs in front of the coal wall of the working face. However, the maximum negative displacement occurs behind the coal wall of the working face, which is also the position where the floor is most likely to be damaged. The main reason is that the "filling body–empty roof area (filling step)–coal wall" support system has been formed. Under the action of concentrated stress, the floor is often prone to damage at this position. In the empty roof area, due to the release of the floor stress, the floor heave is likely to occur.

By fitting numerical simulation results in Figures 12–14 (filling step, maximum displacement of the floor, and advancing distances), it is found that the fitting curves are close to the previous theoretical calculation. It is similar to the maximum deflection curve of the floor. When verifying the previous theoretical calculations, it also shows that the influence of the filling step on the deformation and failure of the floor gradually increases with the increase in the filling step. The influence of the advancing distance on the deformation and failure of the floor increases gradually.

Therefore, there is a "concave" quadratic function relationship between the maximum displacement of the floor and the filling step, and a "convex" quadratic function relationship between the maximum displacement and the advancing distance. The deformation and failure of the floor in the initial stage of the working face will be affected by the filling step and the advancing distance at the same time. When it is advanced to a certain distance, the deformation and failure of the floor will be mainly controlled by the filling step.

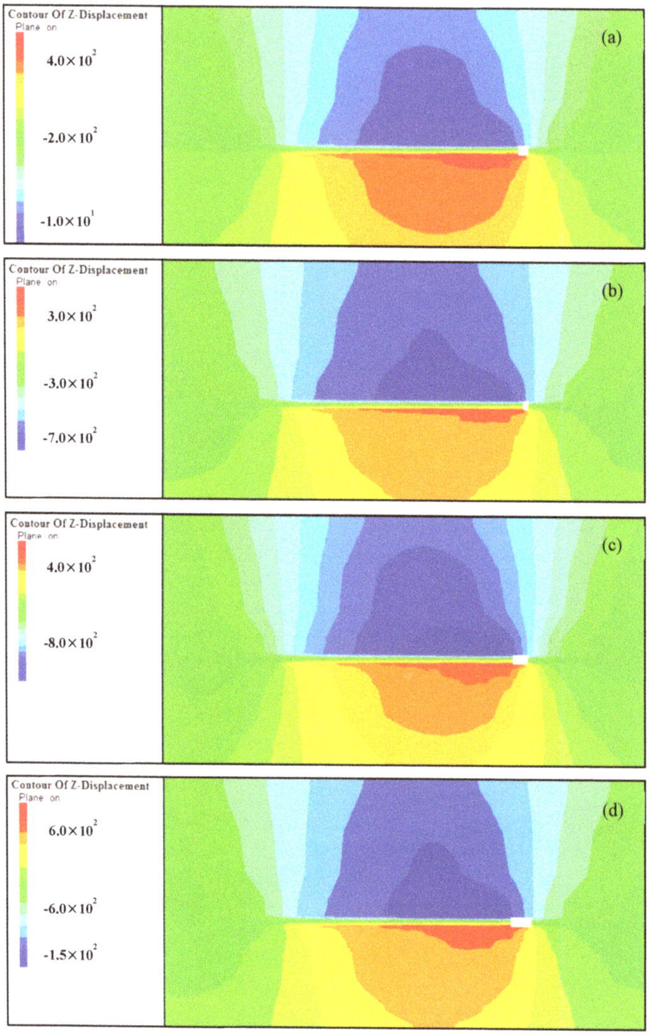

Figure 10. Floor displacement under different filling steps of paste-filling working face. (**a**) Filling step distance 1 m, (**b**) Filling step distance 2 m, (**c**) Filling step distance 3 m, (**d**) Filling step distance 4 m.

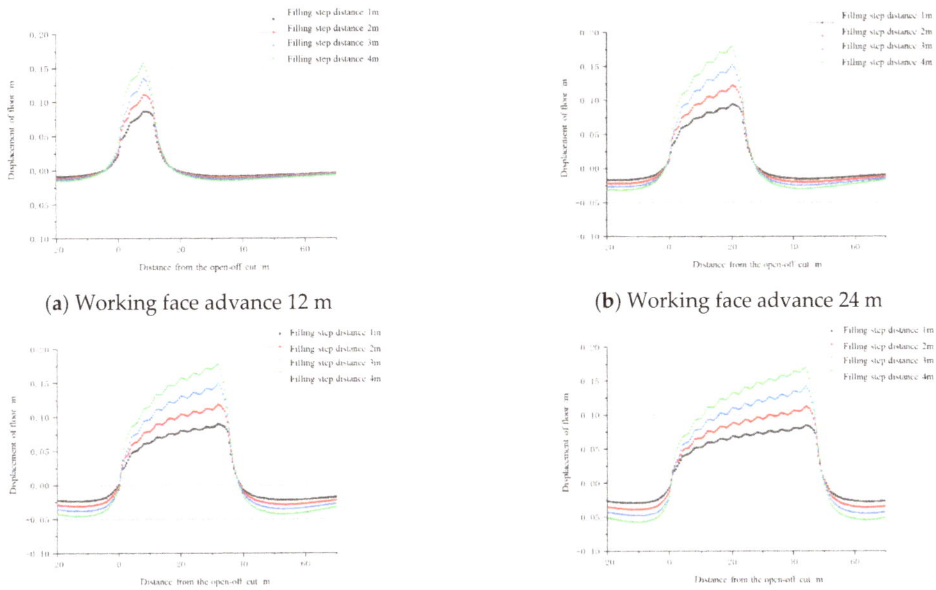

(**a**) Working face advance 12 m
(**b**) Working face advance 24 m
(**c**) Working face advance 36 m
(**d**) Working face advance 48 m

Figure 11. Floor displacement under different filling steps of paste-filling working face.

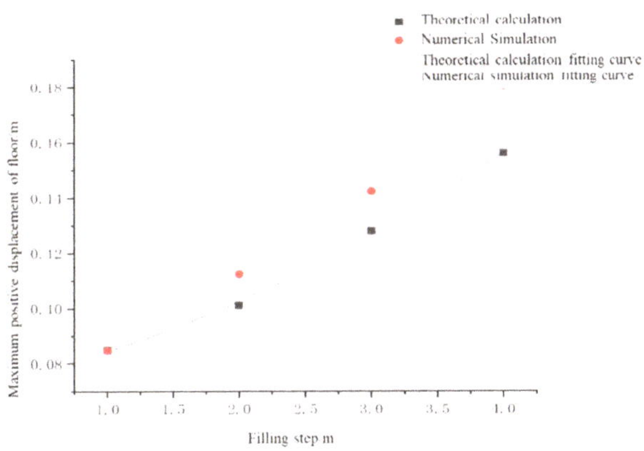

Figure 12. The relationship between different filling steps and the maximum positive displacement of the floor.

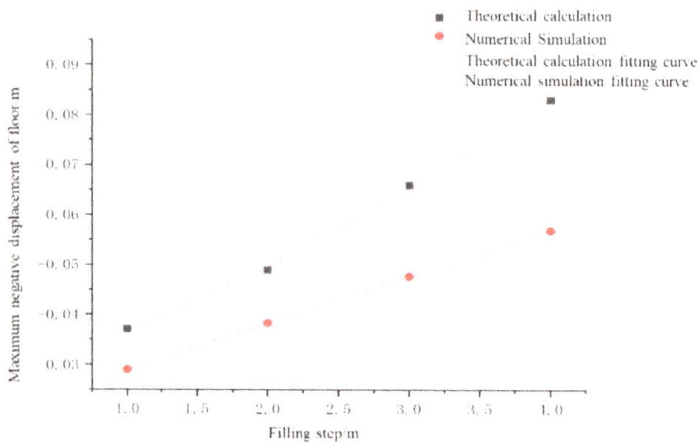

Figure 13. The relationship curve between different filling steps and the maximum negative displacement of the floor.

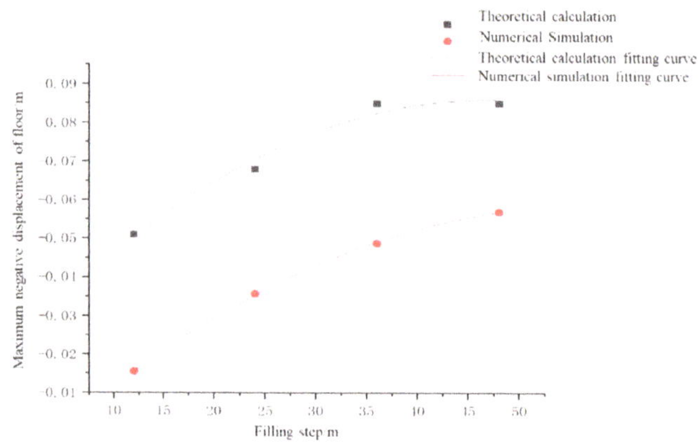

Figure 14. The relationship between different advancing distances and the maximum negative displacement of the floor.

4.3. Destruction Depth Influenced by Filling Step and the Advancing Distance

The filling step distances of the paste-filling working face are 1, 2, 3, and 4 m, and the advancing distances are 12, 24, 36, and 48 m. When the working face is advanced to 48 m, the plastic zone of the floor under different filling steps is shown in Figure 15. When the filling step is 4 m, the plastic zone of the floor under different advancing distances is shown in Figure 16.

The damage to the floor often begins to develop from the front end of the coal wall. Since the paste-filling body has the function of supporting the roof and transferring the pressure, the supporting pressure is greatly reduced, and the damage degree of the floor is extremely controlled. The damage to the floor does not cause the water barrier layer failure, which meets the requirements for safe mining on confined water. It can be seen from Figures 15 and 16 that the floor damage depth gradually increases from 2 to 7 m, the degree of penetration gradually increases, and the confined aquifer plastic failure has not progressed upward. After setting the filling step of the working face to 4 m, and changing the advancing distance to 12 m, 24 m, 36 m, and 48 m, the floor failure depth is maintained at 4 m. Therefore, under certain conditions of filling parameters, the increase in the filling step will cause the vertical damage depth of the floor, and the increase in the advancing distance mainly causes the increase in the horizontal damage range of the floor.

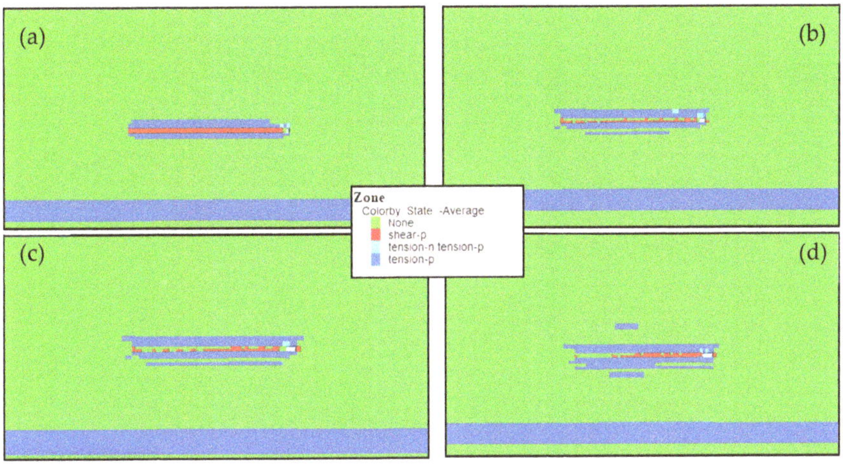

Figure 15. Plastic zone in the paste-filling working face under different filling steps of (**a**) 1 m, (**b**) 2 m, (**c**) 3 m, and (**d**) 4 m.

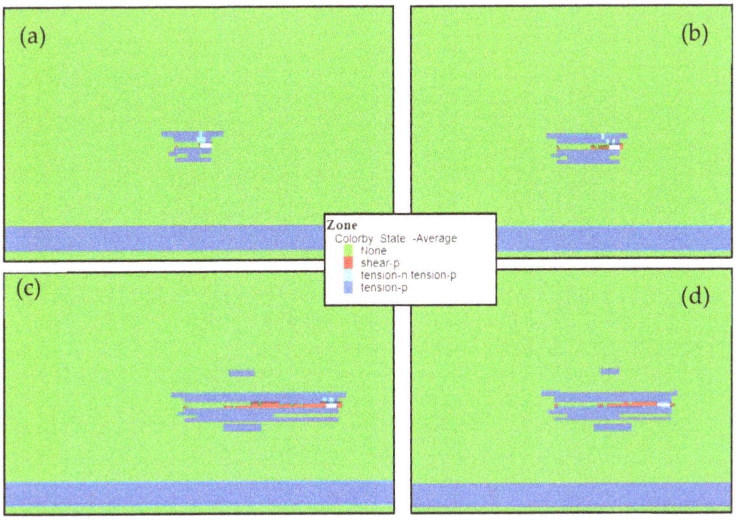

Figure 16. Plastic zone in paste-filling face under different advancing distances: (**a**) 12 m, (**b**) 24 m, (**c**) 36 m. and (**d**) 48 m.

5. Conclusions

(1) Combined with the geological conditions of the 11,607 areas of the working face of Daizhuang Mine, a differential equation for the deflection curve of the floor is established. The maximum deflection of the floor and the advancing distance present a "convex" quadratic function relationship, and as the advancing distance increases, the increase in the displacement of the floor will continue to decrease.

(2) It is verified by numerical simulation that the filling step and the maximum stress of the floor show a linear relationship. During the continuous advancement of the working face, the setting of the filling step has a significant influence on the damage degree of the floor.

(3) With the increase in the advancing distance, the maximum stress of the floor gradually stabilizes. The maximum deflection of the floor and the advancing distance also show a "convex" quadratic function relationship. Therefore, when the working face advances more than a certain distance (36 m in this mine), the longitudinal damage of the floor will not increase with the increase in the advance distance.

Author Contributions: Conceptualization by Q.C.; methodology, X.Y.; investigation, Q.C., J.Q.; writing—original draft preparation, Q.C., J.Q., M.L. and Y.W.; writing—review and editing H.Z., Y.S. and Y.X.; supervision, Q.C.; funding acquisition, Q.C. and H.Z. All authors have read and agreed to the published version of the manuscript.

Funding: This research was supported by the projects of "the Fundamental Research Funds for the Central Universities (2020ZDPY0221, 2021QN1003)", "National Natural Science Foundation of China (52174130, 52104106, 52174089)", and Basic Research Program of Xuzhou (KC21017).

Institutional Review Board Statement: Not applicable.

Informed Consent Statement: Not applicable.

Data Availability Statement: The data presented in this study are available on request from the corresponding author. The data are not publicly available due to privacy.

Acknowledgments: The authors are grateful to Gaohe coal mine.

Conflicts of Interest: The authors declare no conflict of interest.

References

1. Zhang, Y.; Yang, L. A novel dynamic predictive method of water inrush from coal floor based on gated recurrent unit model. *Nat. Hazards* **2021**, *105*, 2027–2043. [CrossRef]
2. Hu, Y.; Li, W.; Wang, Q.; Liu, S.; Wang, Z. Evaluation of water inrush risk from coal seam floors with an AHP-EWM algorithm and GIS. *Environ. Earth Sci.* **2019**, *78*, 290. [CrossRef]
3. Gui, H.; Song, X.; Lin, M. Water-inrush mechanism research mining above karst confined aquifer and applications in North China coalmines. *Arab. J. Geosci.* **2017**, *10*, 180. [CrossRef]
4. Sun, Y.; Li, G.; Zhang, J.; Sun, J.; Huang, J.; Taherdangkoo, R. New Insights of Grouting in Coal Mass: From Small-Scale Experiments to Microstructures. *Sustainability* **2021**, *13*, 9315. [CrossRef]
5. Zhou, D.; Wu, K.; Bai, Z.; Hu, Z.; Li, L.; Xu, Y.; Diao, X. Formation and development mechanism of ground crack caused by coal mining: Effects of overlying key strata. *Bull. Eng. Geol. Environ.* **2019**, *78*, 1025–1044. [CrossRef]
6. Jiang, N.; Wang, C.; Pan, H.; Yin, D.; Ma, J. Modeling study on the influence of the strip filling mining sequence on mining-induced failure. *Energy Sci. Eng.* **2020**, *8*, 2239–2255. [CrossRef]
7. Sun, Y.; Li, G.; Zhang, J.; Huang, J. Rockburst intensity evaluation by a novel systematic and evolved approach: Machine learning booster and application. *Bull. Eng. Geol. Environ.* **2021**, *80*, 8385–8395. [CrossRef]
8. He, J.; Li, W.; Qiao, W. P-H-q evaluation system for risk assessment of water inrush in underground mining in North China coal field, based on rock-breaking theory and water-pressure transmission theory. *Geomat. Nat. Hazards Risk* **2018**, *9*, 524–543. [CrossRef]
9. Liu, W.; Mu, D.; Xie, X.; Yang, L.; Wang, D. Sensitivity Analysis of the Main Factors Controlling Floor Failure Depth and a Risk Evaluation of Floor Water Inrush for an Inclined Coal Seam. *Mine Water Environ.* **2018**, *37*, 636–648. [CrossRef]
10. Liu, S.; Liu, W.; Shen, J. Stress evolution law and failure characteristics of mining floor rock mass above confined water. *Ksce J. Civ. Eng.* **2017**, *21*, 2665–2672. [CrossRef]
11. Sun, W.; Zhang, S.; Guo, W.; Liu, W. Physical Simulation of High-Pressure Water Inrush through the Floor of a Deep Mine. *Mine Water Environ.* **2017**, *36*, 542–549. [CrossRef]
12. Sun, J.; Hu, Y.; Zhao, G. Relationship between water inrush from coal seam floors and main roof weighting. *Int. J. Min. Sci. Technol.* **2017**, *27*, 873–881. [CrossRef]
13. Lu, Y.; Wang, L. Numerical simulation of mining-induced fracture evolution and water flow in coal seam floor above a confined aquifer. *Comput. Geotech.* **2015**, *67*, 157–171. [CrossRef]
14. Yin, S.; Zhang, J.; Liu, D. A study of mine water inrushes by measurements of in situ stress and rock failures. *Nat. Hazards* **2015**, *79*, 1961–1979. [CrossRef]
15. Zhang, W.; Zhang, D.; Qi, D.; Hu, W.; He, Z.; Zhang, W. Floor failure depth of upper coal seam during close coal seams mining and its novel detection method. *Energy Explor. Exploit.* **2018**, *36*, 1265–1278. [CrossRef]
16. Sun, Y.; Li, G.; Zhang, J. Developing hybrid machine learning models for estimating the unconfined compressive strength of jet grouting composite: A comparative study. *Appl. Sci.* **2020**, *10*, 1612. [CrossRef]

17. Apaydin, A.; Korkmaz, N.; Ciftci, D. Water inflow into tunnels: Assessment of the Gerede water transmission tunnel (Turkey) with complex hydrogeology. *Q. J. Eng. Geol. Hydrogeol.* **2019**, *52*, 346–359. [CrossRef]
18. Butscher, C.; Scheidler, S.; Farhadian, H.; Dresmann, H.; Huggenberger, P. Swelling potential of clay-sulfate rocks in tunneling in complex geological settings and impact of hydraulic measures assessed by 3D groundwater modeling. *Eng. Geol.* **2017**, *221*, 143–153. [CrossRef]
19. Golian, M.; Teshnizi, E.S.; Nakhaei, M. Prediction of water inflow to mechanized tunnels during tunnel-boring-machine advance using numerical simulation. *Hydrogeol. J.* **2018**, *26*, 2827–2851. [CrossRef]
20. Sedaghati, Z.; Mikaeil, R.; Bakhtavar, E.; Mohammadnejad, M. Fuzzy Analysis and Risk Management of Water Inrush by Numerical Simulation and FMEA under Uncertainty for Emamzade Hashem Tunnel. *J. Anal. Numer. Methods Min. Eng.* **2019**, *9*, 1–16.
21. Dammyr, Ø.; Nilsen, B.; Gollegger, J. Feasibility of tunnel boring through weakness zones in deep Norwegian subsea tunnels. *Tunn. Undergr. Space Technol.* **2017**, *69*, 133–146. [CrossRef]
22. Ma, K.; Sun, X.Y.; Tang, C.A.; Yuan, F.Z.; Wang, S.J.; Chen, T. Floor water inrush analysis based on mechanical failure characters and microseismic monitoring. *Tunn. Undergr. Space Technol.* **2021**, *108*, 103698. [CrossRef]
23. Meng, X.; Liu, W.; Zhao, J.; Ding, X. In Situ Investigation and Numerical Simulation of the Failure Depth of an Inclined Coal Seam Floor: A Case Study. *Mine Water Environ.* **2019**, *38*, 686–694. [CrossRef]
24. Mo, S.; Ramandi, H.L.; Oh, J.; Masoumi, H.; Canbulat, I.; Hebblewhite, B.; Saydam, S. A new coal mine floor rating system and its application to assess the potential of floor heave. *Int. J. Rock Mech. Min. Sci.* **2020**, *128*, 104241. [CrossRef]
25. Mo, S.; Sheffield, P.; Corbett, P.; Ramandi, H.L.; Oh, J.; Canbulat, I.; Saydam, S. A numerical investigation into floor buckling mechanisms in underground coal mine roadways. *Tunn. Undergr. Space Technol.* **2020**, *103*, 103497. [CrossRef]
26. Zhai, J.; Liu, D.; Li, G.; Wang, F. Floor Failure Evolution Mechanism for a Fully Mechanized Longwall Mining Face above a Confined Aquifer. *Adv. Civ. Eng.* **2019**, *8036928*. [CrossRef]
27. Meng, X.; Liu, W.; Mu, D. Influence Analysis of Mining's Effect on Failure Characteristics of a Coal Seam Floor with Faults: A Numerical Simulation Case Study in the Zhaolou Coal Mine. *Mine Water Environ.* **2018**, *37*, 754–762. [CrossRef]
28. Zhu, S.; Lu, L.; Wu, Y.; Zhang, T. Comprehensive study on the deformation and failure characteristics of a mining-impacted deep double-longwall working face floor. *J. Geophys. Eng.* **2017**, *14*, 641–653. [CrossRef]
29. Sun, Y.; Bi, R.; Chang, Q.; Taherdangkoo, R.; Zhang, J.; Sun, J.; Huang, J.; Li, G. Stability Analysis of Roadway Groups under Multi-Mining Disturbances. *Appl. Sci.* **2021**, *11*, 7953. [CrossRef]
30. Lu, H.; Liang, X.; Shan, N.; Zhang, Y.-K. Study on the Stability of the Coal Seam Floor above a Confined Aquifer Using the Structural System Reliability Method. *Geofluids* **2018**, *9580271*. [CrossRef]
31. Li, G.; Jiang, Z.; Lv, C.; Huang, C.; Chen, G.; Li, M. Instability mechanism and control technology of soft rock roadway affected by mining and high confined water. *Int. J. Min. Sci. Technol.* **2015**, *25*, 573–580. [CrossRef]
32. Sun, J.; Wang, L.; Zhao, G. Stress Distribution and Failure Characteristics for Workface Floor of a Tilted Coal Seam. *Ksce J. Civ. Eng.* **2019**, *23*, 3793–3806. [CrossRef]
33. Sun, Y.; Li, G.; Zhang, J.; Sun, J.; Xu, J. Development of an ensemble intelligent model for assessing the strength of cemented paste backfill. *Adv. Civ. Eng.* **2020**, *1643529*. [CrossRef]

www.ingramcontent.com/pod-product-compliance
Lightning Source LLC
Chambersburg PA
CBHW041016270926
51850CB00024B/684
9783036554419

MDPI
St. Alban-Anlage 66
4052 Basel
Switzerland
Tel. +41 61 683 77 34
Fax +41 61 302 89 18
www.mdpi.com

Processes Editorial Office
E-mail: processes@mdpi.com
www.mdpi.com/journal/processes